A WORLD BANK COUNTRY STUDY

Argentina

Reforms for Price Stability and Growth

The World Bank
Washington, D.C.

First printing April 1990

World Bank Country Studies are among the many reports originally prepared for internal use as part of the continuing analysis by the Bank of the economic and related conditions of its developing member countries and of its dialogues with the governments. Some of the reports are published in this series with the least possible delay for the use of governments and the academic, business and financial, and development communities. The typescript of this paper therefore has not been prepared in accordance with the procedures appropriate to formal printed texts, and the World Bank accepts no responsibility for errors.

The World Bank does not guarantee the accuracy of the data included in this publication and accepts no responsibility whatsoever for any consequence of their use. Any maps that accompany the text have been prepared solely for the convenience of readers; the designations and presentation of material in them do not imply the expression of any opinion whatsoever on the part of the World Bank, its affiliates, or its Board or member countries concerning the legal status of any country, territory, city, or area or of the authorities thereof or concerning the delimitation of its boundaries or its national affiliation.

The complete backlist of publications from the World Bank is shown in the annual *Index of Publications*, which contains an alphabetical title list (with full ordering information) and indexes of subjects, authors, and countries and regions. The latest edition is available free of charge from the Publications Sales Unit, Department F, The World Bank, 1818 H Street, N.W., Washington, D.C. 20433, U.S.A., or from Publications, The World Bank, 66, avenue d'Iéna, 75116 Paris, France.

ISSN: 0253-2123

Library of Congress Cataloging-in-Publication Data

Argentina : reforms for price stability and growth.
p. cm. -- (A World Bank country study)
ISBN 0-8213-1543-9
1. Argentina--Economic conditions--1983- 2. Argentina--Economic policy. 3. Economic stabilization--Argentina. 4. Inflation (Finance)--Argentina. 5. Prices--Government policy--Argentina. I. Series.
HC175.A8665 1990
339.5'0982--dc20 90-35475
CIP

PREFACE

On July 8, 1989, Carlos Saul Menem was inaugurated as President of Argentina. This report was completed on the eve of the transition, and examines the structural problems that contributed to the stagnation of the economy in the 1980s and ultimately culminated in hyperinflation and deep recession. The report is based on a World Bank macroeconomic mission that visited Argentina from January 31 through February 14, 1989 as well as other smaller missions during the final months of the Alfonsin administration. Despite the wide-ranging character of the reform program of the new administration, the intractability of Argentina's pressing economic problems will continue to make the analysis relevant for some time to come.

The mission comprised the following members:

Richard Newfarmer (Mission Leader, Macro, External Finance)
Luca Barbone (Macro, Fiscal and Monetary Policy)
Paul Beckerman (Monetary and Financial Policy)
Luc Everaert (Public Enterprise)
Egbert Gerken (Trade and Industrial Policy)

Other people have made contributions to this report, including James Hicks and David Vetter (provincial government finance), William McGreevey (social security), Dale Gray (energy sector), Steven Oliver (agriculture), and Luis Riveros (labor markets). Roberto Manrique provided research assistance for the external finance sections, and Ann Mitchell prepared the projections and statistical appendix. Diane Bievenour and Alexandra Blackhurst provided secretarial support.

CURRENCY EQUIVALENTS

Currency Unit - Austral (A$)

Exchange Rates Effective August 15, 1989

Official	US$1.00 = A$650
Parallel Exchange Rate (Montevideo)	US$1.00 = A$685/690

FISCAL YEAR

January 1 - December 31

GLOSSARY OF ABBREVIATIONS AND ACRONYMS

ANA	Administracion Nacional de Aduanas	National Customs Administration
AyEE	Agua y Energia Electrica	Water and Electrical Energy Company
BANADE	Banco Nacional de Desarrollo	National Development Bank
BCRA	Banco Central de la Republica Argentina	Central Bank of Argentina
BHN	Banco Hipotecario Nacional	National Housing Bank
BONAVI	Bonos Nacionales de Intereses Variables	Variable Interest Rate Bonds
CEN	Corporacion de Empresas Nacionales	Corporation of National Enterprises
COGASCO	Compañia de Gas Centro-Oeste	Center-Oeste Gas Company
CONADE	Consejo Nacional de Desarrollo	National Development Corporation
CRM	Cuenta de Regulacion Monetaria	Monetary Regulation Account
DEP	Directorio de Empresas Publica	Public Enterprise Board
DGI	Direccion General Impositiva	General Tax Administration
DNPC	Direccion Nacional de Promocion Comercial	National Directorate of Commercial Promotion
ELMA	Empresa de Lineas Maritimas	International Shipping Company
ENTEL	Empresa Nacional de Telecommunicaciones	National Telephone Company
GDE	Gas del Estado	State Gas Company
HISPASAN		Steel Company
HIDRONOR	Hidroelectrica Norpatagonia	North Patagonia Hydroelectric Company
INOS	Instituto Nacional de Obras Sociales	National Institute of Social Insurance Funds
IVA	Impuesto al Valor Agregado	Value-Added Tax (VAT)
JNG	Junta Nacional de Granos	National Grain Board
MCBA	Municipalidad de la Ciudad de Buenos Aires	Municipality of the City of Buenos Aires
NFPS		Nonfinancial Public Sector
OSN	Obras Sanitarias Nacional	National Sewage Company
PEs		Public Enterprises
PRESEX (PEEX)	Programas Especiales de Exportacion	Special Export Program
SEGBA	Servicios Electricos del Gran Buenos Aires	Electric Services of Greater Buenos Aires
SIGEP	Sindicatura General de Empresas Publicas	General Comptroller of Public Enterprises
TAR		Temporary Admission Regime
YCF	Yacimientos Carbonales Fiscales	State Coal Company
YPF	Yacimientos Petroliferos Fiscales	State Oil Company

TABLE OF CONTENTS

PART II - ANNEXES

PUBLIC SECTOR

SECTORAL ISSUES

COUNTRY DATA - ARGENTINA

AREA	POPULATION	DENSITY a/
2766.9 thous. sq.km.	32.0 million (1988) 1.5% annual growth	10.7 per sq.km 16.9 per sq.km of arable land

POPULATION CHARACTERISTICS a/	
Crude Birth Rate (per 1000)	23.6
Crude Death Rate (per 1000)	8.9
Infant Mortality (per 1000 live births)	34.4

HEALTH b/	
Population per physician (thous.)	0.5
Population per hospital bed (thous.)	0.2

INCOME DISTRIBUTION b/	
% of national income, highest quintile	50.3%
% of national income, lowest quintile	4.4%

DISTRIBUTION OF LAND OWNERSHIP	
% owned by top 10% of land owners	..
% owned by smallest 10% of land owne	..

ACCESS TO SAFE WATER (1980)	
% of population - urban	65%
% of population - rural	17%

ACCESS TO ELECTRICITY (1989)	
% of population	95%

NUTRITION a/	
Calorie intake as % of requirements	119.2%
Per capita protein intake (grams per day)	99.7

EDUCATION	
Adult literacy rate % (1980)	95%
Primary school enrollment % a/	100%

GNP PER CAPITA IN 1988 c/ 2537

GROSS NATIONAL PRODUCT IN 1988 d/	US$ Mln. (current prices)	% of GNP	ANNUAL GROWTH RATES (% constant prices) 1970-75	1975-80	1980-85	1988
GNP at market prices	75620.6	100.0	2.9	1.8	-3.4	-0.7
Gross Domestic Investment	10095.4	13.4	1.9	4.4	-16.2	-4.7
Gross National Savings	7456.2	9.9	0.5	2.0	-13.8	29.9
Current Account Balance	-2639.2	-3.5				
Exports of Goods & NFS	10337.3	13.7	-4.7	14.1	5.2	58.6
Imports of Goods & NFS	7849.4	10.4	0.6	13.3	-13.0	-4.5

OUTPUT, LABOR FORCE AND PRODUCTIVITY IN 1988

	Value Added (constant prices) US$ Mln.	% of Total	Labor Force e/ Thousands	%	V.A. Per Worker US$
Agriculture	10897	15.2	1370	12.0	7951.0
Industry	24291	34.0	3586	31.4	6773.6
Services	36356	50.8	6464	56.6	5624.3
Total GDP at Factor Cost	71544	100.0	11421	100.0	6264.4

GOVERNMENT FINANCE f/

	Consolidated Nonfinancial Public Sector g/ Aust. Mln. 1987	% of GDP 1987	% of GDP 1983-87	Central Government Aust. Mln. 1987	% of GDP 1987	% of GDP 1983-87
Current Revenues	54661	30.8	35.1	18588	10.5	9.1
Current Expenditures	59995	33.9	37.8	20527	11.6	8.0
Current Balance	-5334	-3.0	-2.7	-1939	-1.1	1.1
Capital Expenditures	9102	5.1	7.4	551	0.3	0.4
Surplus h/	-13079	-7.4	-9.3	-12723	-7.2	-8.9
External Financing (net)	7411	4.2	1.8	6660	3.8	1.8

a/ For the period 1982-1985.
b/ For the period 1970-1976.
c/ Current US dollars. Estimated using Bank Atlas methodology.
d/ Current US dollar estimates, calculated from data in constant 1970 australes.
e/ Calculated by applying 1980 census shares to 1988 population.
f/ Executed budget estimates in current australes.
g/ Excludes provincial governments
h/ Takes into account all revenues and expenditures.

COUNTRY DATA - ARGENTINA

MONEY, CREDIT AND PRICES	1980	1981	1982	1983	1984	1985	1986	1987	1988
	(Millions of australes; end of period)								
Money and Quasi Money	8.0	16.0	38.3	193.6	1193.5	7624.5	16229	44433	240353
Domestic Bank Credit to Public Sector	1.7	6.4	22.6	76.4	596.8	2845.1	5710	18624	86274
Domestic Bank Credit to Private Sector	8.3	22.2	68.9	290.8	1902.1	8980.6	16495	48819	222373
Money and Quasi Money as % of GDP	22.3	21.0	15.5	15.2	12.2	14.2	17.1	15.9	17.4
Wholesale Price Index (1985=100)	0.067	0.188	0.775	3.961	28.747	133.4	210.6	593.7	3155.9
Annual percentage changes in:									
General Wholesale Price Index	55.8	180.6	312.2	411.1	625.8	364.0	57.9	181.8	431.6
Bank Credit to Public Sector	69.8	282.8	251.7	237.2	681.6	376.7	100.7	226.2	363.2
Bank Credit to Private Sector	108.5	166.0	210.5	321.9	554.1	372.1	83.7	196.0	355.5

BALANCE OF PAYMENTS	1975	1980	1985	1988
	(US$ Millions)			
Exports of Goods, NFS	3704	10765	10242	11301
Imports of Goods, NFS	4518	14024	5891	7789
Resource Balance	-814	-3259	4351	3512
Interest Payments (net)	-460	-947	-4879	-4467
Other Factor Payments (net) a/	-15	-584	-425	-660
Net Current Transfers	5	23	0	0
Balance on Current Account	-1284	-4767	-953	-1615
Direct Investment	..	788	919	1147
Total M< Loans (net)	-12	3400	2786	-252
Disbursements	1018	5809	7564	-
Amortization	1030	2409	4778	-
Other Capital (net) b/	189	-2217	-881	2505
Changes in Gross Reserves (- = increase)	1107	2796	-1871	-1785
Gross Reserves (end year) c/	464	6743	4801	4979
Net Reserves (end year)	-520	6724	-7873	-15429

RATE OF EXCHANGE i/

1980	1988
US$ 1.00 = A$ 0.00018	US$ 1.00 = A$ 8.7703
A$ 1.00 = US$ 5382	A$ 1.00 = US$ 0.1140

MERCHANDISE EXPORTS (Average 1984-1988)	US$ Mln.	% of Total
Agricultural goods d/	3822.9	49.2
Manuf. goods of agric. orig. e/	2004.5	25.8
Manuf. of industrial origin f/	1942.8	25.0
Total Merchandise Exports	7770.2	100.0

EXTERNAL DEBT (as of Dec.31, 1988) g/	US$ Mln.
Total Debt Outstanding & Disbursed (DOD)	58935
IBRD	2265
IDB	1779
IMF	3678
Bilaterals	4470
Bonds	2808
Commercial Banks	43935

DEBT SERVICE RATIO, 1988 h/	43.4%
Interest service ratio (% of exports G&NFS)	26.9%

IBRD/IDA LENDING, DECEMBER 31, 1988 (Mln. US$)	IBRD	IDA
Outstanding & Disbursed	2265	-
Undisbursed	-	-
Outstanding incl. Undisbursed	-	-

a/ Direct investment income plus other factor service income.
b/ Includes short-term capital, net IMF resources, changes in arrears, and valuation adjustments.
c/ Includes valuation and other adjustments.
d/ BCRA categories I, II, and III: livestock and other animal products; agricultural products; fats and oils.
e/ BCRA categories IV, VIII, and XI: i.e. manufactured food, beverages and tobacco; leather, furs and related products; textiles and clothing.
f/ All other manufactured goods categories.
g/ Preliminary estimate.
h/ Amortization and interest payments on medium- and long-term (MLT) debt as a percentage of exports of G&NFS. Excludes arrears; includes rescheduling of debt.
i/ Period average.

ARGENTINA: EXECUTIVE SUMMARY

A. Introduction

1. In his inaugural address of July 8, 1989, President Carlos Menem emphasized that Argentina is facing the most difficult economic situation in its modern history, and promised the country only "sacrifice, work and hope." Inflation had surpassed 100 percent monthly, output was stagnating, unemployment was rising, and the Government was unable to mobilize credit either domestically or abroad. The current situation is the culmination of long-term trends emerging in the 1970s--slow growth in productivity and secular falls in savings and investment. Savings and investment ratios are about half their mid-1970 levels. In spite of efforts at reform in recent years, gross national income per capita is about 23 percent less today than in 1977. Foreign debt, once quite small relative to GDP, has climbed to nearly 100 percent of GDP.

2. The experience of the last two decades indicates that without a permanent reduction in inflation, Argentine savings will not be invested in Argentina. Five stabilization programs in the 1984-89 period have failed, largely because of insufficient adjustment in the public sector. Each time inflation returned and surged to higher levels than on the previous occasion. Lack of social consensus and entrenched business and union power, together with the burden posed by the external debt servicing, conspire to maintain the demands on the public sector beyond its ability to garner resources--thus creating a chronic deficit. The key to controlling inflation is an immediate and sustained reduction in the deficit through a comprehensive reform of the public sector.

3. But controlling inflation may not be sufficient to unleash the enormous productive potential of the country--and thereby provide the basis for increasing real wages over the long term. To reverse declining trends in labor productivity and the productivity of investment, the Government as part of its comprehensive reform must remove price distortions and other policy interventions that have discouraged investment and job creation throughout the country. While achieving price stability is the first order of business, the Government must do so on the basis of a comprehensive public sector reform that makes financial stability permanent and permits sustained increases in productivity and income.

Background

4. The growth of state spending during the 1970s contained three seeds of the crisis that was to become manifest in the 1980s. First, increases in public spending--public expenditures rose from about 25 percent of GDP in 1970-72 to over 38 percent in 1981-83--were not matched by a concomitant expansion of revenues, and so large deficits became commonplace, ranging from 5 to 16 percent of GDP in 1973-83. After 1980, weak tax administration, excessive tax exemptions, low public enterprise prices and falling real revenues associated with inflation initiated a secular deterioration in public revenues relative to GDP that has continued until the present.

5. Deficits in the late 1970s were initially financed through foreign and domestic borrowing and then, with the crisis in the financial system in 1980-82, the Government assumed the foreign debt of the private sector. The build-up in foreign debt during this period created an enormous external and internal transfer problem which hampered efforts at deficit reduction, and has made both foreign and domestic creditors increasingly reluctant to lend to the Government.

6. A third problem associated with the growth of the state was the proliferation of costly subsidies and associated economic distortions: Industrial promotion schemes subsidized domestic industry at a cost to public finances of roughly 5 percent of GDP. Consumers of public services enjoyed subsidies from tariffs that were often below costs. Many borrowers from public banks enjoyed implicit subsidies to housing, industrial investments, and other uses. Only a small portion of these subsidies went to low-income families or the truly needy. At the same time, Government policies raised barriers to competition in several sectors, and other inefficient regulations created additional economic distortions. The need to finance fiscal losses created a highly distorted financial system wherein the Central Bank was used to tap resources from financial intermediaries and channel them to loss-making public banks and the nonfinancial public sector, thus discouraging savings and efficient investment. These policies had the collective effect of directing investment into low productivity areas, and lowering the productivity of the whole economy.

Recent Reform Efforts

7. Inheriting a distorted economy, the Alfonsin administration (1983-89) initiated several reforms that had considerable effect in reducing the external and internal deficits. By keeping the exchange rate generally competitive after the Plan Austral in 1985, the deficit on current account of the balance of payments improved from over 4 percent of GDP in 1983 to under 2 percent in 1988. At the same time, expenditure compression, new tax measures, sporadic improvement in public enterprise pricing, and some adjustment in provincial public sector finance contributed to reducing the deficit of the combined public sector. Deficits of the combined public sector fell from over 20 percent of GDP in 1983 to under 7 percent in 1988.

8. But policies were not always consistent or sufficiently enduring to stabilize the economy. The new taxes, although easy to collect, were often inefficient and seen as temporary; export and energy taxes, for example, were among the most important new measures. Contending political demands prevented coherent action in controlling public finances; 1987 marked a severe setback to the gains in the previous two years, especially in expanding subsidies through the Public Housing Bank (BHN) and Industrial Bank (BANADE). Moreover, insufficient early attention was devoted to the structural problems that were the legacy of the previous decade.

9. The Plan Primavera, initiated in August 1988 against a backdrop of accelerating inflation reaching 30 percent monthly, attempted to remedy some of these problems. By introducing some structural measures--control

(a)

GDP Growth: Long-Term Trend (10 Year Moving Average)

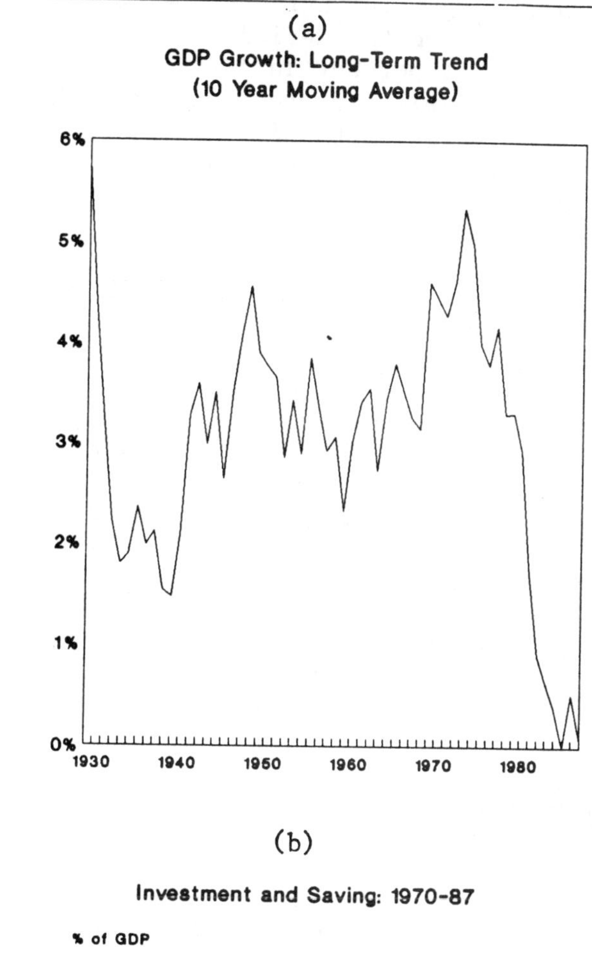

(c)

Origin of Saving: 1970-87

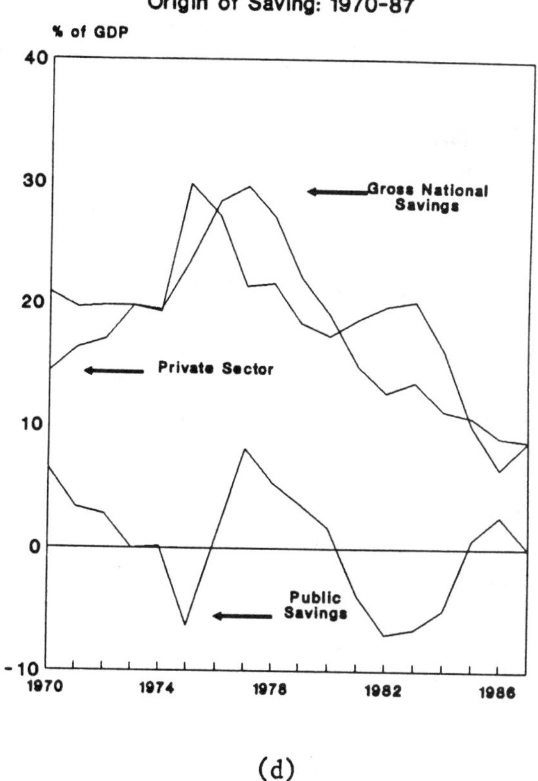

(b)

Investment and Saving: 1970-87

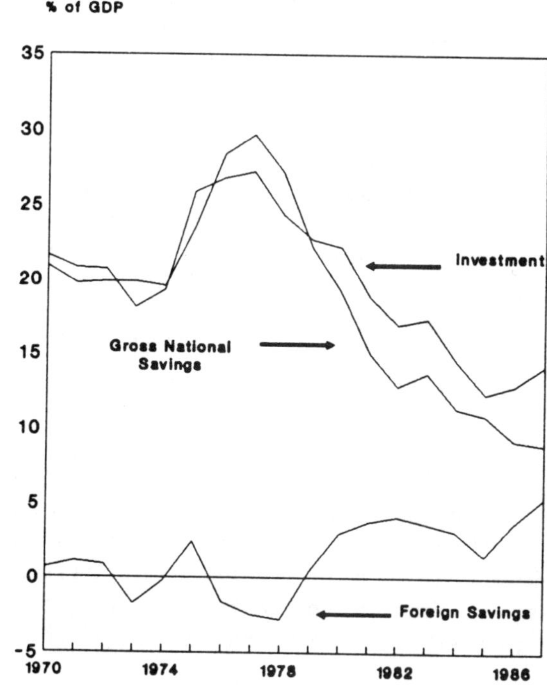

(d)

Public Sector Primary Surplus and External Interest 1970-87

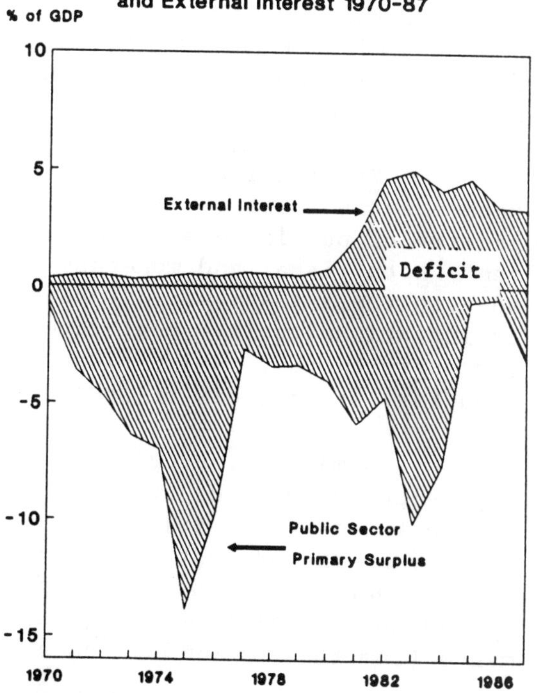

Note: Primary surplus equals public sector balance before interest payments. The deficit shown includes some small portion of private sector foreign interest payments.

of rediscounts and quasi-fiscal expenditures through the Central Bank, a tax reform designed to circumscribe some of the industrial subsidies, and a trade reform to open the economy--the program sought to introduce permanent changes in the state sector, intending to bring down the deficit of the combined public sector from 6.7 percent to under 3 percent in 1989. The trade reform succeeded in lowering the average level of protection and reducing the coverage of quantitative restrictions; financial sector reforms had substantially reduced the flow of rediscounts and maintained the regime of liberalized interest rates initiated in late 1987. By the end of the year inflation had fallen to single digits.

10. However, the political consensus to implement the announced fiscal measures and structural actions affecting the public sector was absent. Underadjustment in the fiscal accounts, as with the Plan Austral, threw the burden of stabilization on monetary policy; the Central Bank had to increase its debt to support the interest- and exchange-rate mix. The private sector, acutely aware that the structural fiscal deficit had not been eliminated, fearing the uncertainty of Argentina's first post-war democratic political transition, and doubting the capacity of the Central Bank to service its debt and support the exchange rate, began shifting its portfolio out of assets denominated in the domestic currency in mid-January 1989. This triggered a self-feeding run on the austral by the end of the month, which the outgoing government was powerless to stop. The free market exchange rate fell from A$16.8 per US dollar in January to A$380 in June. Monthly inflation rose every month to reach 123 percent by June and 203 percent in July.

B. Short-Term Problems and Options

Problems

11. The new economic team taking office on July 8, 1989 confronted four immediate economic problems: hyperinflation, instability in the financial system, price misalignment in the public sector, and the threat of a major recession.

12. **Hyperinflation**. By May, the cash deficit of the Treasury had become extraordinarily large. Widespread tax evasion and falling real tariffs of public enterprises had driven revenues down. But more important, both fiscal and monetary policy had become almost completely endogenous to the inflationary process. The deficit continued to widen because of inflation itself: real revenues were falling and, given low and falling demand for money, public finances were unable to balance through the inflation tax, setting off an inherently unstable process.

13. Similarly, accelerating inflation produced destabilizing monetary expansion through the losses of the Central Bank. This is because interest expenditures on Central Bank liabilities exceed Central Bank interest earnings by an amount that tends to increase with inflation. Interest earnings are dominated by loans to the public banks, which carry a fixed interest spread above inflation, while liabilities are dominated by forced investments from the commercial banks, and are priced at market rates; so higher

nominal interest rates associated with accelerating inflation increase the quasi-fiscal deficit of the Central Bank. Moreover, the lag in interest earnings means that during periods of rising inflation, the quasi-fiscal deficit widens.

14. This situation created a massive imbalance between demand for and supply of money in the first six months of 1989, while the inflationary process left the Central Bank without instruments to deal with it. The collapse of the exchange rate led those who held dollar-denominated deposits in the domestic financial system to shift them abroad, putting pressure on Central Bank reserves and heightening fears about the future exchange rate. Also, the public debt, estimated at 16 percent of GDP, continued to grow explosively because high interest payment obligations by the Central Bank on the forced investments of the financial system were capitalized into new forced investments. Moreover, the Central Bank was being required to finance the payment of maturing government bonds that could not be rolled over. As inflation gathered momentum and confidence waned, the public shifted its portfolio rapidly out of austral-denominated assets in order to avoid the inflation tax and the threat of default. With the demand for money falling, the Central Bank was forced to try to absorb the excess supply, but had to rely on ever higher reserve requirements (i.e. forced investments)--which bear interest and therefore ultimately require monetary expansion--to absorb money. The Government was left with the choice between hyperinflation and financial sector collapse.

15. **Financial System**. The proximate cause of the stress in the financial system was the flight of austral-denominated deposits to dollars and other assets. Withdrawals have put enormous pressure on commercial banks, since the forced deposits of the commercial banks with the Central Bank are not formal reserve requirements and are not automatically released *pari passu* with deposit withdrawals. Moreover, other assets of the banks' portfolio were not performing, as many firms were reportedly unable to service debt with the banks and were receiving rollovers. The Central Bank provided new rediscounts, eased forced investment requirements, and decreed sporadic bank holidays and limits on deposit withdrawals during May and June.

16. **Price Misalignment**. Public enterprise prices were severely misaligned. By end-June, real public enterprise prices were roughly 60 percent below levels prevailing at end-year 1988, the level necessary to generate positive savings. This is the reason why the Government more than doubled real prices as a principal measure in its new program. Realigning these prices will necessarily affect the price level because of their own heavy weight in price indices as well as the fact that so many private sector prices are indexed to these.

17. **Recession**. At the same time, the real economy, already in its second year of recession, was contracting further. The appreciation of the austral in the waning days of the *Plan Primavera* had led to a mini-boom in external travel and import purchases of consumer durables, and the flight from australes to goods in February produced some increase in demand for industrial goods. However, the hyperinflation compounded the effects of

the high interest rates on real sectors. The sharp reduction in real wages in February-June which cut private consumption, the shortage of trade credit and the inability of many business owners to determine what their own prices should be have led to a slowdown in economic activity.

Designing Stabilization Programs: Lessons of the Past

18. The experience of the last five years suggests three principles for the design of a program to stabilize the economy. First, the short-term plan must be built upon a program of structural reforms that provides confidence for investors and the public at large that changes in the conduct of public finances are indeed permanent. Second, the nominal public sector borrowing requirement must be credibly and immediately reduced to levels that can be financed through foreign credit--recognizing that in the immediate weeks ahead, until an agreement with external creditors can be put in place, the predominant foreign source will necessarily be a continual accumulation of arrears. Third, the monetary component of the stabilization package will have to ensure that the public sector will not use the inflation tax to balance its accounts in the future.

19. As the Menem Government implicitly recognized in its July 9 program, past experience argues strongly that the Government take several structural measures prior to--or simultaneously with--putting the stabilization program formally into place. Indeed, the Government began formulating its structural reform program immediately and announced key measures. This strategy inverts the past approach of first trying to close the fiscal gap with *ad hoc* revenue and other measures, and deferring structural measures *until* after stabilization is achieved. Simultaneous efforts at structural reform are necessary because the deficit remains high relative to available domestic and foreign finance, while the demand for austral-denominated assets remains low and volatile. As long as the deficit is seen to be eliminated in sustainable form through structural measures, the private sector will have no faith in the sustainability of the program. Lack of private sector confidence has also resulted in a secular decline in the holdings of financial instruments denominated in domestic currency, and increased the responsiveness of portfolio shifts and money velocity to inflationary expectations. The base of the inflation tax has been narrowed beyond a point where it can help close the accounts of the public sector. Furthermore, early elaboration of a comprehensive medium-term program and successful implementation will credibly convey a sense that price stability will be enduring and that the economy can recover.

20. A companion lesson from past stabilization attempts is that prices of public enterprises cannot be used as a mainstay of stabilization for any length of time. If real prices are allowed to deteriorate, it becomes virtually impossible for the sector to maintain savings at a level consistent with deficit reduction objectives. After the stabilization plans of recent years, falling real prices now raise inflationary expectations since the private sector is aware that tariffs will eventually have to be increased, and may even anticipate these with large price movements of their own.

21. Past experience also indicates that a heterodox program--a social pact on prices, some adjustment in fiscal accounts, and tight monetary policy--entails important pitfalls. The major strength of a heterodox plan, the ability to abruptly bring to a halt inflation through the imposition of wage and prices controls and/or some other form of social consensus, is also its fundamental weakness. In both the Austral and Primavera programs, the magnitude of the reform task was underestimated: The fact that the benefits of the program occurred at its inception removed pressure from the authorities and Congress, who then balked at paying the political price for some of the less popular measures. Once it became evident to the public that the structure and behavior of public finance had not changed fundamentally, inflationary expectations began refueling, the exchange rate-domestic inflation cycle took its course, the positive Olivera-Tanzi effect reversed, and the resulting deterioration in fiscal performance displaced whatever measures of a structural nature may have been under implementation.

Policy Options

22. A thorough stabilization must therefore be predicated upon mutually reenforcing reforms of the public sector, monetary policy, the exchange rate, and external finance. Achieving objectives described below could be attained through various strategies. The key elements of each approach, however, must be the same: up-front fiscal reforms that include difficult-to-reverse structural measures, clear and transparent monetary and exchange rate rules, and an eventual accord with external creditors.

23. **Public Finance Reform**. The cornerstone of the program must be explicit political agreement on a comprehensive reform of public finances. The agreement would have to signal a complete change in policy regime: as such it would have to offer to the public concrete guarantees against reversal. The central component of the package would be a sharply declining fiscal deficit which would remove the need to use the inflation tax; the objective would be an immediate primary surplus to service domestic internal debt and eventually recapitalize the Central Bank. This would be built upon integrated reforms aimed at reducing expenditures, rebuilding the tax base, and reducing the deficits of the public enterprises, social security system, and provinces. This would be enforced by the adoption of a transparent monetary rule stating that future currency emissions would be limited to increases in foreign reserves of the Central Bank. As a consequence, substantial and immediate adjustment of the public sector and public banks would be necessary, so as to reduce the public sector deficit to a level equal to net new borrowings of foreign resources.

24. **Monetary Reform and Policy**. A strong monetary reform that permitted monetary expansion solely as a function of increases in international reserves would provide the public confidence necessary to slow inflation. Monetary policy would be assigned a purely passive role. The policy would be based on an *ex-ante* judgment that the risk that this firm monetary rule might create a prolonged recession, and thus erode public support for the program, is less than the danger that the private sector will not believe the program is sustainable and will maintain its inflationary expectations.

25. Monetary reform would also involve a restructuring of the domestic internal debt, and eventually recapitalizing the Central Bank. This could be done through exchanges of debt instruments with the Central Government. The outstanding stock of forced investments from the commercial banks deposited with the Central Bank would be converted into long-term securities of the Government carrying a fixed nominal interest rate. To service these obligations, the Treasury has to mobilize a sufficient surplus to pay interest and amortization at the rate consistent with the stabilization strategy and targets. The only alternative is to write down the value of these assets, with its adverse implication for future creditworthiness.

26. **Exchange Rate**. The monetary reform (including the new rule on monetary expansion), together with the change in the fiscal policy regime, would allow the Government to maintain a fixed exchange rate against the dollar or a basket of currencies. This would provide the nominal anchor necessary to operate a successful stabilization after hyperinflation. Since international reserves are low, an important element for success is the availability of foreign resources to provide reserves to back the new exchange regime. An early arrangement with the IMF is therefore particularly important.

27. **External Finance**. Garnering foreign support will be difficult. External creditors, like their domestic counterparts, have no alternative but to recognize the limited capacity of the Argentine state to service its obligations. As shown in the projections in Chapter V, it seems unlikely that Argentina can fully service its private commercial debts in the medium term; a strong fiscal program predicated on sharp deficit reduction has extremely limited scope for normal debt servicing in the years immediately ahead. Nonetheless, all external creditors share with the Government a fundamental interest in Argentina's medium-term price stability, growth and restored creditworthiness. Sustained price stability may eventually attract back some of the enormous capital flight that has occurred since 1980. In the short term, however, net transfers to commercial creditors will have to be restricted to a minimum; nonetheless, soon after a comprehensive medium-term program is in place and the stabilization is underway, the Government should initiate negotiations with the commercial banks to explore solutions to Argentina's long-term financing problem so that external transfers can be tailored to the country's capacity to pay at a time when the need for sustained growth is paramount.

C. Main Components of Structural Reform Program: A Comprehensive Reform of the Public Sector

28. Deficit spending of the Argentine public sector has plagued economic management in Argentina for decades. Noninterest expenditures as a share of GDP rose in the late 1970s because of increases in social security and provincial spending, and then rose even more sharply in the early 1980s, driven by continued provincial spending and the South Atlantic War. After the public sector absorbed the private foreign debt, interest expenditures put increasing pressure on public finances; interest payments rose from under 3 percent of current expenditures to over 20 percent.

29. After reaching a peak in 1980, tax revenues began a secular decline that has been offset from time to time only with inefficient and transitory measures, including most notably export taxes, but also taxes on petroleum products, savings, checks, cigarettes as well as compulsory savings. While many of these had the advantage of being readily collectable, they have progressively introduced significant distortions that discourage exports and reduce international competitiveness, discourage savings and channel investment into low productivity areas.

30. At the heart of the inability to contain public finances has been the weak control over the various components of the public sector outside the central administration, including public enterprises, social security and provincial finance as well as the disguised fiscal expenditures through the Central Bank. At the beginning of 1988, the Government adopted a new principle for fiscal policy that enhanced accountability of these governmental sectors: It established clear guidelines for transfers between the sectors and established a rule of sectoral self-sufficiency. At the same time, the Government reduced discretionary expenditures through the Central Bank by slashing rediscounts to public banks and terminating the practice of allowing provincial banks to overdraw their accounts at the Central Bank.

31. These efforts have increased transparency and political awareness of the problem, but have not remedied the deficit of the public sector. For that, a comprehensive reform of the public sector is necessary. This would entail efforts to reduce expenditures, improve taxation, and reform public enterprises, federal provincial financial relationships, and social security. (Structural measures that would markedly enhance the strength of the stabilization program if announced prior to or simultaneously with the program are denoted with an asterisk.)

Expenditure Reductions

32. Expenditures of the nonfinancial public sector have fallen by about 10 percentage points of GDP since 1983. However, present expenditure levels at 30 percent of GDP are still high relative to capacity for mobilizing tax resources as well as relative to other countries at similar levels of per capita income. The process since 1983 has not been part of conscious government policy, but the result of successive marginal contraction imposed by the threat of inflation; consequently, the pace of expenditure reduction has always been too slow to achieve stabilization goals and the process itself has been inefficient. Too often the Government has chosen to contract investment in the hopes that the fiscal crisis would pass, rather than make the difficult cuts in public employment and inefficient programs and subsidies.

33. A major component of a structural reform program would involve expenditure reductions. This would have to focus on employment reduction throughout the public sector. Employment increased by 20 percent in 1983-88 in the central administration and 28 percent in the provinces --despite acute budgetary constraints. Since average public sector wages

have fallen to very low levels, employment reductions should be deep enough to permit some increase in average wages even with some gains in reducing the overall wage bill. The Government should strive to mitigate the hardship imposed by lay-offs through programs of early retirement with full pension, severance pay, and temporary income-maintenance in accordance with workers' length of service.

34. Similarly, the Government must mount a comprehensive effort to identify programs that could be cut to achieve savings and improve the efficiency of the public sector. In the central administration, the health, education and housing budgets should be carefully scrutinized since there is a need for greater efficiency in delivering these services at the same time for additional savings are needed; in education, for example, expenditures for primary and secondary education have been reduced far more than expenditures on higher education, even though the latter entail subsidies to the relatively wealthy and could be supported through a combination of increased direct charges with more scholarships and loans for poor youth who would otherwise qualify for entrance. In transportation, the Government could consider partial divestiture of the port facilities, which are inefficient and costly; greater reliance on toll roads and privately constructed toll roads could increase badly needed investment in this sector as well as reduce the cost to the budget. In agriculture, reforming the National Grain Board (Junta Nacional de Granos) along the lines suggested in Annex Chapter IX would provide some savings. Additional expenditure reductions can be achieved through measures in the public enterprise sector, such as divestitures and employment reduction, and in social security; these are discussed below.

35. These reductions in expenditures as well as other public sector reforms necessarily involve abrupt and disruptive changes in the lives of some Argentines--workers who are laid off from inefficient government programs or enterprises, recipients of subsidies, and consumers of state enterprise products and services whose prices may be increased. While these social costs are relatively small when compared to the alternative of rampant inflation and prolonged stagnation, the Government should make every effort to maintain and strengthen the social support services for the poor--the social "safety net." This means carefully targeting remaining subsidies on low-income groups, improving the efficiency of delivery for existing social services, introducing user charges to recover costs from those who are able to pay, and improving collection and financial management of earmarked revenues. It also means improving the efficiency of regulations governing the private sector in the provision of these services. Finally, it entails channelling that component of shared revenues destined to service the poor to those provinces with a larger number of low-income families.1/ These efforts could substantially mitigate the human costs of stabilization and reform.

1/ While these ideas are not developed in this report, detailed suggestions to strengthen the social safety net are presented in the World Bank's report, Argentina: Social Sectors in Crisis, June 1988.

The Tax System

36. The Argentine tax system has become increasingly deficient. Taxes have fallen and new taxes have been neither efficient nor equitable. The system has shown a relatively low buoyancy, largely because of the increased use of the tax system to promote regional and sectoral industrial development. Ad hoc taxes have been imposed on several occasions; the Government has repeatedly granted tax amnesties, with lower revenue results. Meanwhile, the administration of taxes, beset by lack of resources and excessive variability in legislation and management, deteriorated substantially.

37. Important changes were introduced through a tax package in December 1988. Although Congress rejected a proposal to generalize the value-added tax (VAT), the Government introduced changes in direct taxes that modified the industrial promotion system; if sustained and implemented they could lead to a significant reduction of tax avoidance and evasion and of the economic distortions generated by industrial promotion. Still, these efforts did not go far enough.

38. A comprehensive tax reform is therefore necessary to overhaul the tax system. It should be guided by principles such as those outlined in the Law of Economics Emergency introduced as part of the July 9 program: namely, that all subsidies through the state be eliminated, except for those directed at poverty alleviation. A program of structural reform in tax regime would include:

* (a) A reform of the value-added tax that would lower the rate and broaden the base as well as revamp the income tax to broaden its coverage and enforce its application. (This reform would encompass some of the other measures described below.)

* (b) The permanent suspension of the general and the provincial industrial promotion regime to replace the proposed temporary and partial suspension (laws 21.608, 22.021, 22.702 and 22.973).

* (c) The phase-out of the tax exemption regime for Tierra del Fuego. Some temporary financial transfers to the provincial budget could help compensate workers; benefits in the form of tariff exemptions could be retained if the Government wished to convert the island to an export processing zone.

* (d) Acquired rights of beneficiaries under the above industrial promotion could be capitalized via the issuance of securities; these should be paid on the basis of audited statements of projected production (i.e., the "theoretical cost") made at the time of original application for benefits.

* (e) The permanent abolition of promotional schemes for exports --including the PEEX program and the export subsidies. The competitive exchange rate already provides sufficient incentive for exports and the fiscal savings are substantial.

39. Measures that would at the same time promote a more efficient utilization of resources, foster public enterprise reform and increase public sector revenues in the area of taxation of the energy sector, would include:

(a) The imposition of VAT on all oil products; a simplification of excise taxes to be replaced by only one ad valorem tax, fixed on the supply price at the plant.

(b) The state oil company, YPF, should be subject to income taxes and, if public finances require additional transfers, the Government should use its power to pay dividends as the means to extract profits from the sector.

40. Trade taxes distort investment and, therefore, the move away from them should be pursued with determination.

(a) The exemptions from import tariffs on capital goods and other imports should be abolished. This would reduce the dispersion of effective protection rates, increase revenues, and increase employment.

* (b) Product-specific export taxes should be rapidly replaced with a general tax reform that includes the agricultural sector. In the interim, export tax payments should be credited against value added tax obligations of producers.

41. Solving the problems of the tax administration will involve a determined medium-term effort on the part of the Government, as well as the implementation of measures in the tax system to simplify its administration. This will involve among other things:

* (a) Measures to increase General Tax Administration's (DGI) internal technical ability through improved organization, data processing, and better personnel procedures.

* (b) Strengthen administration measures that would reduce the complexity of the tax system, through the abolition of low-yield nationally administered taxes.

* (c) Abolishing the simplified VAT system and approving laws pending in Congress that would substantially increase DGI's powers, including meaningful and appropriate sanctions for noncompliance by taxpayers.

Public Sector Enterprises

42. Public sector enterprises have contributed substantially to the overall deficit of the public sector--about half on average during the 1980s. This poor performance is the result of sporadic reliance on public enterprise price adjustments to achieve (temporary) macroeconomic stability as well as the use of the sector as a source of resources to channel

subsidies to the private sector. The Government has pursued pricing policies that have oscillated between providing sufficient resources to cover costs and compressing prices to achieve anti-inflation objectives as in the Austral and Primavera Plans. Legal constraints have been imposed on public procurement through the buy-Argentina law (Compre Argentino), resulting in inflated costs and massive subsidies to private industry and other suppliers. Similarly, entrenched unions have exploited their political power to saddle the sector with excessive employment, especially in the railways. Finally, pricing distortions have been accentuated by the use of the energy sector as tax collector to cross-subsidize losses in the railways, social security system, and provinces.

43. To deal with these problems, the Argentine Government has pursued a dual strategy. The first element consisted of clearly delineating the rules for the transfer policy to the sector. This was based upon limiting Treasury contributions to cover the servicing of medium-term commercial bank external debt and what is due on account of special funds. The second element of the strategy was the full or partial divestiture of selected enterprises. The most important efforts were the proposed partial privatization of ENTEL (the state telephone company) and of Aerolineas Argentinas, although these have not yet been carried out. Several smaller privatizations were concluded for smaller government equity holdings owned through Fabricaciones Militares, the armed forces' holding company. Through the Petroplan and the Plan Houston, the Government has relaxed the monopoly of the state oil company (YPF) on potentially oil-rich areas so as to permit private participation in oil exploration. However, additional measures to increase the role of the private sector are needed to increase further revenues for the Treasury as well as expand production.

44. These efforts permitted a substantial reduction of the value of transfers from the Central Government in 1988, to only 1.1 percent of GDP. However, insulating the central administration from the public enterprise sector has required an internal cross-subsidy system, whereby the profit-making enterprises (essentially YPF) have provided funds to finance the loss-making ones (essentially the railways). The continued viability of this system depends on guaranteeing sufficiently high real prices and reducing the deficit of loss-making enterprises. Strains have already appeared, as the Government has a perennial conflict between anti-inflationary targets and revenue needs; other unresolved issues (such as royalty payments to the provinces from the oil company that involve a strong subsidy element) also threaten the ability to generate sufficient resources. But the more fundamental problem with this approach is that it leads to greater distortions in prices and discourages investment in one of the country's highest return sectors, oil and gas, in order to maintain consumption in the railways, social security, and provincial governments.

45. A medium-term program of structural reforms would contain several urgent elements, which if enacted would markedly improve the basis for stabilization and growth:

* (a) **Prices**. The Government needs to maintain the post-July level of real prices of output, which implies future adjustments that keep pace with increases in costs. By 1990, tariffs in

the sector need to be sufficiently high relative to costs so as to generate a surplus on their noninterest current account equal to about 3 percent of GDP, thereby covering about half of their investment after interest expenses. Improvements over present levels could come from a combination of either expenditure reductions or real tariff increases.

* (b) As it adjusts prices, the Government should formulate the new pricing structure to reflect an efficient resource allocation, at least in terms of producer prices.

(c) **Royalties**. The issue of royalty payments to the provinces on petroleum and gas entails a heavy implicit tax on the sector to support provincial finances. The current reference price on which these royalties are paid dates from the early 1980s and is too high; it should be reduced to international levels as soon as possible.

(d) **Budgetary Control**. The Government needs to enhance budgetary control over the public enterprises. The respective roles of the successor to DEP (if any), the Ministry of Economy, the SIGEP and the various Secretariats in the Ministry of Public Works should be clearly delineated. The collective budget for the public enterprises, the Ministry of Economy's cash and commitment budgets and the financial statements should all be drawn up within the framework of a uniform information system.

(e) **Institutional Framework**. The market structure and regulatory framework of each sector requires definition so that enterprises are subjected either to the discipline of the market or well-defined regulations regarding pricing. For those enterprises designated to function as regulated monopolies, the regulatory framework should be clearly established in the law so as to permit maximum managerial discretion within a carefully elaborated legal framework, thus minimizing the imposition of political and noncommercial objectives.

(f) **Enterprise Restructuring**. Without enacting internal reforms within each enterprise to induce managerial responsibility, accountability, and autonomy, changes in the regulatory and competitive environment will not have their full desired effects. The Government should therefore continue and accelerate restructuring programs currently under discussion to strengthen management, improve personnel policies, and enhance financial controls. Management could be strengthened by restricting political appointments to boards of directors and/or a few senior-level posts as well as providing improved salaries, attention to qualifications in appointments and promotions, and sound training. These efforts could permit considerable reduction in expenditures and improvements in long-term efficiency.

* (g) **Procurement**. The Compre Argentino law should be modified to allow foreign competition in the bidding for contracts of the public sector, restricting the advantages of domestic firms to the level of the ad valorem tariff. This would effectively abolish the implicit subsidy system to the private sector and reduce the operating costs of the public enterprises.

* (h) **Employment Rationalization**. Redefining the role of public enterprises surfaces the need to examine employment levels and ways to reduce public employment without causing affected workers unnecessary hardships. The Government should examine programs of early retirement; in the railways alone, one-third of the workforce is over the age of 55. It should also consider programs of severance pay and/or programs of income maintenance for laid-off workers in accordance to their tenure with the enterprise.

(i) **Cross-Subsidies**. The Government needs to redefine the mechanisms and degree of cross-subsidization among the public enterprises and needs to establish a transparent scheme for resource transfers. If the Government wants to subsidize selected loss-making enterprises, it should transfer these to the budget so that these expenditures are annually subjected to the budgetary review process. Highest priority should be given to reducing the demand for expenditures in Ferrocarriles, which absorbs most of the cross-subsidies and accumulates losses of nearly one percent of GDP annually.

* (j) **Divestiture**. The Government should accelerate planned privatizations and review for possible inclusion remaining enterprises and activities that could be privatized. In the process, it should establish clearly defined legal procedures to ensure an unbiased selection of private investors and competitive bidding.

Federal-Provincial Relationships

46. Prior to the new coparticipation (i.e. revenue sharing) law passed in late 1987, the distribution of federally-collected revenues had become subject to political discretion. This was the culmination of years of unclear rules over revenue sharing. In addition, provincial governments had for years relied on provincially-owned public banks to fund short-term deficits; the provincial banks could in turn appeal to the Central Bank for rediscounts whose repayment would subsequently be capitalized. The system of irregular and highly politicized transfers invited provincial administrations to increase spending with minimal concern for revenues. It is no surprise that provinces have been the largest single source of increases in noninterest expenditures of the consolidated public expenditures. No less important, provinces have made too little effort in improving their own revenue collections, which have languished over the last decade.

47. The new coparticipation law that took effect in 1988 was designed to change this incentive framework. It increased the provincial share of coparticipated revenues to 57.5 percent, and set clear limits on discretionary contributions from the Treasury to a maximum of 1 percent of GDP. The law also provided a political mechanism for secondary distribution of the resources. The approval of the law was one of the crucial elements for the success of the "separation" strategy pursued by the previous administration. This agreement was also reinforced by the closing of the rediscount window for the provincial banks at the Central Bank, which had been an important source of deficit finance, particularly in 1987. However, the arrangement has proved tenuous. The provinces, unable or unwilling to increase revenues and reduce expenditures, requested and were granted additional funding twice in 1988.

48. A program of structural reform would transform the provinces into agents of development which generate surpluses that they can invest wisely to increase total provincial product, thereby augmenting their future tax revenues. Specifically, this would include:

* (a) Improving the information on provincial public finance. A first step in improving public sector finances for general macroeconomic planning should be to require the provinces to report existing data on their finances to the Central Government. Provincial governments should be required by law to report to the Treasury expenditures on a cash and commitment basis every quarter as a condition for receiving their share of revenues under the coparticipation agreements. The Treasury should establish an efficient information system for the collection and analysis of the provincial budget data. Fourth, information on the provincial banks' net asset position with provincial governments should be reported regularly to the Central Bank;

* (b) The Government should declare its intention to maintain the current revenue sharing law, halt any further transfers, and oppose any additional appropriations in Congress during the budget year;

(c) The Federal Government should work with the provinces to increase their own revenues, reduce expenditures and increase their efficiency. Revenue collection and personnel policy is of highest priority in the provincial governments. The Government could design an incentives system as part of future revenue sharing laws or investment financing from external credit sources to encourage these provincial efforts; and

* (d) Disallowing Central Bank financing to provincial banks as a means of financing the provincial governments, including reintroduction of overdraft facilities and rediscounts.

Finances of the Social Security System

49. The social security system has had a strategic role in defining the shape of Argentina's public finances. The aging of the population has caught up with the system; the ratio of the economically active population to the number of retired beneficiaries fell by 13 percent between 1970 and 1985, and is expected to fall even further. At the same time, the pension program had virtually no earnings from invested capital, since the surpluses of the early years of operation of the system had been consumed by the inflationary episodes of the 1960s and 1970s. To make matters worse, the heavy burden of wage taxes--55 percent of net wages--has led to widespread evasion and under-reporting; moreover, inadequate attention to revenue record-keeping has permitted widespread abuse of potential benefits.

50. The initial reaction to the crisis consisted of arbitrary reduction of benefits and increasing recourse to transfers by the central administration. From their legal level of between 70 and 85 percent of salary, payments have fallen to under 40 percent of salary in recent years. In 1987-88, the Government undertook several revenue measures, with the dual objective of gradually restoring benefits to their legal level and insulating the central administration from transfers to the social security system. The measures were comprised of increases in wage-related contributions to the social security system and the earmarking of taxes on certain goods provided by public enterprises (gasoline, telephone, electricity).

51. The additional revenues allowed the gradual closing of the financial gap in the social security accounts in the very near term. However, projections of the medium-term deficit at current benefit levels imply that either contributions must keep rising, or that the revenues from the earmarked taxes on goods must rise in real terms, or both. However, the scope for either of these measures is virtually nonexistent because wage taxes for social security and other programs are already very high. A continued increase in earmarked revenues would imply a continued undesirable increase in relative prices of the goods to whose prices the taxes are linked. Adjustments aimed at mobilizing more resources for the current system--such as increasing formal-sector employment and reducing the current transfer of 10 percent of social security income to the health insurance fund for retired persons--offer no hope for increasing system revenues.

52. A program of structural reforms might focus on constructing a different pattern of benefits that would sharply reduce pension obligations at higher income levels to substitute for entitlements that will be unsustainable in the future. Benefits could be reduced in a way that could actually increase both the efficiency and equity of the benefits package:

(a) Present low retirement ages--age 55 for women and 60 for men--drive the system towards deficit, even at high quota rates. There is no alternative to reducing benefits. One option is to increase the retirement age by 5 or 10 years so that years of contribution would rise relative to years of receiving benefits; alternatively, the Government could offer substantially lower benefits for an optional retirement at the earlier ages;

(b) Reduce the rate of salary replacement from 82 to about 40 percent; alternatively, the government could permit a low, basic rate of salary replacement with voluntary purchase of additional coverage. This should be implemented with a transition period to account for acquired rights; and

(c) Use an extended salary base for calculating pension rights in lieu of the last salary--for example, a 10 year real average.

D. Structural Reform Program: Monetary Policy and Finance

53. For decades, the financial system was accustomed to sequestering private financial savings to fund the public sector deficit. As part of this process, the Central Bank became over time a vehicle for channelling subsidies from private commercial banks to the general government and to the inefficient public banks. The Central Bank used forced investments as a way of tapping into depositors' funds. These were then channelled to: (i) the general government via payment of expiring public sector bonds and balancing of a short-term float to the social security system administered through the commercial banks; and (ii) via rediscounts to money-losing public banks, most notably the National Mortgage Bank (BHN), the National Development Bank (BANADE), and to provincial banks. These institutions in turn channelled substantial subsidies to middle-class loan recipients, industries and other activities as well as some provincial governments. Losses in these intermediation activities have become the quasi-fiscal deficit of the Central Bank.2/

54. The consequences of this structure have been twofold. First, the scope of the Central Bank to pursue anti-cyclical stabilization has been tightly constrained by the need to secure financing for the nonfinancial public sector deficit and its own deficit. By adding a source of expansion to the monetary base, the operating losses of the Central Bank complicate the difficulties of managing monetary aggregates in the presence of exceptionally narrow austral-denominated financial markets. Second, the financing requirements of its own quasi-fiscal deficit have added to the burden on domestic savings produced by the public sector deficit. The resources drawn from the commercial banks have gone to finance relatively inefficient public-sector activities, leaving the private sector starved of resources essential for working-capital and commercial finance. The scarcity of such finance has contributed to Argentina's disappointing real economic performance over the past 15 years.

55. Both the interest rate rules for Central Bank assets and the inherent quality of those assets have contributed to the Central Bank's deficit. The external component is the interest paid on external obligations (incurred as part of the 1981-87 debt refinancings) less interest

---

2/ This is defined as comprising the operating loss (or profit) of the Central Bank on account of intermediation operations, plus any adjustments necessary to take into account the fact that some rediscounts issued to the financial system and other entities may never be recoverable, i.e., they represent subsidies that should be part of the NFPS budgetary allocation.

earned on gross international reserves. The domestic component is the loss (or profit) on account of domestic operations. Most Central Bank assets are rediscounts to failed financial institutions or to public banks such as BHN and BANADE, which are unable to service the debt; the interest on these assets has been set at inflation plus some slight mark-up. Nonmonetary domestic liabilities are dominated by forced investments from commercial banks, which pay market interest rates. During periods of high real interest rates, the domestic component of the quasi-fiscal deficit tends to widen; for example, in December 1988, the average rate of interest on assets was 8.2 percent per month while that on liabilities was 11.4 percent. These losses also include provisions for bad or nonperforming loans.

56. A program of structural reform to end inflation and revitalize the financial system must limit the power of the Government to fund its deficit through capturing resources from the financial system. The best way to do this after the failed experiences of the Plan Austral and Plan Primavera is to limit the powers of the Central Bank to create money as well as sever the ties of intermediated cross-subsidies between the private commercial banks and the public banks. While this would depend heavily on the choice of stabilization program, the most effective structural reforms are those implicit in the program outlined above, including:

* (a) Strict limitations on the power of the Central Bank to create money through reforms that would tie money creation to increases in international reserves.

* (b) Making the Board of Directors fully independent of the Government (i.e., removable only through an impeachment process for misconduct, and thus able to deny commercial-bank and government credit requests without fear of dismissal).

* (c) Recapitalizing the Central Bank through exchanging existing liabilities for new longer-term liabilities; this would end the Central Bank's current intermediation nexus that creates the quasi-fiscal deficit.

* (d) The external component of the quasi-fiscal deficit could be substantially reduced by transferring the foreign liabilities associated with the 1983-87 commercial debt restructurings (including the "new money" operations) to the Central Government. While this has no effect on the combined public sector deficit, it does make the operations of the Central Bank and nonfinancial public sector more transparent, and facilitates the conversion of the Central Bank into a stronger monetary authority.

And for the financial system:

(e) The net flow of all Central Bank credits to the BHN, BANADE, and other public banks should cease immediately. This, together with the reforms of the Central Bank, would sever the intermediation relationship between the Central Bank and the public banks. It would also imply that restructuring plans already contemplated for these banks would have to be accelerated or the institutions would have to be closed.

E. Structural Reform Program: Trade and Industry

57. Argentina's postwar experience of persistent macroeconomic instability and secular decline was preceded by the closure of its economy to foreign trade. Prior to the Great Depression the share of imports in GDP kept close to 50 percent. This indicator of openness fell to 5 percent at the end of World War II, and again in the mid-1950s. Since then, the import share has remained near 10 percent. Exports as a share of GDP, while fluctuating due to weather and domestic demand conditions, have followed the general downward trend of imports.

58. Inward-oriented trade and industrial policies initiated during the 1930s were maintained and intensified in the 1950s, a time when other countries then at Argentina's level of development were removing their external trade barriers and taking advantage of the rapid expansion of international trade. Successive governments opted for import-substitution through import restrictions hoping to foster investment and productivity growth through accelerated domestic industrialization. Instead, the economy became more dependent on selected machinery imports for industrial growth and on agricultural commodities for export growth. Recurrent balance of payments crises frequently cut short industrial expansion and led to recession because imports vital during the second stage of import substitutions for expansion became unavailable. Total factor productivity in the nonagricultural sectors (excluding government) has grown at less than two thirds of the rate achieved in the agricultural sector, and during the 1970s and 1980s, total factor productivity in industry has actually declined. While the share of industry in GDP has grown substantially in neighboring economies over the last decades, the industrial sector in Argentina remains at about the same level as four decades ago.

59. The closure of the Argentine economy, instead of reducing macroeconomic instability, has contributed to the secular increase in the inflation rate and to the instability of the real exchange rate. The share of exports and imports in GDP turned out to be less important for the country's exposure to trade shocks than the mechanisms that enable the open economy to cope with them. A price elastic aggregate import demand and capital flows tend to absorb a good part of the shock impact on the real exchange rate and thus on relative prices, while the discipline of intense competition keeps relative price changes from turning into a source of inflation. These mechanisms have been weakened. Increasing the import share of non-substitutable primary inputs and intermediates rendered aggregate import demand prices inelastic; intermittent balance of payments crises impaired the country's access to external credit; and the creation of sheltered domestic markets fostered a pricing behavior that has added an inertial component to inflation. Moreover, pricing in entrenched oligopolies undisciplined by competition has probably acted to accelerate inflationary impulses through expectation-based mark-up pricing.

60. Also, trade and industrial policies undermined stability by adding to the deficit. Trade tax revenues have been an unstable part of the government budget. Revenues foregone, through industrial promotion incentives and overcharging on public sector purchases resulting from the "buy national" obligation of public procurement, have added up to 5 percent of GDP to the deficit.

61. Beginning in early 1987, the Government began to transform the consensus that the import-substitution strategy had run its course into an active reform policy. To have started and continued the reform is a major achievement because of the long-standing resistance of powerful vested interests. The new reform emphasized free trade status for exporters early in the reform process, coupled with a phase-out of various specific export promotion measures; a negotiated sector-by-sector approach to import liberalization; and an attempt, though not always successful, to maintain a competitive exchange rate for trade transactions. As a consequence, the average production-weighted tariff was lowered from 43 to 28 percent; quota coverage was reduced from 30 to 18 percent; and discretionary import licensing procedures were provisionally circumscribed.

62. A program of structural reform that would support sustained price stability and growth therefore requires that trade and industry reform not be postponed. In addition to the termination of the industrial promotion regime and of export taxation cited above, a program for structural reform would include:

(a) The replacement of all quantitative restrictions that were exempted from the 1987-88 reform with *ad valorem* tariffs, and a subsequent tariff adjustment to bring the rates within the general tariff band.

(b) The removal of specific tariffs; only for seasonal products should specific tariffs at non-prohibitive levels be considered.

(c) A narrowing of the tariff band and reduction in average tariff levels; the band should be narrowed as soon as possible from 0-40 percent to 10-40 percent; a program should be announced to reduce the band to 10-20 percent in two years to reach an average of about 15 percent.

(d) Removal of export license requirements.

F. New Sources of Growth

63. As President Menem indicated in his inaugural address, there is every reason to believe that the sacrifice and work implicit in an ambitious program of deep-seated reforms would be rewarded with economic recovery and sustained growth. Argentina's productive potential in abundant resources, agricultural lands, and skilled labor is perhaps unsurpassed in Latin America. A sound program of consistent reforms could not only put people back to work, but create new jobs at a much higher rate than during the last decade.

64. With macroeconomic stability resulting from a comprehensive reform of the public sector and a program of structural reforms that removes price and other distortions, it is entirely possible to foresee a rather rapid recovery from the recession of 1988-89. Output could conservatively be estimated to expand at rates in excess of 3.0 percent annually for the 1990-94 period. If the Government's reform program were especially strong --extending to trade, finance, and the real sectors (energy, agriculture and industry)--it is entirely possible that foreign savings of Argentines would be attracted back, allowing for even more rapid rates of growth. Argentine holdings abroad are estimated to be about US$45-60 billion--about the same size as its foreign debt; if only the annual income on these assets were to return, these flows could provide savings to fund a much more rapid and enduring recovery.

65. Stabilizing the economy--the most immediate priority--will permit the recuperation of domestic savings and investment. If stability is to be sustained, savings--led by the public sector during the initial phase of recovery--must roughly double from their present extremely low rates by 1994 to finance increased domestic investment in productive activities. Private savings could well be a driving force financing growth in the financial system to channel savings to highly productive investments and inspire confidence among private savers and investors that their efforts will be rewarded.

66. The private sector would be a leading force during this period, increasing investment in response to new growth opportunities, exports and efficient import substitution. Such improvements in both saving and investment depend critically upon private sector confidence in the macro and sectoral policy framework and on the stability of these policies over time.

67. Exports would become one leading sector, based on industrial goods and nontraditional and processed agricultural goods. Assuming a consistently competitive exchange rate, there is also some room for the growth of agricultural exports beyond the growth of consumption in industrialized countries, as Argentina could readily regain the market share it has lost in recent years due to domestic supply factors. Nontraditional exports, led by new capacity in chemicals, plastics, machinery and transport products, could grow at over 7 percent annually for the period. Industry too could be expected to grow in response to new demand for exports and efficient import-substitution activities. Finally, new investment would also provide a strong impetus to growth. If these possibilities come to pass, they could indeed mean that the sacrifice and work of today would unleash the abundant productive potential of the Argentina of tomorrow.

CHAPTER I: STATE-LED GROWTH AND INFLATION

A. Background

1.01 State-led growth in Argentina has now come irretrievably to an end in a wave of macroeconomic instability and inflation. In the late 1940s, the State began to expand its role in the economy--to become ever more important as a net borrower of funds, investor, and source of subsidy for favored activities and interest groups. Subsidies took the form of tax exemptions and hidden transfers through the financial system. These eroded the tax base and with other factors compelled the state to rely increasingly on borrowing and the inflation tax. Large and chronic fiscal deficits became a central feature of the Argentine economy, and during the mid-1970s and early 1980s they exceeded 10 percent of GDP. State-led growth could therefore be sustained only through foreign borrowing and money creation.

1.02 The other dimensions of this state-led model were high protection for domestic industry and concentrated formal labor markets. High nominal tariffs, import prohibitions, quantitative restrictions, and opaque subsidies to industry (such as the tax exemptions in industrial promotion legislation and buy-Argentina policy of state enterprises) insulated the sector from international competition. Not only did the resulting price structure channel investment into activities with low productivity, it also reduced the importance of trade, enhanced the oligopolistic price setting power of industry, and made the economy more vulnerable to external price shocks. The highly organized labor sector succeeded in capturing some of the rent from the industrial system--through demands that the public sector increase employment as well as wages. This was possible as long as the economy was expanding. As the economy slipped into long-term stagnation, the system produced conflicts over income shares with sporadic destabilizing effects on the price level. These policy interventions and structural characteristics of the economy led to declining trends in the productivity of labor and investment that would heighten the country's vulnerability to financial instability.

B. Growth, Investment and Savings

1.03 The state-led model began to exhaust itself in the mid-1970s when the long-run growth path of the economy--as measured in the 10 year moving average of GDP growth--turned sharply downward (Figure 1.1). Investment, traditionally 20-23 percent of constant price GDP, began a sustained fall to 11-13 percent. While investment by the public sector has stabilized at low levels only slightly less than the early 1970s, the private sector has recovered to only 50-60 percent of those earlier levels. The investment rates of the economy are now so low that they are barely sufficient to replace depreciating capital stock, portending low growth for some years to come.

1.04 The savings performance of the economy also reflects serious structural problems. National savings rates have fallen from about 20 percent of GDP in the early 1970s to under 10 percent. Trends in public and private savings provide a clue to the erosion in national savings. Public savings fell steadily from 1970 to 1975, and then again from 1977 to 1982, with only partial recovery thereafter. Public savings before interest

Figure 1.1
Investment Savings

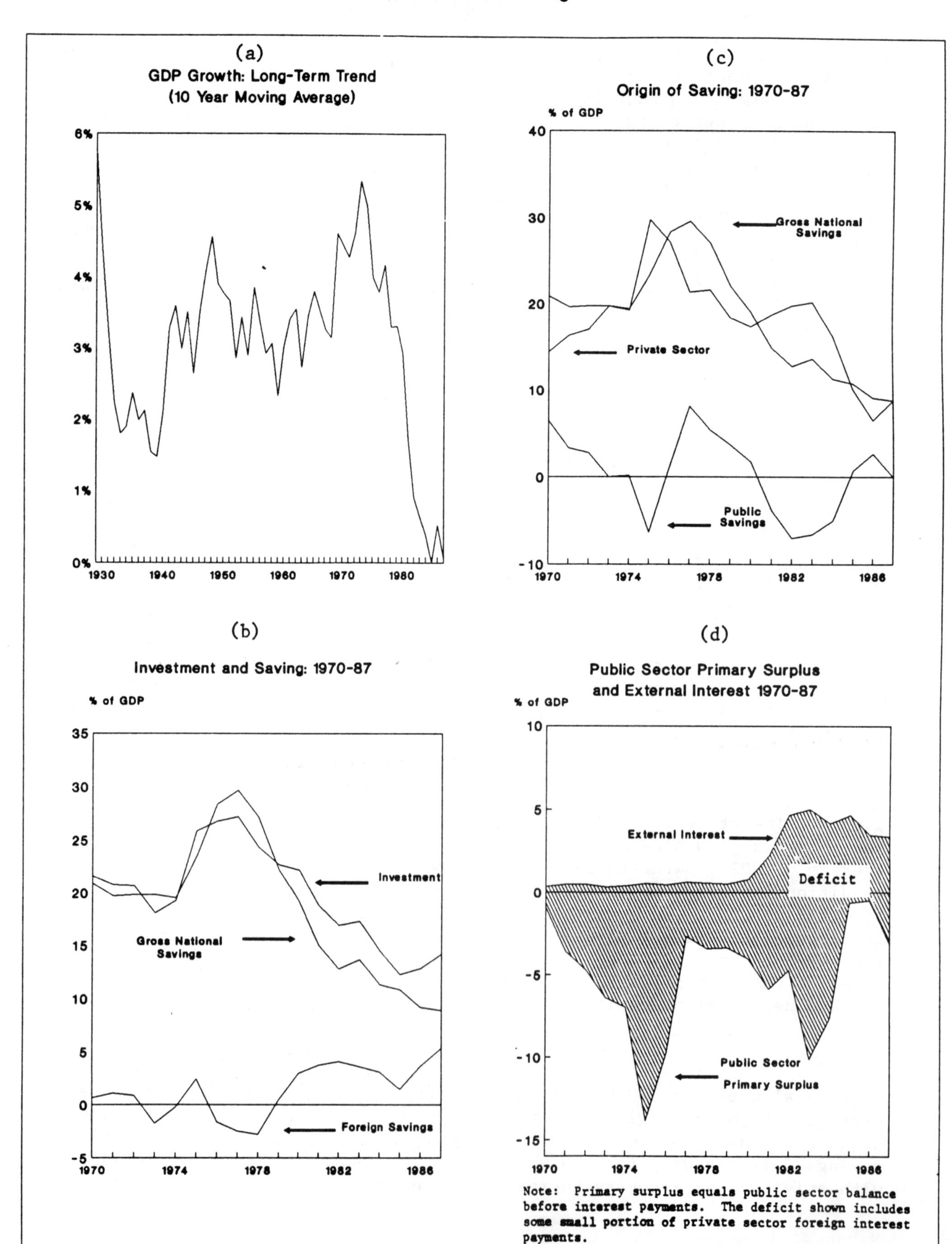

Note: Primary surplus equals public sector balance before interest payments. The deficit shown includes some small portion of private sector foreign interest payments.

payments had, by 1987, recovered virtually to the same level as those of 1970-72, about 5 percent of GDP, but these were still insufficient to generate a primary surplus. This underscores both the difficulty imposed upon Argentina's public sector because of the heavy weight of its external debt as well as its lack of sustained progress in controlling the non-interest deficit. Private sector savings continue to be a major problem; they are a little more than half their early 1970s levels. Economic conditions have led Argentines to save vast amounts abroad and, aside from understating private savings in national accounts, these resources abroad are unavailable for job-creating investment in Argentina.

1.05 The adjustment process which depressed national savings and investment in the 1980s entailed two related problems: adjusting to abruptly increasing external transfers which diverted domestic savings into foreign factor payments, and the struggle to control severe fiscal imbalances. The fiscal imbalance was attributable in part to the internal transfer problem associated with suddenly increasing debt service, but even more was a reflection of the inadequate control of public finances that had plagued the economy in the 1970s and indeed worsened in the 1980s. As the decade wore on, macroeconomic instability--manifest in endemic inflation--and policy-induced distortions have become the main impediments to the recovery of savings and investment.

C. Adjustment and External Transfers

1.06 The origins of the external component of the adjustment problem are to be found in the macroeconomic policies of the late 1970s. Beginning in 1976, macroeconomic policy relied on an appreciated exchange rate to combat inflation rather than fiscal adjustment. The Government liberalized the capital account in end-1976, and monetary policy kept real interest rates high to attract foreign capital inflows. The policy was continued with the "tablita" experiment beginning in January 1979 which, in effect, produced assured dollar rates of return in the domestic market and thus provided even greater incentive for massive foreign borrowing. The shift in relative prices through the sustained real appreciation of the exchange rate encouraged import growth while exports stagnated. The current account swung into deficit in 1979 and reached 7 percent of GDP in 1980-81. The foreign debt rose from US$8.2 billion in 1976 to US$27.2 billion in 1980 and to US$45 billion by 1983 (Figure 1.2). Whereas use of foreign savings had been relatively low in the early 1970s--and indeed was substantially negative in five of six years between 1973-78--macroeconomic management after 1978 discouraged domestic savings and the economy became dependent on external credit. The lack of consistency between monetary targets and the fiscal deficit doomed the economy to an unsustainable external position.

1.07 As the end of the Videla government approached in March 1981, the realization grew that the "tablita" experiment could not be sustained; private net foreign borrowing slowed and then turned into capital flight in 1980. The Government tried to sustain its exchange rate policy through accelerated public borrowing in 1980-81, but the run on the peso became severe. Private capital outflows and reserve losses between 1980-83 totalled US$23 billion. These outflows implied that 61 percent of

Figure 1.2

(a)

Argentina: Real Exchange Rate
1987 = 100

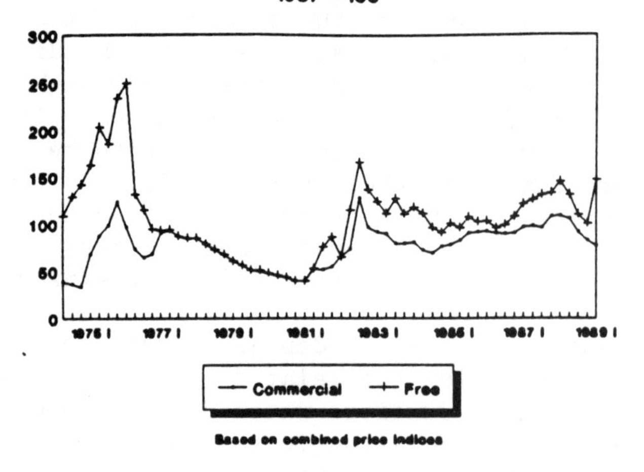

(b)

Trade Balance and Current Account
1970-1988

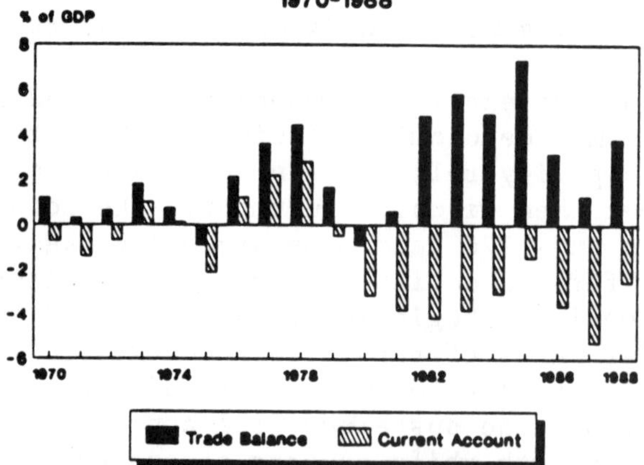

(c)

External Debt, 1975-1988
(US$ Million)

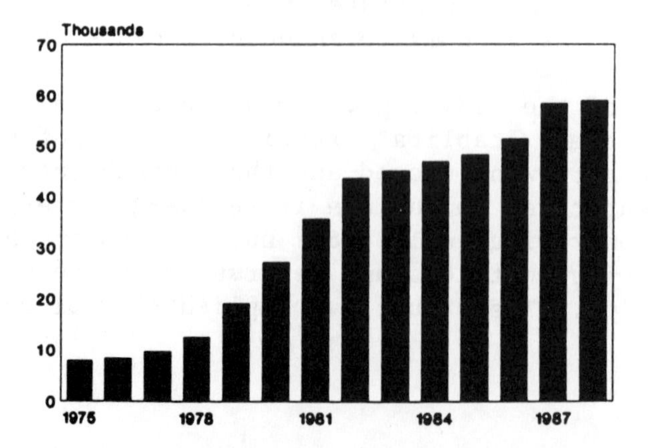

resources supplied to the Argentine economy between 1979-83 financed capital flight rather than the transfer of real resources.1/ Only 39 percent of borrowing financed the transfer of real goods and services in the form of the current account deficit.

1.08 The collapse of the 1978-81 model meant that the foreign debt could no longer be rolled over but had to be serviced, which required the country to reduce expenditures relative to income to generate the foreign exchange necessary to service the debt. The Government had no choice but to begin what would be a series of devaluations. The devaluations imposed an immediate real wealth loss on those with foreign liabilities, amounting to 15 percent of GDP in 1981 and 51 percent in 1982.2/ However, the public sector did not (or could not) reduce spending because, among other reasons, the South Atlantic war broke out, and it had to shoulder the increased interest burden of the public sector. This meant an explosion in the deficit financed primarily through the inflation tax.3/

1.09 For the private sector, the devaluations had an immediate effect in reducing spending, and drove many firms, long accustomed to low or even negative real interest rates and therefore heavily leveraged, into default. The Government in mid-1982 took action to relieve the pressure on private debtors by "liquifying" debts through inflation. This took the form of setting nominal asset and liability rates at levels significantly below the rate of inflation and depreciating the exchange rate.

1.10 In retrospect, the policy course of the early 1980s involved a high cost that has cast its shadow over the entire decade. Rather than adjusting quickly to the reluctance of private markets to lend voluntarily into Argentina's policy environment in 1980, the Government chose to defend the appreciated peso through massive foreign borrowing. Rather than maintaining an arm's length distance in the private sector's bilateral negotiation over the terms of the private debt, Argentina essentially absorbed the private liabilities and used inflation to impose losses on creditors (including the publicly owned banks). Rather than adjusting public expenditures to accommodate higher interest payments, the Government financed accelerated spending through inflation, shattering any possibility of immediate recovery.

1/ Total sources of funds in 1979-83 included increases in external debt (US$32.6 billion), decreases in reserves (US$2.7 billion) and foreign direct investment (US$2.4 billion). This total supply of funds of US$37.7 billion was used to finance current account deficits of US$14.8 billion and increases in net foreign assets abroad worth US$22.9 billion.

2/ H. Reisen and A. Van Trotsenburg, Developing Country Debt: The Budgetary Transfer Problem, Paris: OECD, 1988, p. 60

3/ External shocks, while secondary in influence to domestic policy, also played a role in the form of rising real interest rates. Between 1979 and 1980, Argentina's interest bill quickly rose from 1 percent of GDP to 5 percent, and peaked at 8.5 percent in 1983.

D. Public Sector Deficits and Inflation

1.11 Public sector deficits were common in the 1950s and 1960s, but they became particularly large during the mid-1970s and early 1980s. Although their causes differed (as discussed below), their economic consequences were similar in that they eventually led to inflation.

1.12 The inflationary consequences of the deficit depend on the financing of the public sector and the willingness of the public to hold assets denominated in local currency, including money. The deficit of the public sector (including the Central Bank) can be financed through foreign borrowing, borrowing from the private sector, and the inflation tax. As long as foreign and domestic creditors are willing to finance the full deficit through holding Government paper, deficits need not create inflation.4/ However, at no time in the past two decades has this condition held, and so the Government has had to rely on compulsory borrowing from the financial system (in the form of high reserve requirements or forced investments with the Central Bank) and the inflation tax to close the public accounts.5/

1.13 A given level of unfinanced deficit can cause a higher rate of inflation when the public shifts its real balances out of austral-denominated assets in the financial system, reducing the base against which the inflation tax can be applied; this implies inflation (the tax rate) will rise in order to bring the public accounts into balance. Since demand for money balances falls as fears of inflation rise, the process is inherently unstable. The process becomes explosive when the demand for austral assets falls so low that increasing inflation rates reduces revenues from the inflation tax.

1.14 The link over the last two decades between public sector deficits and inflation has become more direct (Figure 1.3). In contrast to the period until the late 1940s--when price increases in Argentina were positively correlated with demand pressures emanating from high levels of economic activity, low rates of unemployment, and expansionary monetary policies--in nine of the recessions since 1949, economic downturns have been positively correlated with accelerations in the inflation rate. Inflation since 1970 shows a strong upward trend with three particularly severe episodes, 1974-76, 1982-85, and 1988-89. Each inflation episode has left the country on a higher inflation plateau than the previous one and with an apparent greater potential for instability.

4/ Strictly speaking, the Government can run a noninflationary deficit financed through seigniorage equal to the growth rate of the economy, provided money demand is stable. This is a relatively less important consideration in the Argentine context.

5/ The inflation tax is the inflation rate times the stock of net liabilities of the Central Bank (excluding credit to the Government).

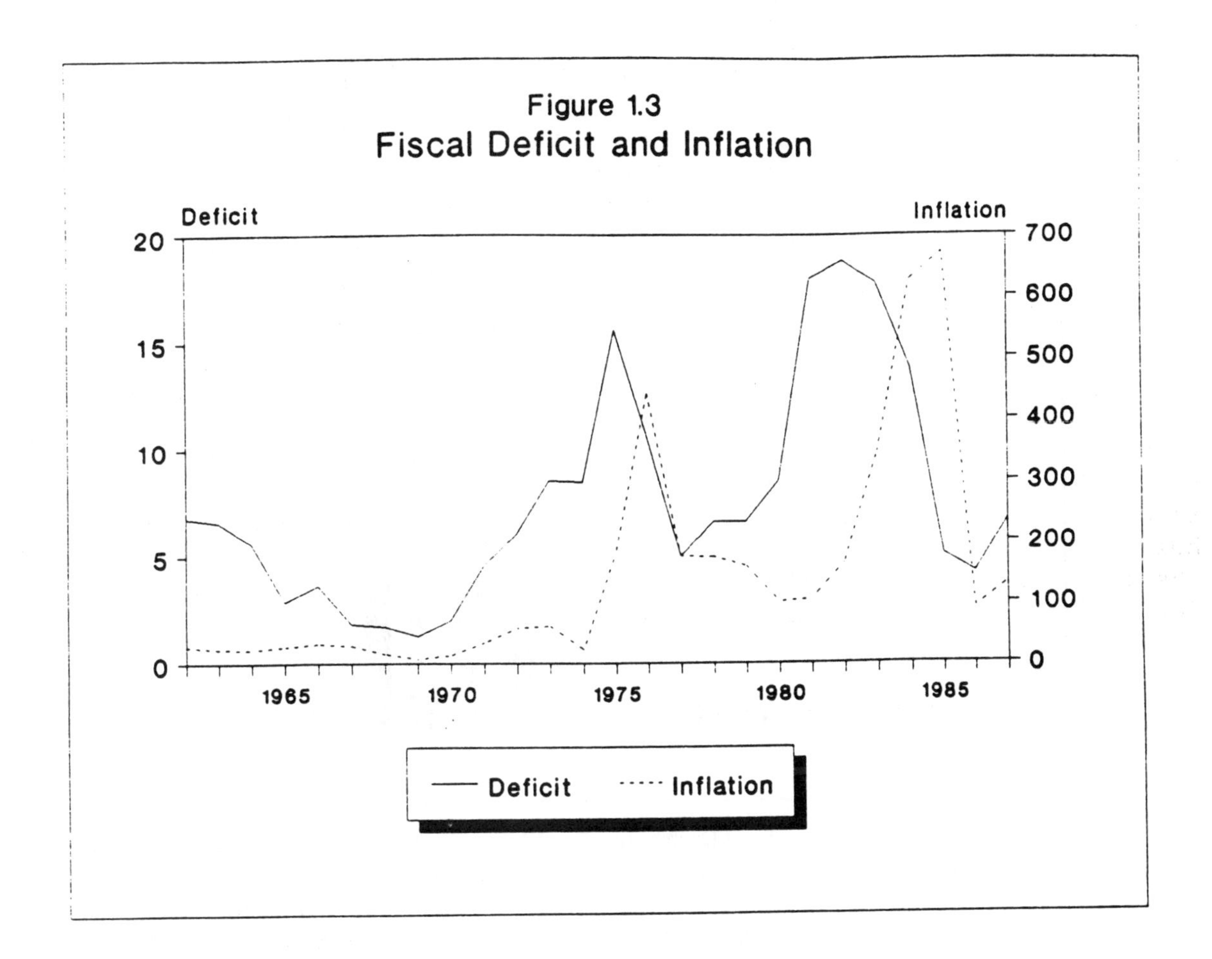

Figure 1.3
Fiscal Deficit and Inflation

The 1974-76 Episode and its Aftermath

1.15 High deficits emerged in 1973-75 after the second Peron administration sought to increase real wages, growth and public investment without undertaking concomitant revenue measures. In spite of renewed growth in 1973-74, public sector deficits became progressively worse and exceeded 15 percent of GDP in 1975. The Government resorted to Central Bank financing (Figure 1.4), and the inflationary impact was severe.

1.16 The new military government in 1976 sharply reduced public sector employment and the wage bill, improved public enterprise prices, and thus increased public sector savings. But because investment continued at very high levels and public enterprise accounts were deteriorating, the overall deficit of the public sector--while reduced by half relative to 1974-75--remained quite large, one-third higher than 1970-72. The deficit then grew steadily after 1977.

Figure 1.4

Financing Deficits

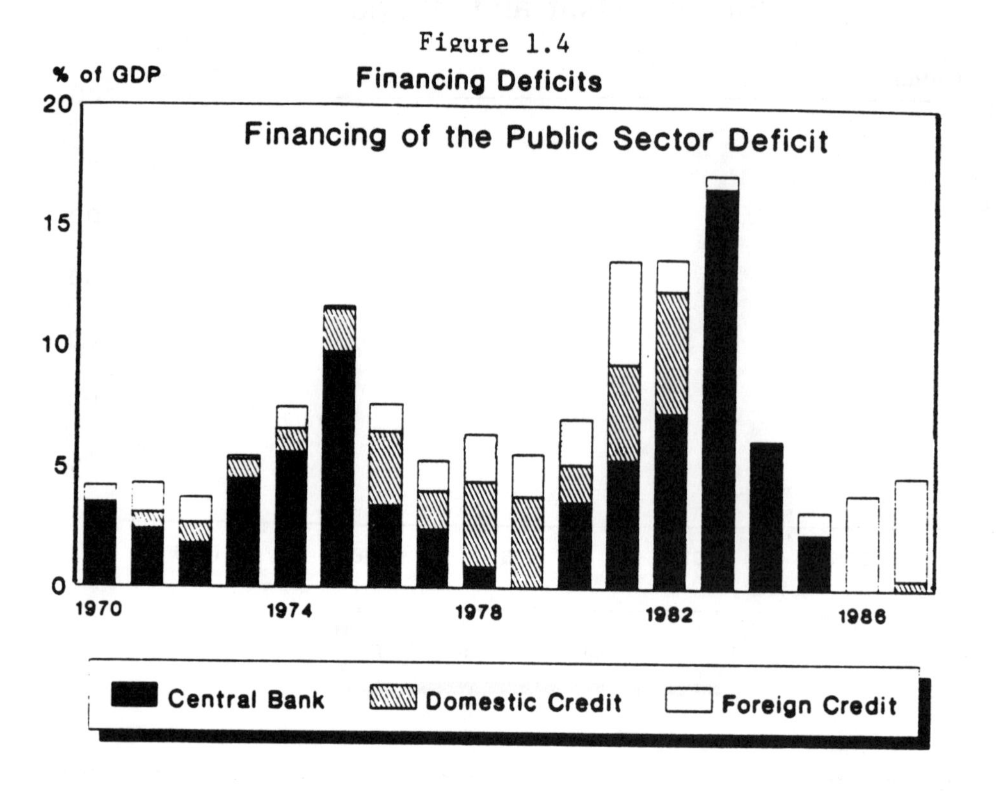

Figure 1.5

Falling Money Demand

RATIO OF M1 TO GDP

(Percent)

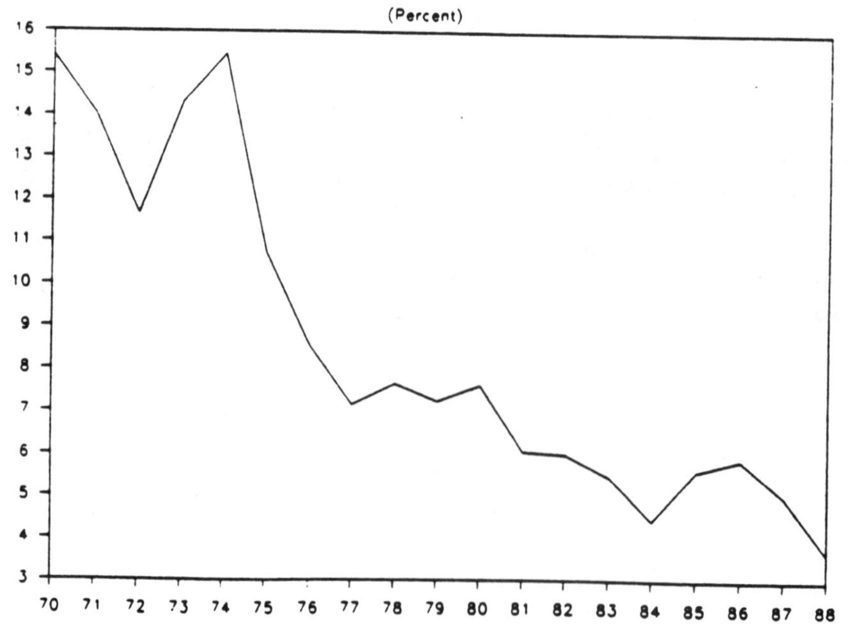

1.17 The availability of foreign and domestic finance explains why the Government had some scope for deficit expansion without immediately negative inflationary consequences (Figure 1.4). Borrowing domestically to finance deficits was viable as long as the public was willing to hold austral-denominated assets and the risk premium demanded for government paper was not excessive. The Government also relied on high reserve requirements in the financial system--about 45 percent--to ensure that Central Bank deficits would be financed. The appreciating exchange rate and, after 1979, official assurances to the private sector that the peso would not be devalued, encouraged the private sector to borrow abroad and then relend to the Government. This minimized the need for Central Bank finance.

1.18 The Government also benefited from relatively stable money demand. Even though the 1974-75 inflation had driven down real balances to a lower plateau, money demand remained around 7-8 percent of GDP during the period (Figure 1.5). This meant that seigniorage increases occurred on a stable base--and one that was high relative to later years--and so the Government could collect the inflation tax with less risk of unleashing an unstable inflationary process.

The 1982-85 Episode

1.19 After 1980, the internal transfer problem associated with the foreign debt drove up the overall public sector deficit. The Government took steps beginning in mid-1981 that led to a progressive takeover of the private external debt.6/ From 1979-80, when the public sector held about 53 percent of the total external debt, the public sector's share of total external debt rose to 88 percent by 1986 (Figure 1.2 (c)). By the end of this period, all but short-term commercial credit lines had been transferred to the public sector.

6/ To encourage refinancing of existing private debt as well as new foreign borrowing at a time when the macroeconomy was deteriorating, the Government--at the urging of foreign creditors and domestic borrowers--extended foreign exchange insurance worth US$5 billion to cover private renegotiation of external debt service payments and new inflows. Shortly thereafter, successive real devaluations more than doubled the real exchange rate, and the private sector's real debt was reduced at the expense of capital losses to the Central Bank. The exchange insurance program, initially terminated at end-1981, was renewed in July 1982 and then applied to more than US$10 billion. The average exchange rate for contracts coming due at end-1982 was 10 percent of the prevailing official rate. Since the Government did not have the foreign exchange to service the debt, the Central Bank assumed responsibility in November 1982 for all payment of the debt contracted under the program; as collateral, foreign creditors received a new government bond (BONOD) and promissory notes. These events transformed the responsibility for external liabilities in relatively short order. As described in Chapter IV, this became the origin of the external component of the Central Bank's quasi-fiscal deficit.

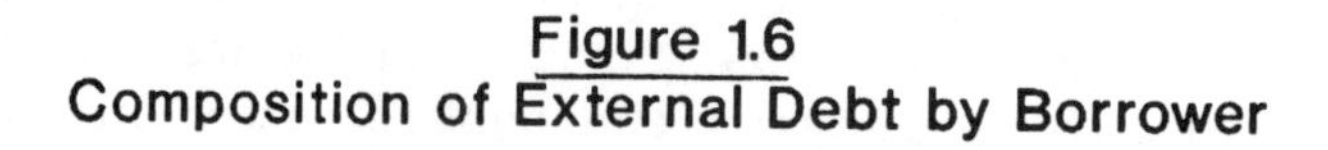

Figure 1.6
Composition of External Debt by Borrower

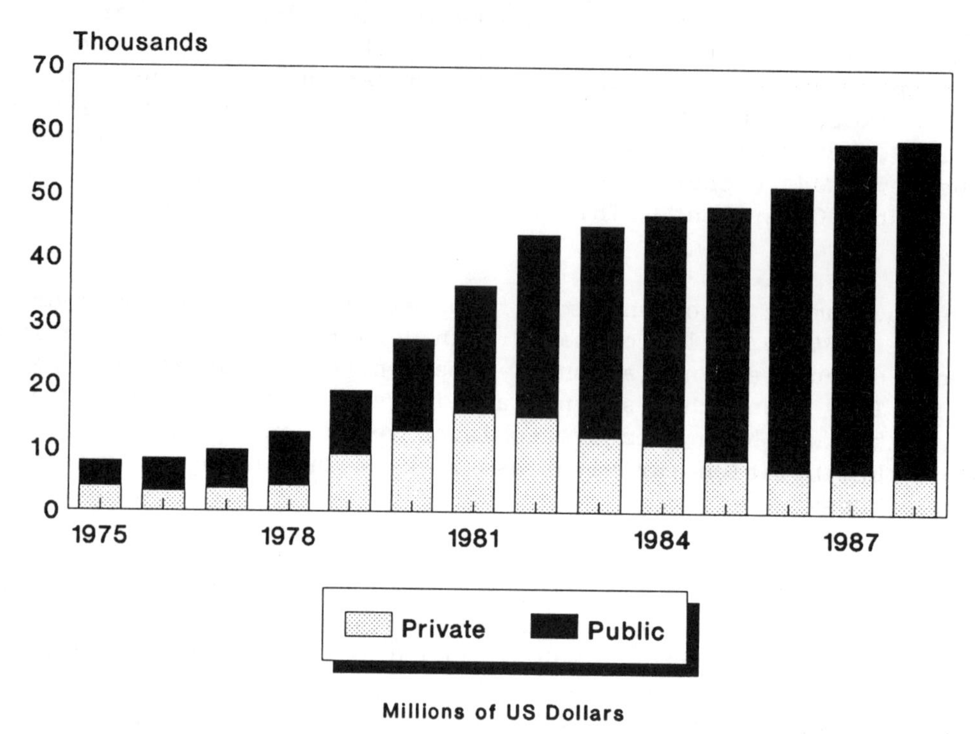

Millions of US Dollars

1.20 Interest payments for the economy increased substantially with the increase in the real domestic debt stock and high rates after 1979. Total interest payments rose from less than 1 percent of GDP in 1970-72, to 3.1 percent in 1978-79, to over 7 percent in 1980-83. By 1983, interest payments had come to comprise 15 percent of current expenditures of the public sector. The new debt burden meant that the Government had to generate a primary surplus (i.e. overall balance before interest payments) equivalent to the value of interest service, or seek refinancing of the difference and/or an increase in its net debt.

1.21 But the Government was unable to generate any surplus with which to pay its external debt service. As shown in Figure 1.1 (d), the primary deficit of the nonfinancial public sector actually worsened from 1980 to 1983 because of the effects of inflation in reducing real tax revenues (the Olivera-Tanzi effect), lack of buoyancy in the tax system associated with the tax expenditures in the industrial promotions law (see Chapters III and Annex Chapter VII), the inability to impose reductions on wages, and lack of control on expenditures as well as exceptional expenditures for the South Atlantic war. Moreover, the prices of public enterprises, used as a tool for reducing inflation and garnering foreign loans after mid-1978, continued to be low. To make matters worse, the budget process control continued to erode.

1.22 The Government, in the absence of adjustment in its fiscal accounts, and without financing alternatives, relied almost exclusively on Central Bank financing in 1982-83. This set off the second major wave of inflation. As with the previous episode, the rapid inflation drove down the demand for money as economic agents adjusted money balances and shifted their portfolios to escape the inflation tax (Figure 1.5). This meant that the inflationary process became inherently more unstable since even small fiscal deficits could now set off an explosive process wherein the inflation tax would not close the financing gap. This situation endowed the new constitutional government with a new, lower plateau of money demand that averaged between 3-4 percent of GDP, and eroded the base for a stable inflation tax. It also left economic agents with a skepticism about government instruments, high inflationary expectations and, with an open capital account, the capacity to adjust quickly to changes in expected inflation.

1984-89

1.23 Although inheriting an economy shackled with a huge foreign debt, large balance of payments and public sector deficits, and a budgetary process in disarray, the Alfonsin administration initiated several reforms that have had considerable effect on both external and internal disequilibria. By keeping the exchange rate broadly competitive since 1985, the current account of the balance of payments improved from an average of over 5 percent of GDP in 1980-83 to under 2 percent in 1988. The Government also made discernible progress in reducing the nonfinancial public sector deficit. Deficits of the nonfinancial public sector fell from over 16 percent of GDP in 1983 to about 5 percent in 1988. New tax measures, sporadic improvements in public enterprise pricing, and some adjustment in provincial public sector finance contributed to the improvement.

1.24 But policies were not always consistent, sufficiently forceful, or enduring to stabilize the economy. Contending political demands prevented sufficient action on the fiscal deficit relative to available financing, and 1987 particularly marked a severe setback to the gains in the previous two years. Moreover, insufficient attention was devoted to the structural problems that were the legacy of the previous decades: excessive private sector demands on public expenditures, a tax system badly eroded by past inflations and poor administration, many public enterprises that were overstaffed and inadequately managed, inadequate definition of taxing and spending responsibilities between the various sectors of government, increases in public sector employment (especially in the provinces), inappropriate use of the Central Bank as a vehicle to capture and spend financial resources, and a closed economy that channeled resources toward inefficient economic activity. In these circumstances, price stability proved elusive, and efforts to stabilize the economy ultimately failed with the hyperinflation of 1989, the subject of Chapter II.

CHAPTER II: STABILIZATION EFFORTS AND EMERGENCE OF HYPERINFLATION

2.01 Argentina attempted four stabilization programs in 1984-89. Each time inflation returned and surged to higher levels than on previous occasions. The failure of the latest of these, the Plan Primavera, illustrates the difficulties of stabilizing the economy without deep-seated reform of the public sector. The collapse of the Plan left the economy spiralling toward hyperinflation, a crisis in the financial system and deep recession. While the new program of July 9 has abated the worst aspects of the crisis, only rapid implementation of comprehensive structural measures to reform the public sector can return the economy to a path of sustained growth.

A. Stabilization Efforts from 1984 to 1987

The Plan Austral

2.02 After an ill-fated attempt to restore real wages and growth through expansionary fiscal policies in early 1984, and a tentative stabilization effort beginning in October 1984, the Government announced a comprehensive initiative in mid-June 1985, the Plan Austral. The program's objective was to break the "inertial" pressures arising from the effects of inflationary expectations on contracts and on asset demands. The measures included a wage-price freeze, tightened fiscal and monetary policy, and a fixed exchange rate following a steep devaluation.

2.03 The Plan Austral succeeded in establishing temporary price stability. Monthly inflation fell from about 30 percent to about 2 percent over the second half of 1985. The economy achieved some remonetization as the M1/GDP ratio rose from 4 to 8 percent in 1986. Partly because the fall in inflationary expectations increased the willingness to hold money and partly because of the monetary restriction, real interest rates turned sharply positive. Real GDP contracted 4.4 percent in 1985, but a rebound in consumption and investment in the fourth quarter of 1985 and renewed confidence and rapid expansion of monetary aggregates led to a strong recovery the following year. Investment rose in real terms in 1986 after five years of contraction, and real GDP rose 5.4 percent. The trade surplus reached US$4.6 billion due to improved agricultural production and prices, as well as to lower imports resulting from the stabilization program. The program paved the way for Argentina to reschedule its external debt with commercial and official creditors in August 1985.

2.04 The fundamental sources of instability in public finances, however, were not decisively addressed. While the cash deficit of the nonfinancial public sector was reduced from about 7 percent of GDP in the first half of 1985 to almost zero in the third quarter of 1986, this was due almost exclusively to the operation of the positive Olivera-Tanzi effect and the use of temporary measures (such as the forced savings scheme) rather than to fundamental reform. Increases in the real prices of public enterprises relieved the pressure of their deficit on the Treasury, but only for a limited time, as inflation eventually overtook them. This cycle was repeated in 1988-89. The perceived lack of fundamental improvement in the fiscal accounts contributed to a progressive loss of confidence and eventual refueling of inflationary expectations. This was

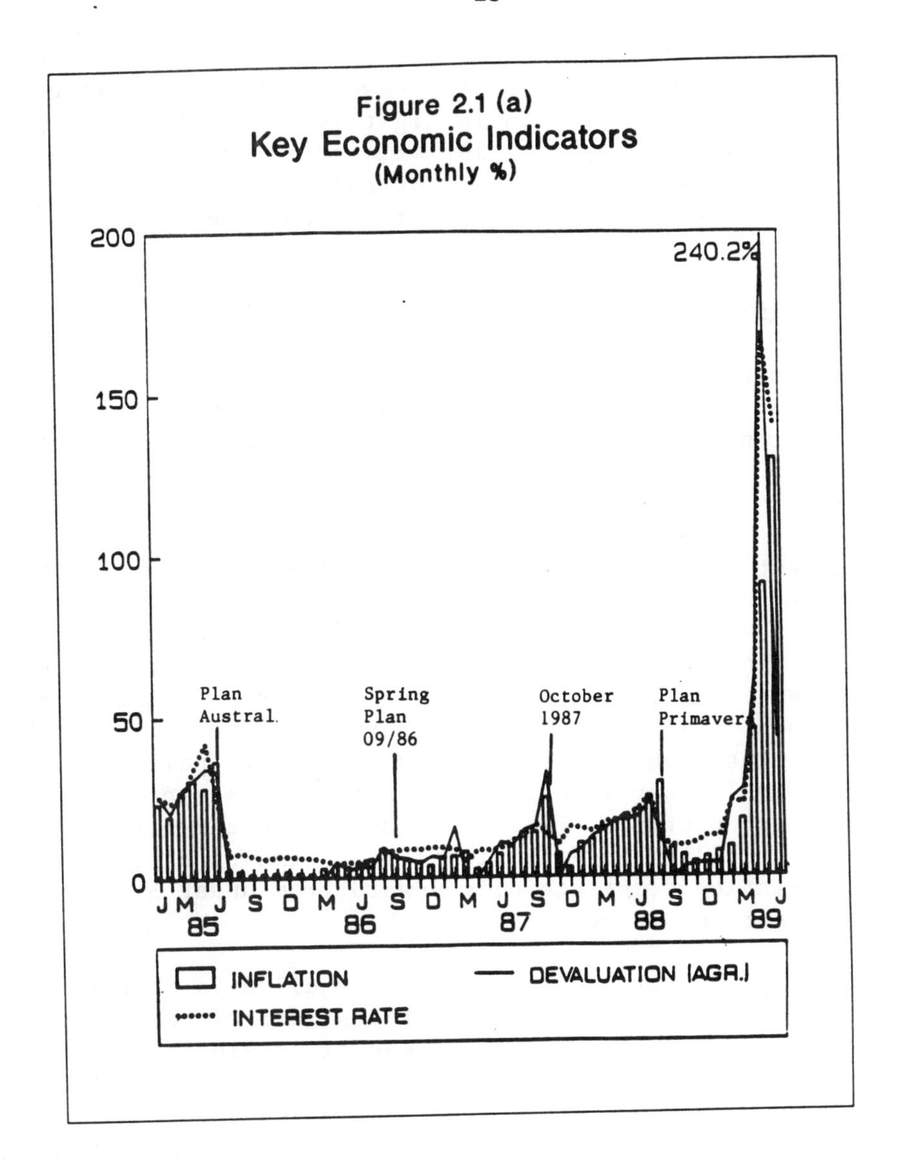

Figure 2.1 (a)
Key Economic Indicators
(Monthly %)

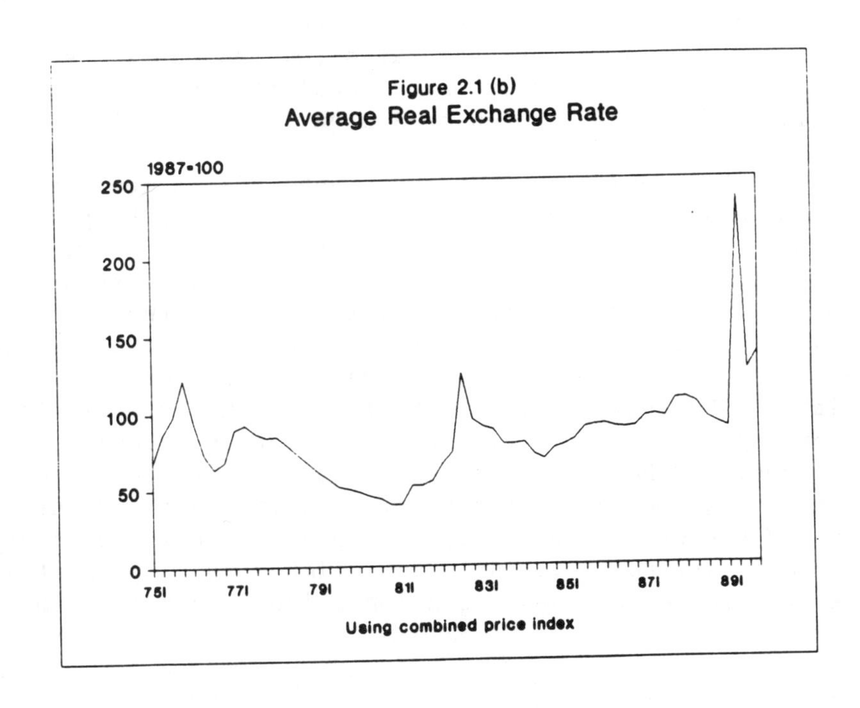

Figure 2.1 (b)
Average Real Exchange Rate

compounded by a large money-supply increase in the second semester of 1985 that proved inconsistent with single digit monthly inflation in 1986. Other contributing factors included the inflationary effects of adjusting relative prices that had been frozen "out of balance" as the price freeze was lifted, as well as some wage-push inflation, especially in July 1986.

2.05 These factors, together with the strong demand recovery in 1986, once again re-ignited inflation. By August, monthly inflation had crept up toward 8 percent. The Government put in place a new program based on a tight monetary program and price guidelines, but took no action on the fiscal situation, which was seen as essentially sound. But despite tight money and higher nominal interest rates, real wages and production remained relatively high. Both fiscal and monetary accounts deteriorated in the fourth quarter.

2.06 The fact that the large fiscal imbalance evident in the last quarter of 1986 remained in the early months of 1987 cast a pall over the credibility of the program and threatened the sustainability of the tight money policy. Many firms, perhaps anticipating the reimposition of price controls, appear to have overadjusted their prices in January and February. With the approach of legislative and gubernatorial elections scheduled for September 1987, the Government announced a new wage policy in early May that permitted public wages to increase while private prices would be subject to government control. Fiscal policy during the first 9 months of 1987 was substantially more expansionary than the same period the year before, and inflation crept steadily upward from under 3 percent in April to over 25 percent in October.

2.07 Demand pressures and financial disequilibria therefore remained high in 1987 because of the increasing fiscal deficit over the period leading up to September. The spending of several provincial governments, municipalities and certain public enterprises far exceeded their budgets and more than consumed the central administration's surplus. In addition, the Central Bank's quasi-fiscal expenditures sharply increased in the form of rediscounts through public banks that subsidized middle class housing, industry and other favored sectors. To make matters worse, falling world grain prices and reduced export tax rates lowered public sector revenues. The demands of the public sector on national savings accelerated, as the combined deficit for 1987 mounted to a high level of 8.3 percent of GDP.

2.08 In October 1987, the Government announced a new emergency effort to control inflation. In contrast to the September 1986 plan, the program included some new _ad hoc_ fiscal measures, but lacked a serious attempt to reduce Government expenditures; the program also reinstated a price freeze. It succeeded in bringing down monthly inflation for the next two months. However, with the price thaw in the first quarter of 1988, inflation began to accelerate, and by June 1988 the situation appeared to have become uncontrollable. Revenues continued to fall because of the effects of inflation on lagged tax collections, tax exemptions eroding the value added and other taxes, and because of an apparent increase in tax evasion. Accelerating inflation in the first semester led to a decline in public confidence and the demand for money fell precipitously, putting further pressure on monetary policy. Inflation climbed to 30 percent in August 1988. The severity of the new inflation episode compelled the Government to act.

B. The Plan Primavera: August 1988-February 1989

2.09 On August 3, 1988, the Government announced a new heterodox policy package, the Plan Primavera. The Plan, building on the experience gained during the execution of the Austral Plan, envisaged a series of revenue and expenditure measures that would be expected to reduce the demands of the consolidated public sector on the austral-denominated financial system to levels compatible with the inflation target and the desired accumulation of international reserves. At the same time structural reforms in the public sector, trade and finance were intended to buttress the credibility of the program by making changes that would be seen as permanent improvements.

2.10 Three elements were crucial to the achievement of the targets of the Plan: (i) a rapid decrease in the rate of inflation, along the lines of an agreement on the price guidelines negotiated with the industrialists; (ii) additional measures to boost government and public enterprise revenues above what could be expected from the operation of the positive Olivera-Tanzi effect, thus compensating for the fall in the inflation tax that would accompany success of the program; and (iii) the availability of sufficient external credit. Given these conditions, the economic plan was designed to avoid excessive strains on monetary policy, and to be consistent with relatively low real interest rates of 1-1.5 percent per month.

2.11 The heterodox character of the plan was reflected in its emphasis on four nominal anchors. After a large initial price adjustment, industrial groups agreed to hold monthly price increases to no more than 3.5 percent in September, to zero in October, and 4 percent through February 1989. The Government announced that public enterprise prices, after an increase of 30 percent just before the program was announced, would be kept unchanged until end-September; a 4 percent increase was then made in October and monthly increases after that time were to have kept pace with inflation. The Central Bank devalued the official rate of the austral by 11.4 percent at the start of the program, and froze the rate through September 30; a further devaluation of 3 percent took place in October, with a 4 percent monthly trajectory envisaged thereafter. Wages of the central administration were also frozen, but, in contrast to the Plan Austral, wages of the private sector were to have been determined through collective bargaining.

2.12 Stabilization implies a loss of revenues accruing to the consolidated public sector from the inflation tax (provided that inflation is not extremely high).1/ Any credible and consistent effort at stabili-

---

1/ If inflation is above the level that makes the elasticity of the demand for money unitary with respect to inflation, then a reduction in inflation increases the revenue from the inflation tax. For Argentina, the revenue maximizing point for inflation (the unitary elasticity point) is between 20-25 percent per month, a level that was surpassed, during 1988 only in August. These models cannot be precise because the parameters governing the relationship between the demand for money and inflation are unstable. See F. Desmond McCarthy and Alfredo E. Thorne, "Argentina: Problems for Achieving Macro Stability", LAC Internal Discussion Paper, World Bank: January 1988; and R. Fernandez "Inflación y economia del Estado," Buenos Aires: CEMA, 1986.

zation must involve a corresponding increase in tax revenues. The planned reduction in inflation--from a monthly average of 18.7 percent during the first eight months of 1988 to an average of 5 percent during 1989--implied that net revenues from the inflation tax (after accounting for the net exposure of the Central Bank vis-à-vis the financial system) would fall from an average of almost 5 percent of GDP during the first three quarters of 1988 to about 3-3.5 percent during 1989, depending on the strength of the remonetization of the economy during the period. The implied loss in nontax revenues of the public sector required that the tax revenue effort be increased, expenditures contained, and/or external resources mobilized for the public sector.

2.13 The 1989 budget was presented to Congress on September 29, 1988. Together with supplemental measures and an effective exchange rate tax, the budget contained targets intended to reduce the combined deficit from 4.6 percent of GDP in 1988 to 2.4 percent in 1989. This result would have implied a reduction of over two-thirds from the extremely high 1987 level (8.3 percent). This would also have produced a reduction in the non-interest deficit of the nonfinancial public sector (including public enterprises, but excluding provincial governments) from the expected 1988 outcome of 0.9 percent of GDP to a surplus of about 1.9 percent. The turnaround in public finances was to be achieved through an increase in total revenues relative to 1988 (excluding public enterprises) and a slight increase in expenditures (including the financing requirements of public enterprises). Revenue measures (including changes in the VAT legislation and modifications in the industrial promotion system) were contemplated to take effect in January 1989.

2.14 External interest payments of the nonfinancial public sector and Central Bank were expected to absorb about 4.5 percent of GDP; domestic interest payments (after correction for the inflationary component) would increase the financing needs of the consolidated public sector by a further 0.6 percent of GDP.

2.15 The Government also placed all imports except oil as well as about half of industrial exports in the free exchange market. By buying foreign exchange from exporters at the official rate, the Government intended to realize an exchange rate gain when it sold this exchange to importers at the free market rate. It also introduced a weekly auction market at the Central Bank designed to set an upper limit on the movement of the free rate. This effective exchange rate tax was expected to yield revenues of about 0.5 percent of GDP on an annual basis in both 1988 and 1989 and thus reduce the quasi-fiscal deficit. The tax was made possible because of an unexpected rise in export prices of 20-50 percent in late June and July 1988. The Government committed itself to preventing the spread between the official and free exchange rates from going beyond 25 percent, and to unifying exchange markets gradually beginning April 1, 1989.

2.16 To finance these needs, the public sector was expected to increase its external indebtedness by about 1.4 percent of GDP, equivalent to one third of external interest payments coming due in 1989. It was also expected to be able to roll over the maturing stock of government debt. These financing levels would have been sufficient to guarantee that the combined public sector would not require resources from domestic financial markets in 1989.2/

C. Performance Under the Program (August 1988-February 1989)

2.17 The Plan Primavera brought down monthly inflation and the Government continued to implement some structural reforms intended to complement the anti-inflation program, most notably a substantial tariff reform and reduction in protection. Inflation fell for three consecutive months following the announcement of the program in August to a low of 4.8 percent in November (combined index), and remained in single digits through January. Part of the initial reduction in inflation was due to the substantial difference between list prices of industries (from which the wholesale price index is computed) and actual prices; the gap was initially large because entrepreneurs correctly guessed that price controls were in the offing and marked-up prices in late July and early August.3/

2.18 The fall in the rate of inflation, coupled with the public's belief that both the price and exchange rate guidelines were sustainable in the short run, did in fact result in a strong increase in the demand for all monetary aggregates soon after the announcement of the Plan in August 1988. After hitting a floor of about 3 percent of GDP, M1 is estimated to have reached about 4.2 percent of GDP in January 1989; increasingly higher real interest rates also led to even brisker growth of broader aggregates, with M4 reaching 18.1 percent of GDP by January 1989.

2.19 Signs of strain in the execution of the program began to appear late in 1988. In the absence of quick fiscal action, monetary policy shouldered the burden of the stabilization and of maintaining the conditions for relatively low inflation rates. Given that no controls were imposed on the free exchange rate market, monetary policy was targeted on the free exchange rate to maintain it within a narrow band above the official rate, thus providing a nominal anchor that could allow an unwinding of inertial inflation. However, real interest rates became extremely high, as the public kept increasing its risk premium for holding australes and the Central Bank was bent on preventing incipient capital outflows that would otherwise bid up the dollar rate, and threaten the anti-inflation program.

2.20 The generally restrictive stance of monetary policy over the period is evident in the fact that the major source of expansion of the monetary base in the third and fourth quarters of 1988 was the external sector. In the absence of purchases of dollars by the Government to pay for external interest, and with the demand for dollars for imports largely satisfied through private dollar inflows, the surplus in the trade balance resulted in a massive creation of base money. The second largest source of creation of money was credit to the Central Government, mainly through the

---

2/ Annex 2 presents a detailed discussion on the consistency framework described here, including the integration of the public, monetary, balance of payments, and quasi-fiscal accounts of the Central Bank.

3/ One indicator was that the WPI consistently outpaced the CPI for the second quarter of 1988, a tendency which was subsequently reversed. As costs rose, the "colchon" (wedge between list prices and actual prices) was consumed, and the price guidelines came under pressure.

regular purchase by the Central Bank of government bonds coming to maturity. The August program had envisaged a complete roll-over of these bonds, but this assumption proved unrealistic. The flow of credit to the financial sector through the rediscount mechanisms was kept well within the bounds anticipated.

2.21 Monetary authorities sterilized the base money creation through the active use of forced investments and additional reserve requirements, together with the continued immobilization of credit due to the automatic capitalization of interest on Central Bank liabilities. This led to a progressive squeeze on credit and to higher interest rates, which rose from being slightly negative in real terms in June-August 1988 to an average of 4.7 percent per month in the fourth quarter. While the tightness of policy was sufficient to quell a mini-run on the austral in November, the higher level of real interest rates that followed inexorably led to a major crisis.

2.22 The program began to unravel in late December. Inflation rose in December to 6.3 percent and continued to rise thereafter. The first ominous signs began to appear when wage settlements, following a series of bitter strikes by public sector employees in November-December, were substantially higher than anticipated in the program. Signs of loss of control over public finances also surfaced: transfers to the provinces had to be increased above the budgeted levels because provincial spending did not adjust as quickly as anticipated to the new revenue-sharing law; also shared revenues were less than anticipated because of an apparent increase in tax evasion, making the adjustment more difficult for the provinces and central administration alike. Public enterprise tariff adjustments were kept below the rate of inflation longer than originally envisaged. The Government decided to advance a substantial increase in pensions from July to January 1989. Finally, the tax reform, presented to Congress in December with the intention of improving revenues by 1.5 percent of GDP, was approved with less than half that amount. Congress did not accept the proposed widening of the base of the value added tax, but insisted that it be lowered further to 14 percent (it had been lowered from 18 percent to 15 percent at the outset of the Primavera to elicit support of industrialists for the price agreements); the executive acquiesced to gain the support of key interest groups for a reform restricting tax expenditures in the industrial promotion law. By year-end it had become apparent that, without new policy initiatives, it would be difficult for the Government in 1989 to hold the combined deficit to less than 5 percent of GDP, more than twice the 2.4 percent target in the program.

2.23 Events outside the control of Government compounded the problem. The country suffered from an extensive drought which reduced production and exports. The drought caused severe and prolonged shortages in power supply throughout the country, as hydropower generation, which accounts for a third of total generation, fell. Also, Argentina's presidential campaign caused nervousness in financial markets about the course of future economic policy. Finally, discussions with external creditors remained deadlocked. All of this contributed to increased economic uncertainty.

2.24 The uncertainty and continuing imbalances in macroeconomic fundamentals precipitated a run on the austral in the final days of January. The high real interest rates necessary to make austral-denominated assets attractive produced extremely rapid asset growth (through interest capitalization)--growth at a faster rate than the austral equivalent of US dollar reserves (see Figure 2.1). Eventually the domestic asset stock capable of conversion into foreign denominated assets reached a point at which devaluation became unavoidable, and speculators tried to escape australes before the collapse. The real exchange rate appreciation of 11 percent relative to July 1988 and the climbing interest rates had created an explosive policy combination; sagging confidence translated itself into an accelerating demand for dollars at the Central Bank auction window in the last ten days of January.

2.25 At first, the Central Bank reacted with even tougher talk and actions than it had at end-November when a similar run transpired. As the demand for dollars reached US$300 million per day, it announced a new forced investment on January 30 that absorbed about A$7,000 million (4 percent of all deposits)--with the result that the monthly call rate rose from about 12 percent to 19 percent in the same period. This pushed the demand for dollars back down to US$100 million on February 1-2, and US$31 million on February 3. The fact that the Government had sold over US$900 million in the first 34 days of the year, together with the forced investments and increased reserve requirements, exerted a tremendous contractionary pressure on money markets. But these measures were not enough, as holders of austral-denominated assets had decided that it was time to shift their portfolios irrespective of the dollar price. On February 6, the Government decided to cut the parallel exchange rate loose by halting the dollar auctions of the Central Bank, thus ending the Plan Primavera. The stage was set for the deep crisis that would lead the country to hyperinflation in the next few months.

D. Recent Developments: Emergence of Hyperinflation

2.26 The demise of the Plan Primavera in February gave way to several futile efforts by the Government to maintain control over inflation through the exchange rate. On February 6, the Government enacted a package of measures centered around changing the exchange rate regime from a dual exchange rate system to a three-tiered system. This regime was short-lived. As the parallel rate rose, exporters withheld their export earnings from the Central Bank or even preferred to store their exports and await a better rate. The progressive drying up of export earnings drove the free rate higher and added to the incentive to withhold future exports. The exchange gap with the commercial rate rose from a level of about 20 percent in the Plan Primavera to between 70-100 percent. Monthly interest rates on call money loans, meanwhile, hovered between 15-20 percent (about 5-6 percent real). By end-March, the run on the austral had become so severe that trade lines and sight deposits at the Central Bank had fallen and threatened depletion of liquid reserves.

2.27 On April 1 the Government unified and floated the exchange rate, enacting an effective 40 percent real devaluation. In a last effort to control inflation and raise revenues, authorities enacted an export tax with rates that were subsequently changed several times. 4/ Throughout April and May the austral entered a free fall and the effective real devaluation was more than 100 percent; the Argentine currency reached its lowest real level since the waning months of the second Peronist administration in early 1976 (Figure 2.1).

2.28 The fiscal situation became desperate. Revenues fell sharply in real terms during March and April (Figure 2.2). Only part of the fall appears attributable to the Olivera-Tanzi effect; the remainder reflected the weakness of tax administration and enforcement as well as widespread expectations of a tax amnesty. (The Government in early May did enact a partial amnesty that included overdue taxes up to April 30, 1989.) Tariffs of public enterprises fell sharply in real terms until May. While a cumulative adjustment of 68 percent during May recouped some of the losses, prices were still substantially lower than at the outset of the Plan Primavera (Figure 2.2).

2.29 Soon after the election on May 25, the Government announced a new package that comprised fixing the exchange rate, limitations in deposit withdrawals, and some new taxes. The latter included adjustments in the export tax (now 30 percent on agriculture and 20 percent on industrial products), a 4 percent emergency sales tax on the first sale transaction for selected agricultural products and measures to improve tax collection (including sanctions for nonpayment) as well as some expenditure cuts. In addition, on June 3, the Congress passed the Government's proposed legislation providing for a temporary suspension of 25 percent of the industrial promotion benefits, a 3 percent across-the-board export tax, a permanent 5 percent agricultural sales tax on first transactions, indexation of VAT and excise taxes, payment of reimbursements of export subsidies with a new bond (TIFISO), a tax on foreign exchange transactions and several lesser measures. These measures failed to calm markets, however. Massive portfolio shifts to dollar-denominated and fixed assets, precipitously falling revenues to the public sector and exploding nominal interest rates, propelled the momentum of hyperinflation.

4/ The initial rate was a 100 percent surcharge on all export dollar receipts above A$36/US$ on all exports, with the expectation that the rate would normalize in the range of A$40/US$. Unfortunately, the rate accelerated to much higher levels, averaging A$67 for April and topping A$100/US$ by the end of the month. On May 1, the surcharge was converted to a 20 percent tax on all exports. At end-May this was changed to 30 percent on agricultural exports and 20 percent on all other goods.

FISCAL DEFICIT AND ITS CAUSES

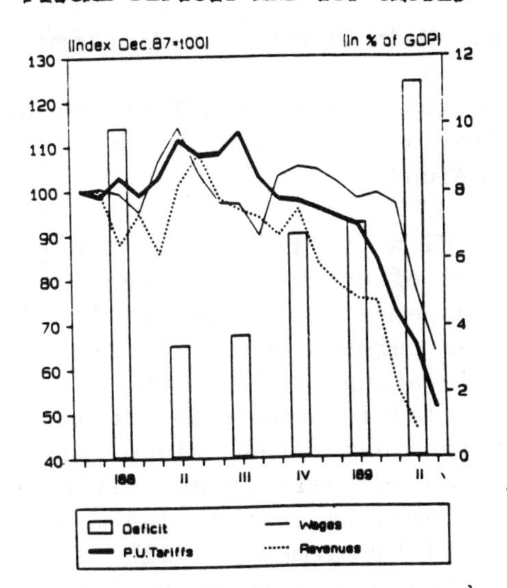

Figure 2.2
(a)

TAX COLLECTION
(Dec. 1987=100)

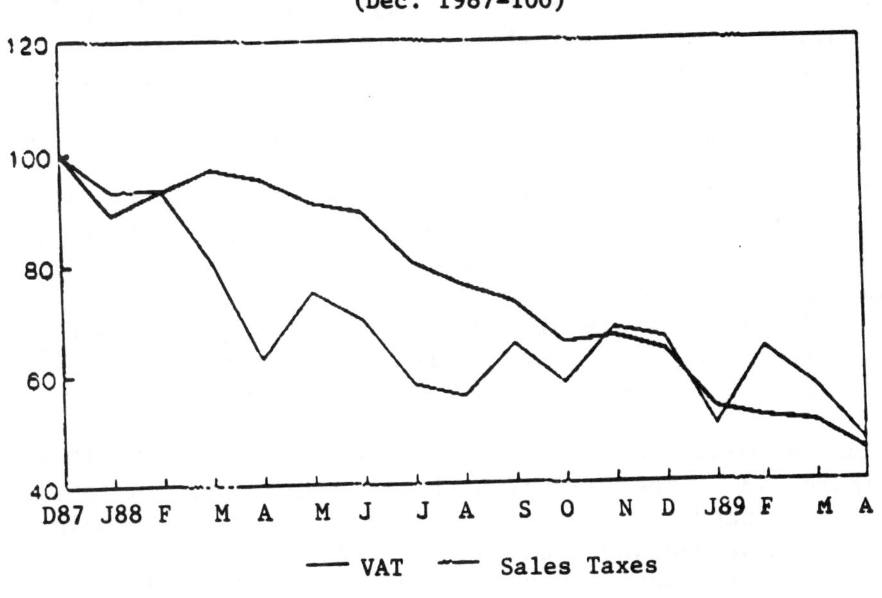

Figure 2.2
(b)

PUBLIC UTILITY TARIFFS

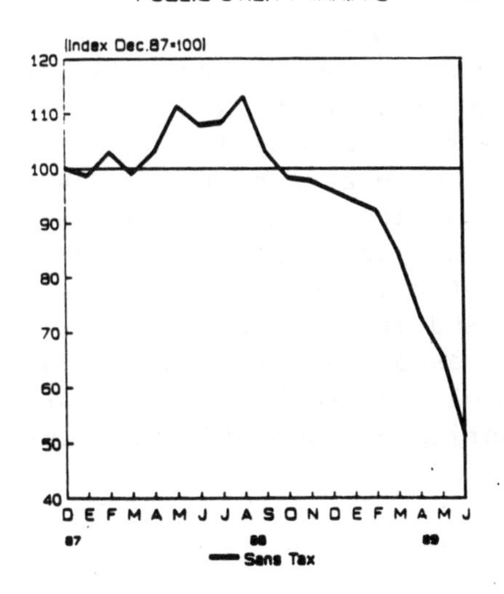

Figure 2.2
(c)

E. Main Short-Term Problems

2.30 The new economic team confronted four immediate economic problems: hyperinflation, instability in the financial system, price misalignment and the threat of major recession.

2.31 **Hyperinflation**. With the failure of the Plan Primavera, the Government was left without effective instruments to deal with the crisis. The already precarious management of fiscal policy during 1988 collapsed with the failure of the tax reform and the Government could do nothing in the face of its own deteriorating credibility. Monetary policy was powerless in the face of the inelastic demand for dollars and to some degree captured by its own high interest rates, since very high market rates increased the Central Bank's own deficit and ultimately led to money creation to cover losses. As normally restrictive measures were taken--for example, increases in forced investments--high real interest rates to be paid on those liabilities provoked a monetary disequilibrium, and the resulting uncapitalized portion (the smaller portion to be sure) increased the monetary base. The loss of control over monetary policy and the evaporation of confidence meant that there could be no control over the nominal exchange rate. Finally, price controls had completely lost any effectiveness--as seen in the closing days of April when the Government reimposed them in a last futile attempt to control the price explosion.

2.32 Inflation was driven on the fiscal side by the inability of public finances to balance through the inflation tax, and the instability produced by the inflation-deficit relationship. The deficit continued to widen because of inflation itself: real revenues were falling and extraordinarily high real interest rates further widened the quasi-fiscal deficit of the Central Bank. Furthermore, low public enterprise prices continued to put pressure on the deficit.

2.33 Inflation was driven on the monetary side by the free fall in money demand (Figure 2.3). The Central Bank was being required to finance the payment on index-linked bonds (equivalent to half the outstanding money supply) which were coming due and could not be rolled over. The internal public debt, estimated at roughly 16 percent of GDP, continued to grow because high real interest rates paid by the Central Bank on the forced deposits of the financial system are capitalized.

2.34 Both fiscal and monetary policy became completely endogenous to the inflationary process. Accelerating inflation reduced real tax revenues because of time lags in collection, and this increased the deficit of the nonfinancial public sector. Similarly, higher inflation produced destabilizing monetary expansion because of the wedge between income on assets and interest expenditures on liabilities of the Central Bank; the former are dominated by loans to the public banks, which carry a fixed interest spread above inflation, while the latter are dominated by forced investments from the commercial banks, and are priced at market rates. Therefore, higher nominal interest rates associated with accelerating inflation endogenously increases the deficit of the Central Bank. Moreover, the lag in interest earnings means that during periods of rising inflation, the quasi-fiscal deficit widens.

Figure 2.3
Monetary Variables

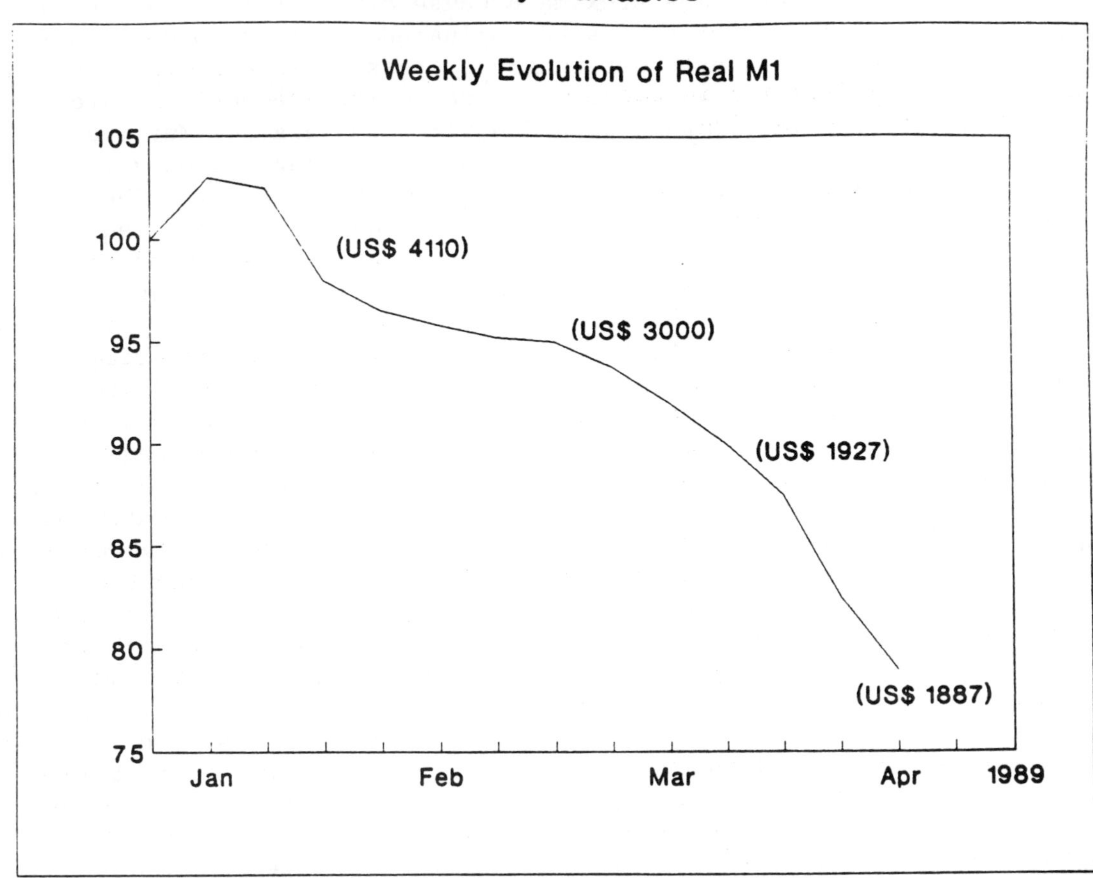

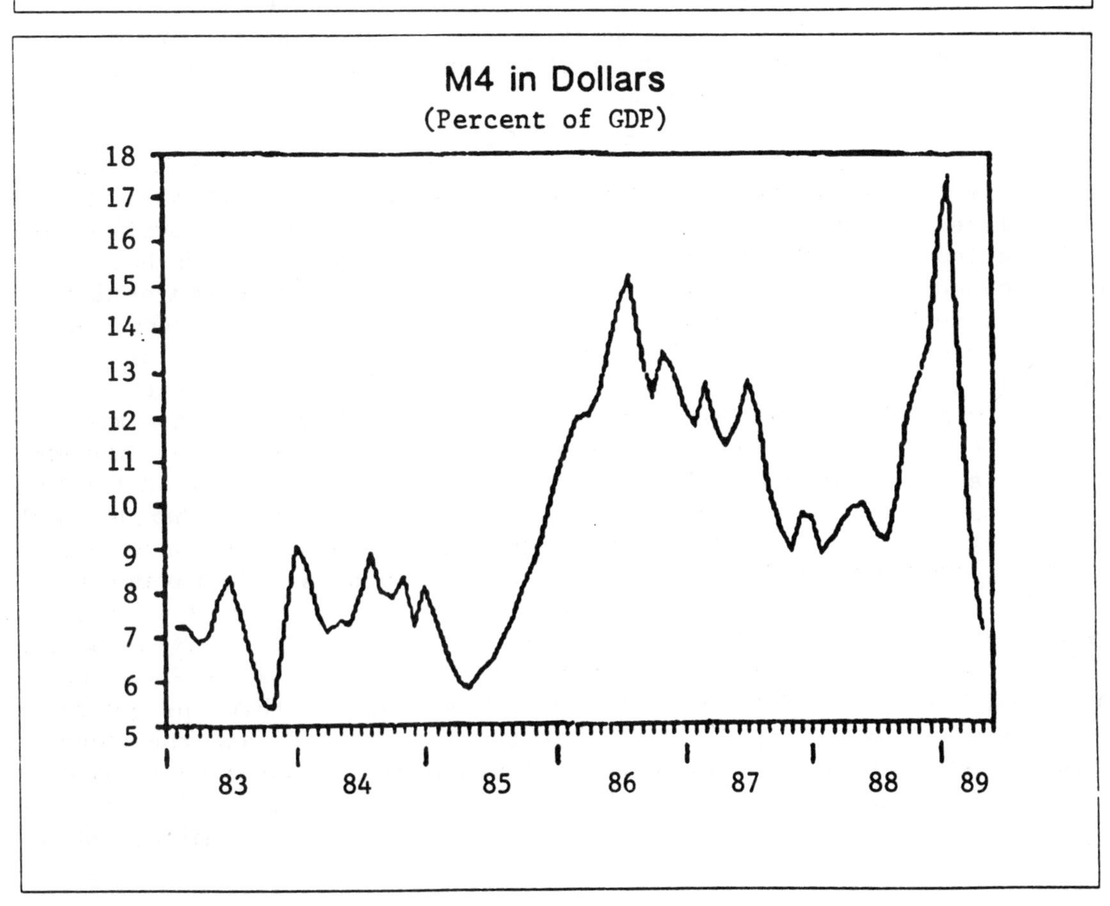

2.35 **Financial System**. The proximate cause of the stress in the financial system was the flight of austral-denominated deposits to dollars and other assets. Withdrawals have put enormous pressure on commercial banks since the forced deposits are not formal reserve requirements and are not automatically released pari passu with deposit withdrawals. Moreover, other assets of the bank's portfolio were not performing, as many firms were reportedly no longer able to service debt with the banks. The Central Bank has had to increase rediscounts and ease forced investment requirements to support liquidity of the banking system as well as impose frequent bank holidays and limits on deposit withdrawals during May and June.

2.36 **Price Misalignment**. Public enterprise prices were severely misaligned. By end-June, real public enterprise prices were roughly 60 percent below levels prevailing at end-year 1988, the level necessary to generate positive savings. This is the reason why the Government more than doubled real prices as a principal measure in its new program. Realigning these prices will necessarily affect the general price level because of their own heavy weight in price indices as well as the fact that so many private sector prices are indexed to these.

2.37 **Recession**. At the same time, the real economy, already in its third year of recession, was contracting further. The real appreciation of the austral in the waning days of the Plan Primavera had led to a mini-boom in external travel and import purchases of consumer durables; the collapse of the program produced the flight from australs to goods in February and led to a short-lived boom for industry. However, the inflation in June compounded the effects of the high interest rates on the real sectors. The sharp reduction in real wages in February-May cut private consumption, and the shortage of trade credit and the inability of many store owners to determine their prices also led to a slowdown in economic activity.

F. Short-Term Options

2.38 The current situation does offer an important opportunity which, if capitalized upon, could ease the path towards sustained recovery. Inertial inflation common to the pre-Austral and Primavera programs has in fact been mitigated as contracts have been shortened to very limited periods; backwards indexation no longer carries inflation forward, and accounts are increasingly carried on a dollar basis, so that price movements are substantially linked to movements in the austral/dollar exchange rate. Formidable challenges, however, also make the Argentine situation more difficult to handle than in similar hyperinflationary episodes. In particular, the almost complete indexation to market interest rates of the public debt implies that hyperinflation will not, per se, erase the internal debt, but rather that it may increase its burden on public finances unless explicit action with high social and political costs is undertaken.

Policy Priorities and Sequencing

2.39 The experience of the last five years suggests three principles for the design of a package to stabilize the economy. First, the short-term plan must be built upon a program of structural reforms that provide confidence for investors and the public at large that changes in the conduct of public finances are indeed permanent. Second, the nominal public sector borrowing requirement must be credibly and immediately reduced to levels that can be financed through foreign credit--recognizing that in the short term the predominant foreign source will be arrears. Third, the monetary component of the stabilization package will have to ensure that the public sector will not use the inflation tax to balance its accounts in the future.

2.40 Past experience argues strongly that the Government should take structural measures simultaneously with--or prior to--putting the stabilization program formally into place. The Government implicitly recognized this with its July 9 program, which contained several structural reforms. Ideally, the new program will be progressively broadened to include the Government's approach toward tax policy and subsidies, public enterprises, revenue sharing with provincial governments, tax subsidies to industry, policies toward the financial sector and trade policy. Policy options for such a program--with an indication as to sequence and priority--are outlined in the Executive Summary and elaborated in subsequent chapters.

2.41 This inverts the usual approach of first trying to close the fiscal gap with *ad hoc* revenue and other measures, and deferring structural measures until after stabilization is achieved. A major lesson of the last five years is that many structural measures, especially in the public sector, must precede--or at least proceed simultaneously with--stabilization measures. This is because the structural deficit remains high relative to available domestic and foreign finance and demand for austral-denominated assets remains low and volatile. As long as a chronic deficit is present, the private sector will have no faith in the sustainability of the program. Early formulation of the comprehensive medium-term program and initial policy steps will ensure that stabilization measures are consistent with efficiency objectives and medium-term objectives of increased growth, savings and investment. Only in that way will reforms credibly convey a sense that price stability will be enduring and that the economy can recover.

2.42 A companion lesson from past stabilization attempts is that prices of public enterprises cannot be used as a mainstay of stabilization for any length of time. If real prices are allowed to deteriorate, it becomes virtually impossible for the sector to maintain savings at a level consistent with deficit reduction objectives. After the stabilization plans of recent years, falling real prices now raise inflationary expectations since the private sector is aware that tariffs will eventually have to be increased, and may even anticipate these with large price movements of their own.

Program Options: Moderate-Inflation Option

2.43 Policymakers might attempt to formulate another program aimed at reducing inflation to under 5 percent per month. Recent stabilization failures have probably precluded the moderate-inflation option. The moderately high inflation equilibrium of the early 1980s is now unstable. Despite the lower financial disequilibria generated by the public sector in the second half of the 1980s compared to the previous ten years, it is apparent that the economy could function with short-term cycles of easy and then tight monetary policy only because it was traversing a relatively narrow course between deep recession and hyperinflation. That path is now closed.

2.44 This is because the structural deficit remains high relative to available domestic and foreign finance and demand for austral-denominated assets remains low and volatile. Lack of social consensus and entrenched business and union power, together with the burden posed by the external debt servicing, conspire to maintain the demands on the public sector beyond its ability to garner resources, thus creating a structural deficit. This has resulted in the secular decline in the holdings of financial instruments denominated in domestic currency, and the responsiveness of portfolio shifts and money velocity to inflationary expectations. This narrows the base of the inflation tax beyond a point where it can help close the accounts of the public sector.

2.45 Using a conventional heterodox program--a social pact on prices, some adjustment in fiscal accounts, and tight monetary policy--is also now precluded. The major strength of a heterodox plan, the ability to abruptly bring to a halt inflation through the imposition of wage and price controls and/or some other form of social consensus, is also its fundamental weakness. In both the Austral and Primavera plans fundamental fiscal reform was recognized as a necessary condition for the ultimate success of stabilization. In both cases, however, the magnitude of the reform task was underestimated, and the authorities and congress balked at the prospect of paying the political price for some of the less popular measures. This was made doubly difficult because of the sequencing of the gains from the Olivera-Tanzi effects, which provide the temporary illusion of a politically cost free adjustment. Once it became evident that the process had not changed fundamentally, inflationary expectations rose, the exchange rate-domestic inflation cycle took its course, the virtuous Olivera-Tanzi effect reversed, and the resulting deterioration in fiscal performance displaced whatever measures of a structural nature may have been under implementation.

2.46 If this route is pursued by the incoming Government, any apparent success will likely prove purely transitory, and the hyperinflationary process will continue its course until lasting stabilization, pursued along more conventional lines, as suggested below, is achieved. Moreover, false starts at stabilization will make ultimate stabilization more difficult to achieve by any existing government, since credibility would be eroded.

Program Options: Bold Reform

2.47 A more radical approach to stabilization would be predicated upon mutually reenforcing reforms of the public sector, monetary policy, the exchange rate, and external finance. The objectives described below could be attained through various strategies, including a currency reform. The key elements of each approach, however, must be the same: up-front fiscal reforms that include difficult-to-reverse structural measures, clear and transparent monetary and exchange rate rules, and an eventual accord with external creditors.

2.48 **Public Finances Reform**. The cornerstone of the program must be explicit political agreement on a comprehensive reform of public finances. The agreement would have to signal a change in policy regime: as such it would have to offer to the public concrete guarantees against reversal. The main component of the package would be sharply declining fiscal deficit target, which would remove the need for using the inflation tax; this would be predicated upon an integrated package of reforms aimed at reducing expenditures, rebuilding the tax base, and redressing the deficits of the various components of the public sector. This would be enforced by the adoption of a transparent monetary rule, discussed below, stating that future currency emissions would be limited to increases in foreign reserves of the Central Bank. As a consequence, substantial and immediate adjustment of the public sector and public banks would be necessary, so as to reduce its deficit to a level equal to net new borrowings of foreign resources.

2.49 The objective would be to reduce the deficit immediately to a level that requires no financing from domestic sources. This will necessarily imply recourse to inefficient taxes--such as export and energy taxes--in the short run. However, structural reforms are necessary to replace these. As discussed in Chapter III, they should include expenditure reductions and fiscal reforms aimed at strengthening tax administration, widening the base of the VAT and reducing tax subsidies on a permanent basis through the industrial promotion schemes and Tierra del Fuego regime, and buy-Argentina legislation. In addition, reducing the deficit would require additional fiscal discipline in the provinces and continued institutionalization of financial relationships between the provinces and the national government. It would also require institutional changes in the public enterprises designed to subject them to the discipline of the market, divorce price setting from macroeconomic policy and make spending more efficient; sustained improvement must be predicated upon adequate and continued price-cost margins as well as some reduction in employment so that wages could increase without increasing the wage bill. Similarly, the social security system requires major changes in order to make it financially viable over the medium term. Finally, reforms of the financial sector are necessary to disengage the Central Bank's operating performance from price instability. The Government's July 9 program has already put many of these reforms on the agenda.

2.50 **Monetary Reform and Policy**. A strong monetary reform that permits monetary expansion solely as function of increases in international reserves would provide the public confidence necessary to slow inflation. Monetary policy would be assigned a purely passive role, and the monetary authority would concentrate on its role as supervisor of the rebuilt financial system. In contrast to the Austral Plan, monetary policy would not be designed to temper the usual appearance of high interest rates in the wake of any stabilization. The monetary rule that domestic increases be solely a result of Central Bank reserve increases would allow for sufficient liquidity to the private sector as it seeks to rebuild depleted balances. The policy would be based on an ex-ante judgment that the risk that this firm monetary rule might create a prolonged recession and thus erode public support for the program is less than the danger that the private sector will not believe the program is sustainable and will therefore maintain their inflationary expectations.

2.51 Both the Plans Austral and Primavera experienced high real interest rates in the wake of stabilization. However, in the Primavera, real interest rates were driven far too high to maintain the austral appreciation in the face of waning confidence of the public over the extent of changes in fundamentals; an important lesson is that the program must be credible.

2.52 Monetary reform would also involve a restructuring of the domestic internal debt, and, as a consequence, of the corresponding assets of the public. This could be done through exchanges of debt instruments with the Central Government. The outstanding stock of forced investments would be converted into long-term securities of the Central Government carrying a fixed nominal interest rate. To service these obligations, the Treasury would have to mobilize a sufficient surplus to pay interest and amortization at the rate consistent with the stabilization strategy and targets. The only alternative is to write down the value of these assets, with its adverse implication for future creditworthiness. This policy should be woven together with the other reforms of the financial system including the recapitalization of the Central Bank (see Chapters IV and Annex Chapter IV).

2.53 **Exchange Rate**. The monetary reform in the wake of the change in fiscal policy regime would allow the Government to maintain a fixed exchange rate against the dollar or a basket of currencies. This would provide the nominal anchor necessary to operate a successful stabilization after hyperinflation.5/ Since international reserves are low, an important element for success is the availability of foreign resources to provide reserves to back the new exchange rate regime. Strict adherence to the rule guarantees that new additions to the money stock are automatically backed by foreign reserves. Since the initial stock of high-powered money is limited, this will guarantee that as remonetization proceeds virtually the entire money supply will be backed by foreign reserves. An early arrangement with the IMF is therefore particularly important to provide initial reserves.

2.54 **External Finance**. An important element for success is the availability of important foreign support. In contrast to the Plan Primavera, the relative success of the Plan Austral as well as the more enduring successes in other countries in permanently stopping hyperinflation was partially attributable to the willingness of external creditors to support the program. In addition to providing badly needed reserves, a foreign financing program enhances its credibility by providing much-needed finance to the public sector, by eliminating the uncertainty stemming from irregular relations with external creditors and by spreading confidence in the private sector that would lead to an adequate supply of regular trade finance.

5/ See R. Dornbusch and S. Fischer, "Stopping Hyperinflation: Past and Present", Weltwirtschaftliches Archiv, April 1986, Vol. 1, p. 47. Also M. A. Kiguel and N. Liviatan, "Inflationary Rigidities and Orthodox Stabilization Policies: Lessons from Latin America," World Bank Economic Review, Sept. 1988, pp. 273-298.

2.55 Garnering foreign support will be difficult. As shown in the projections in Chapter V, it seems unlikely that Argentina can fully service its private commercial debts in the medium-term; a strong fiscal program predicated on sharp deficit reduction simply has extremely limited scope for renewed debt servicing in the years immediately ahead. External creditors, like their domestic counterparts, have no alternative but to recognize the limited capacity of the state to service its obligations. Nonetheless, all external creditors and the Government have a fundamental interest in Argentina's medium-term price stability, growth and restored creditworthiness. These facts suggest that, while commercial arrears are likely to build in the short term, after the medium-term program is in place and the stabilization is underway, the Government should initiate discussions with private creditors with the objective of regularizing financial relations.

CHAPTER III: FISCAL POLICY AND PUBLIC FINANCE

A. Introduction

3.01 Insufficient control of the main components of the public sector has traditionally hampered fiscal policy in Argentina. The need to increase mobilization of domestic resources to service the external debt--partly a consequence of the 1982 decision to nationalize the privately-held external debt--complicated this task. From representing less than 2 percent of current expenditures in the 1970s, external interest payments jumped substantially in the 1980s, and have since oscillated around 15 percent of current expenditures. This placed severe stress on the other components of expenditures and increased the burdens on the tax system. The adjustment process therefore entailed large fiscal deficits, a worrisome increase in inefficiency in the provision of services and the management of public enterprises, and the use of the Central Bank as a conduit for concealed subsidies outside the budget process.

3.02 The unraveling of the Plan Austral showed that a sustained decrease in inflation is impossible without a reversal of fundamental imbalances in the public sector. More recently, the Government began pursuing structural reforms coupled with macroeconomic stabilization in an effort to provide a durable solution to public sector problems. At the beginning of 1988, the Government adopted a principle for fiscal policy that improved control over public finances: clear guidelines for the transfers to the provincial governments, the social security system, public enterprises and the application, within those guidelines, of the principle of budgetary self-sufficiency. At the same time, considerable efforts were made to rein in the sources of expansion of the deficit of the Central Bank (the quasi-fiscal deficit), culminating in the presentation to Congress of the annual budget of subsidies through the Central Bank. These reforms, however, have yet to resolve the structural imbalances in public finances.

B. The Public Sector: Struggling with the Deficit

Background: The Pre-1984 Period

3.03 The central role of the public sector accounts in determining Argentina's macroeconomic performance and steering the course of structural developments derives from its importance in relation to total economic activity and the pervasiveness of the public presence across economic activities. Total expenditures of general government were equivalent to about a third of GDP in 1987; including outlays of public enterprises, the size of the nonfinancial public sector (NFPS) exceeded 44 percent of GDP (Figure 3.1).1/ This is only a partial indicator of the importance of the

---

1/ The data in Figure 3.1 are estimated on a budget basis, and includes revenues and expenditures of the provinces. Quarterly estimates of revenues and expenditures (with the exception of the provincial governments) are available on a cash basis, and are discussed in the section on recent performance. The budget estimates calculate interest payments in real terms and therefore exclude from the computation of interest payments the part attributable to the reduction in real value of the debt due to the rate of inflation.

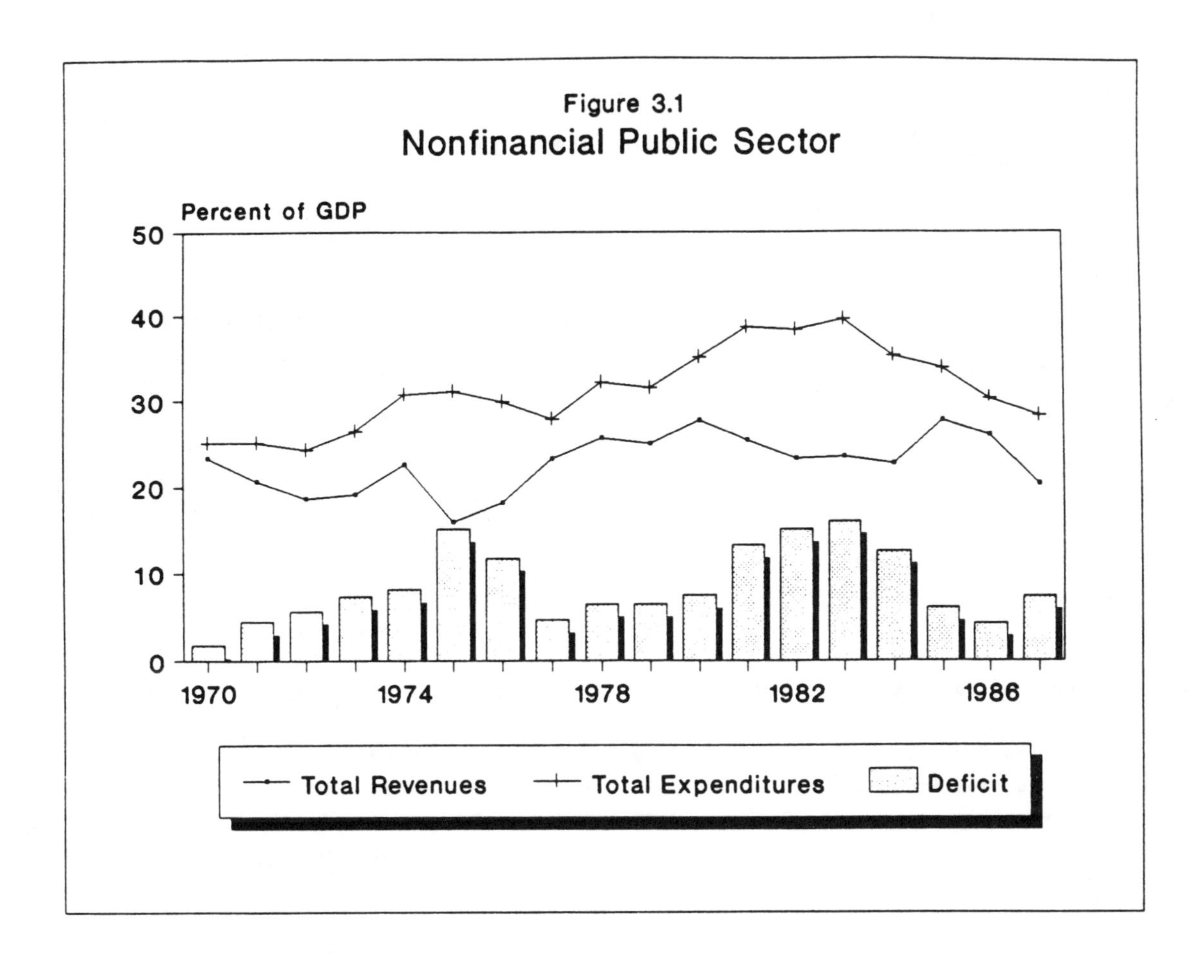

public sector; indeed its sphere of action extends to many economic sectors, through a system of public enterprises that have long held substantial monopolistic positions in steelmaking, transportation, telecommunications, and a variety of services.

3.04 The overall public sector has been prone to budget deficits, which have absorbed a substantial share of national savings (Table 3.1). Since 1972, overall financing requirements of the NFPS have been lower than 5 percent of GDP only in 1977 and 1986. The upheaval of the early 1970s coincided with a rapid increase in the deficit, which ballooned to 15 percent of GDP in 1975. Substantial progress was made in the following two years, but starting in 1977 a steady widening of NFPS financing requirements began, culminating in a peak of over 16 percent of GDP in 1983. This fragile financial position was the result of contrasting trends in revenues and expenditures, brought about by frequent changes in policies, and by a general inability to exercise an effective fiscal control.

Table 3.1: ARGENTINA - NONFINANCIAL PUBLIC SECTOR ACCOUNTS, 1970-1987
(Percentage of GDP)

	Average 1970-74	Average 1975-80	1981	1982	1983	1984	1985	1986	1987 a/
Current Revenues	20.58	22.20	24.35	22.54	23.14	22.39	26.92	25.18	19.67
Tax Revenues	17.61	18.90	20.34	18.73	18.55	18.15	22.01	21.93	17.62
Non-tax Revenues	2.98	3.30	4.00	3.82	4.58	4.24	4.91	3.25	2.05
Current Expenditures	18.59	20.36	27.16	26.05	27.44	26.20	25.43	23.74	22.63
Personnel	9.35	8.73	9.58	7.66	9.76	10.58	9.69	9.06	4.17
Goods and Services	2.29	2.93	2.99	3.47	3.78	2.87	3.26	3.06	2.19
Interest on Debt	0.55	1.52	3.88	6.28	3.50	2.93	2.93	2.33	2.69
Domestic	0.35	1.35	2.75	3.86	0.46	0.49	0.23	0.26	0.78
Foreign b/	0.20	0.17	1.13	2.42	3.04	2.44	2.70	2.07	1.91
Current & Capital Transfers	6.40	7.19	10.71	8.64	10.40	9.82	9.55	9.29	13.57
Social Security Payments	4.28	4.34	6.08	4.86	6.08	5.56	5.59	5.52	4.96
Other	2.13	2.85	4.63	3.78	4.33	4.26	3.95	3.77	8.61
General Government Savings	2.00	1.83	-2.82	-3.51	-4.31	-3.81	1.49	1.44	-2.40
Public Enterprise Savings	0.54	0.29	-1.91	-3.87	-2.57	-1.44	-1.51	0.45	-0.61
Noninterest Current Account	0.94	1.25	1.61	0.22	-0.10	0.59	1.01	1.94	0.90
Interest Payments	0.40	0.96	3.52	4.09	2.46	2.03	2.52	1.48	1.51
Domestic	0.19	0.55	2.42	1.85	0.48	0.30	0.55	0.07	0.05
Foreign	0.21	0.42	1.10	2.24	1.98	1.72	1.97	1.42	1.46
Public Sector Savings	2.54	2.13	-4.72	-7.38	-6.87	-5.25	-0.03	1.90	-3.01
Capital Revenues c/	0.30	0.50	1.11	0.83	0.48	0.43	1.00	0.91	0.77
Capital Expenditures (Gross)	8.31	11.29	9.65	8.56	9.69	7.82	7.06	7.11	5.14
Public Sector Financing Requirements	5.48	8.67	13.26	15.11	16.08	12.64	6.09	4.30	7.38
Memo: Primary Surplus	-4.53	-6.19	-5.86	-4.74	-10.12	-7.68	-0.64	-0.49	-3.18
Central Bank Foreign Interest	..	..	..	0.06	0.45	1.56	1.72	1.20	1.10

Source: Ministry of Economy.
a/ Excludes provincial governments.
b/ Does not include interest on foreign debt by public enterprises or the Central Bank.
c/ Includes forced savings.

August 1989

3.05 Expenditures substantially increased throughout the 1970s and early 1980s, with only a moderate pause between 1975 and 1978. The share of the consolidated public sector in total output rose from about 25 percent in the early 1970s to almost 40 percent in 1983, fueled by the outlays associated with the South Atlantic War and the increasing deficit in the public enterprises. While many countries experienced a tendency towards a larger public sector, the post-war growth of the state was particularly marked in Argentina, and was a corollary of the import-substitution development model.

3.06 The tax system, on the other hand, was unable to provide an increase in revenues that could finance rising expenditures. Tax collection virtually collapsed in 1975, in part because of the political upheaval of the time and because of the concomitant explosion in the rate of inflation, which sharply reduced the real value of taxes collected with a delay from the occurrence of the taxable event. Overall revenues fell by almost 7 percent of GDP, a third in real terms. In the following years revenues steadily recovered, and, buoyed by a series of reform measures (the most important of which was the adoption and progressive generalization of the VAT between 1974-80), reached an unprecedented 26.8 percent of GDP in 1980. However, the 1980s have witnessed a steady erosion of government revenues that has generally continued despite several reform initiatives starting in 1985.

Progress in Reducing the Deficit

3.07 The restoration of the constitutional government in December 1983 marked a reversal of the trend towards bigger government and higher deficits--a process, however, that has been marred by setbacks and that is still insufficient to provide a stable macroeconomic environment. Total expenditures declined in relation to GDP in every year after 1983, with the exception of 1987, and are estimated to have fallen by nearly 10 percentage points of GDP to 30.3 percent of GDP in 1988, the lowest level in over a decade. However, public finances have remained fundamentally weak, and the reductions in expenditures have not resulted in a durable fall of the deficit to a sustainable level, chiefly because of the continued deterioration of tax revenues. Although the overall financing requirements fell from the unsustainable levels of the early 1980s to 4.3 percent of GDP by 1986--helped in this by the early success of the Plan Austral in lowering inflation and reducing the loss due to the delayed collection of taxes (the Olivera-Tanzi effect)--progress has been substantially more limited afterward.

3.08 Despite the implementation of repeated stabilization plans, the borrowing requirements of the consolidated public sector widened considerably in 1987, spurred in part by the spending concomitant with the fall elections. The deficit rose to 7.4 percent of GDP (on a budget basis), at a time when the quasi-fiscal deficit of the Central Bank also surged. As discussed in Chapter II, the fiscal performance in 1988 improved somewhat because of much stricter control of expenditures made possible by the application of the separation principle. Nonetheless, the reduction in the deficit was modest because of the sharp decline in tax revenues, and 1988 ended with an overall deficit of an estimated 6 percent of GDP on a budget basis.

3.09 The lack of clear rules applying to the relationships between the different bodies of government is one of the main reasons for the repeated crises in public finances. A major step forward towards a consolidation of fiscal policy was made with the decision in 1988 to set limits on the transfers from the central administration to the other three components of government--the provincial governments, the social security system, and public enterprises--after taking measures aimed at providing sufficient amounts of resources to each of them. However, this arrangement is extremely fragile and rife with derived inefficiencies. A more fundamental resolution of the structural shortcomings of the public sector will be needed in the years ahead. In addition to reducing expenditures (discussed in conjunction with the subcomponents of government), profound reform of the public sector should address the inadequacy of the tax system, the relationships between provincial and central governments, the social security system, and the control of public enterprises.

C. The Tax System: Problems and Reforms

Tax Revenues: Lack of Buoyancy and Growth of Distortions

3.10 The ideal function of the tax system is the collection of taxes, with due regard paid to economic efficiency and equity in taxation. The Argentine tax system has become increasingly deficient by all of these standards. The system has shown a relatively low buoyancy, and has been unable to ensure a stream of revenues that could reconcile desired domestic spending with the external interest obligations.

3.11 This, in turn, is attributable largely to the increased use of the tax system to promote regional and sectoral industrial development. As a result, revenues have lagged, horizontal and vertical equity has suffered greatly, and investment decisions are now subjected to arbitrary distortions in the structure of incentives. For these reasons, tax reform must remain an important item in the agenda for structural reform of the public sector.

3.12 While high and variable rates of inflation have undoubtedly contributed to a worsening of the tax performance, the lack of buoyancy is also attributable to the progressive extension of tax expenditures and of tax evasion. Table 3.2 shows that the "theoretical" fiscal cost of the industrial promotion law (excluding the special regime applicable to Tierra del Fuego) had grown to 2.6 percent of GDP over the period 1980 to 1987. This represents only the tax exemptions agreed to when the promoted project was approved (based on a forecast of output over the life of the project). Since the promotion law sets no limit on production levels, there exists an incentive to expand production to benefit from subsidies beyond the "theoretical" limits; thus, actual tax expenditures are almost certainly much larger. In addition, other tax expenditures have entailed large losses to the Treasury. Tax subsidies to selected imports, exports, Tierra del Fuego, and holders of government bonds cost the Treasury another 2.6 percent of GDP (Table 3.3).

3.13 With the major taxes being eroded by the special regimes, the Government increasingly looked for additional revenues in areas where collection was easy, but at the cost of increasing distortions. Thus,

Table 3.2: ARGENTINA - THE THEORETICAL COST OF INDUSTRIAL PROMOTION
(Percent of GDP)

	1980	1981	1982	1983	1984	1985	1986	1987	1988a/
Administered by SICE	0.67	0.65	0.32	0.86	0.59	0.62	0.80	0.77	0.10
Administered by four Provinces b/	0.01	0.03	0.04	0.10	0.25	0.88	1.39	1.82	2.15
Total	0.67	0.67	0.37	0.96	0.84	1.50	2.19	2.60	2.25

Source: Secretaria de Hacienda

a/ Includes only projects approved through 1987.
b/ Catamarca, La Rioja, San Juan, San Luis.

Table 3.3: ARGENTINA - OTHER TAX EXPENDITURES, 1988
(Percent of GDP)

	Percent
Exemption from Import Duties	1.1
Nontraditional Export Subsidies	0.6
Tax Rebates	0.4
Income Tax Exemptions (10 percent)	0.1
Exemption of Subsidies from Income Tax	0.1
Subsidies to Tierra del Fuego	0.8
Exemption from Income Tax for BONEX, Interest	0.1
TOTAL	2.6

Source: Gonzalez Cano (1988)

taxation on energy-related products has become progressively more important,[2] contributing to distortions in relative prices, lack of transparency in the finances of the energy sector, and jeopardizing at times anti-inflationary objectives. Ad hoc taxes have been imposed on several occasions. Also, the Government has repeatedly granted tax amnesties, with lower revenue results. Meanwhile, the administration of taxes, beset by lack of resources and excessive variability in legislation and management, deteriorated substantially.

Reforming the Tax System

3.14 A tax reform program was launched in 1985 as an integral part of the Plan Austral. Its principal aim was to increase revenues, but also to introduce changes in the structure of taxes to make the tax system more equitable, essentially through greater reliance on direct taxes. The reform also included a series of measures to increase the powers of the DGI. The relevant laws and regulations were approved between September 1985 and September 1986, with the exception of a proposed national land tax, which failed to win Congressional support. In addition, the new co-participation law discussed below was approved in December 1987. The following were the main innovations introduced: (i) Two laws substantially increased--at least in theory--the anti-evasion powers of the General Tax Administration (DGI), by establishing nominative equity shares and limiting banking and financial secrecy. These measures, advanced by international comparison, had a limited initial effect because of the administrative bottlenecks in the tax agency; (ii) The reform modified the VAT through a unification of the rates at 18 percent, that came at the expense of a reduction of the base, and introduced a simplified system for small taxpayers, a "patent" assessed on the basis of objective indicators of economic activity; and (iii) The strengthening of direct taxes included several changes to the income tax,[3] an important adjustment in the inflation adjustment method, a strengthening of the capital gains tax, an increase in the marginal rates for the tax on capital and net wealth, a reduction in the taxable minimum, and the inclusion of shares in the base for the personal wealth tax (with credit given for the taxes paid by enterprises).

3.15 Also, a "forced savings" scheme was also introduced, which in effect amounted to a one-time tax (as the interest rate it supposedly carried was extremely low). The Government repeated the tax in 1988.

2/ Taxes collected by the energy sector now amount to about 20 percent of total revenues, an unusually large proportion of revenues in a country that does not have large energy exports.

3/ Personal deductions were decreased, and their indexation system modified; deduction of interest was limited; the concept of corporate income was extended; profits of limited partnerships were attributed to the partners. The period for carry-over of tax breaks was reduced from 10 to 5 years.

3.16 On balance, the reform fell considerably short of its objectives. The erosion of the tax base due to the system of fiscal incentives to industry increased, rather than decreased, because of separate legislative action. The fragmentation of the tax system was not reduced, and the revamping of the income tax turned out to be a timid one: the additional 1987 yield was limited to about 1.5 percent of GDP. More importantly, the system for inflation indexation may have in fact contributed to a further reduction of the tax liability of corporations.

3.17 Nor did the reform produce a durable increase in revenues (Table 3.4). After the strong recovery in 1985 (which was mostly due to the positive Olivera-Tanzi effect caused by the sharp reduction in the rate of inflation), total tax revenues fell in 1986 and 1987, and collapsed again in 1988. In terms of the yield of the taxes in the reform, disaggregated data are only available up to 1986; while the income tax doubled over the all-time low of 1984, it still remains at 1.5 percent of GDP. Only marginal gains were registered for the wealth taxes, and although VAT collections were higher than in earlier years, in 1986 they stood at 3.1 percent of GDP, a far cry from the high of 4.7 reached in 1981.

3.18 The Government reacted to the deteriorating fiscal situation in 1987 and early 1988 through the adoption of several ad hoc revenue measures. In March 1988 the tax rates on bank checks were increased; surcharges on gasoline, fuels and telephone were approved in early 1988 and earmarked for the finances of the social security system and of the provinces; selected excises (particularly on cigarettes) were increased; the forced savings scheme was extended and moved forward to meet the Treasury's obligations at end-1988.

3.19 Important changes were introduced through a tax package in December 1988. Although Congress rejected a proposal to generalize the VAT and the Government introduced changes to direct taxes that modified the industrial promotion system, these changes, if sustained and implemented, could lead to a significant reduction of tax avoidance and evasion, and of the economic distortions generated by the promotion. The reform abolished the blanket tax exemption given to promoted firms, replacing it with a non-negotiable tax bond equivalent to the subsidy calculated at the moment of approval of the project. While this move leaves the value of the tax expenditures untouched, it would make it impossible to extend tax exemptions beyond limits that are known.

Issues in Tax Administration

3.20 Another major cause of the problems of the tax system is the inability to effectively administer it.4/ Several agencies are charged with collecting national taxes: the General Tax Administration (DGI), the Social Security Administration, the Customs Service, and other minor

---

4/ These issues are treated in depth in the forthcoming World Bank Study, Tax Policy for Stabilization and Economic Recovery.

TABLE 3.4: ARGENTINA - TAX REVENUES BY SOURCE, 1970-1988
(Percent of GDP)

	Average 1970-74	Average 1975-79	Average 1980-84	1985	1986	1987	1988
National Tax Revenue	14.7	15.0	17.3	19.8	19.8	16.9	14.1
Income Taxes	1.9	1.3	1.3	1.0	1.3	1.6	1.2
Property Taxes	0.5	0.4	0.9	0.9	0.9	1.0	0.9
Capital Tax	0.0	0.4	0.8	0.6	0.6	0.6	0.6
Net Worth Tax	0.0	0.0	0.0	0.0	0.1	0.1	0.1
Other	0.5	0.0	0.1	0.2	0.2	0.2	0.2
Sales and Excise Taxes	5.3	6.7	9.1	9.0	9.4	7.5	6.2
Value Added Tax	1.7	3.0	4.1	3.2	3.3	3.2	1.8
Unified Excise Tax	1.3	1.1	1.6	1.5	1.7	1.6	1.2
Tax on Bank Drafts	0.0	0.1	0.1	0.4	0.6	0.4	0.9
Oils and Fuel Tax	1.3	1.3	2.2	2.8	2.8	1.7	0.8
Stamp Duty	0.4	0.3	0.3	0.2	0.3	0.2	0.1
Foreign Exchange Transaction Tax	0.1	0.1	0.1	0.1	0.1	0.1	0.1
Other	0.6	0.8	0.7	0.8	0.7	0.2	1.3
Foreign Trade Taxes	2.0	1.8	1.9	3.2	2.5	1.9	1.2
Import Taxes	1.0	1.0	1.1	1.0	1.3	1.5	0.9
Export Taxes	1.0	0.7	0.7	2.1	1.1	0.3	0.2
Other Trade Taxes	0.0	0.1	0.1	0.1	0.1	0.1	0.1
Social Security Taxes	4.7	4.6	3.9	5.0	5.0	4.7	3.2
Other Taxes	0.3	0.0	0.1	0.0	0.0	0.2	0.7
Forced Saving	0	0	0	0.7	0.6	0.0	0.7
Provincial Tax Revenue by Source	2.2	2.5	3.5	3.4	3.8	4.9	..
Property Taxes	0.4	0.5	0.8	0.9	0.9	..	..
Gross Income Tax	0.9	1.3	1.8	1.7	2.1	..	..
Automotive License Tax	0.2	0.2	0.3	0.4	0.3	..	..
Stamp Duty	0.5	0.4	0.4	0.3	0.5	..	..
Other Taxes	0.3	0.2	0.2	0.1	0.1	..	..
Total Tax Revenue	17.0	17.5	20.7	23.2	23.6	21.8	14.1

Source: Secretaria de Hacienda; IBRD estimates.

bodies. The DGI is, by far, the most important of the agencies, both with regard to the amount of revenue collected and to its auditing and prosecuting powers. It operates over the entire national territory, and its main functions are the assessment of tax liabilities and the collection of revenues. It collects approximately 60 percent of total national revenues, corresponding in recent years to roughly 10 percent of GDP. It administers the income and capital taxes, the VAT on domestic products, the national excise taxes, and other minor taxes. Tax revenues are distributed to the National Treasury, the provinces, several special funds and, more recently, to the social security system.

3.21 One of the reasons for the inability of DGI to increase collection substantially--despite the numerous new revenue measures over the years--is that the tax system is excessively fragmented. Also it has been upset by great variability in legislation, confused or contradictory application of the law and of regulation, and often a disregard for the administrative viability of proposed tax changes. Tax amnesties, granted repeatedly in the attempt to raise revenues, have inhibited DGI's pursuit of evaders by the DGI, and created expectations of further immunity from prosecution of evasion. An additional constraint is the need of DGI to interact with other public bodies in its operations. The tax tribunal and the accounts tribunal deal with DGI's institutional processes, the first acting on taxpayers' appeals on DGI decisions, and the second verifying DGI's performance and behavior. However, the legal powers of the DGI *vis-à-vis* the taxpayers are limited, and extremely lengthy procedures can follow an unfavorable audit before a final injunction is issued. This also has contributed to the impression of insufficient willingness to prosecute tax evasion.

3.22 The effects of this unfavorable external environment were compounded by increasing strains on DGI's resources and the absence of forward-looking planning. During the 1970s, the number of registered taxpayers increased dramatically as a result of the introduction and generalization of the VAT and of other legislative modifications, as have the complex features of incentive schemes such as the industrial promotion law, which would require increased auditing ability. Resources available for these purposes, however, have failed to match the increasing demands posed by the evolution of the tax system and the increase in the number of tax payers. The total workforce of DGI has been relatively stable at less than 11,000 for the past few years (about 6,000 personnel assigned to operations in Buenos Aires, and the rest distributed among the provinces). Of the total, only 1,300 are tax inspectors, handling 1.6 million registered taxpayers. Many problems hamper the ability of the agency to retain and motivate qualified personnel. The DGI is organized as an autarkic agency, i.e., with a certain degree of independence from general public sector rules. This allows the agency to offer somewhat more attractive working conditions than other government bodies. However, the managerial scope in these matters is severely limited by existing union contracts that mandate an extremely flat compensation curve. As a result, private auditing firms have routinely utilized the DGI as a pool for qualified manpower recruitment.

3.23 The internal ability of the agency to process and utilize information has also been hampered by the lack of long-range planning and the failure to introduce more modern systems of information gathering and processing. DGI entered the 1980s without a clearly defined strategy to cope with increasing demands on its resources and with obsolete internal operating and processing systems. Compounding these problems was a long-standing tradition of politicalization of key managerial positions and a very high turnover in the Director General position: the current administration appointed four directors in the first two years of its office, and no director has ever had a tenure of three-and-a-half years.

D. Federal-Provincial Relationships

3.24 The rules governing the revenue sharing between provinces and the central administration have often been controversial and subject to arbitrary changes. According to the Constitution, all revenues collected at the national level should be subject to sharing ("co-participation"), with the exception of those taxes that are specifically earmarked for some purpose. In practice, this general principle needs to be supplemented with a resolution of two other important issues: the proportion of revenues to be shared (the primary distribution), and the rule for the sharing among the different provinces (the secondary distribution). Both of these problems have represented perennial bones of contention between the central administration and the provinces.

3.25 During the military rule, the transparency of the primary and secondary distribution of resources decreased drastically, particularly as a result of the 1980 tax reform, which essentially resulted in a sharp decrease in revenues available for sharing, while the expenditure responsibilities of the provinces were being increased. The co-participation law was allowed to expire at the end of 1985, and it was not until the end of 1987 that a new co-participation law was approved to apply to 1988. Meanwhile, the distribution of resources from the Treasury to the provinces continued to be effected on an arbitrary basis.

3.26 The new law increased the provincial share of co-participated revenues to 57.5 percent, and set clear limits to discretionary contributions from the Treasury at a maximum of 1 percent of GDP. The law also provided a political mechanism for the secondary distribution of the resources.

3.27 The approval of the law was one of the crucial elements for the success of the "separation" strategy pursued by the Government. By setting clear limits to the extent of transfers, and decreasing the uncertainty regarding the amount of resources available for each province, the law was supposed to establish rules of the game that would compel the provinces to spend within their resources limit. This was also reinforced by the closing of the rediscount window for the provincial banks at the Central Bank, that had allowed the provinces to finance their deficit particularly in 1987.

3.28 However, the law has fallen short of its objectives. Although transparency has--to a certain extent--been restored to the relationships between different levels of government, financial problems have not been resolved. Provinces continue to generate deficits because they too have experienced declining revenues associated with the erosion of co-participated taxes; also their expenditures have failed to adjust (Table 3.4). The Central Government relied increasingly on non-shared revenues (for example, the forced savings schemes); thus, 1988 marked one of the worst years in collection of co-participated taxes, and, despite their increased share, the actual revenues going to the provinces decreased sharply in real terms. Changes made to the excise tax on combustibles resulted in the abolition of revenues subject to co-participation, exacerbating their plight.

3.29 The provinces, unable or unwilling to reduce expenditures, requested once again additional funding. This was granted twice in 1988, first in March through the approval of a specifically earmarked surcharge on cigarettes, and subsequently in December, through the granting of an additional A$3 billion transfer (theoretically repayable to the Central Government in two years). The imbalance in provincial finances is likely to continue in the future unless action is undertaken.

E. Finances of the Social Security System

3.30 The social security system has had a strategic role in shaping Argentina's public finances, and represents one of the main points of uncertainty regarding the future. The problems of the social security system derive from the imbalance between benefits and the resources available to finance them. From the late 1940s to the late 1960s, the social security system provided large surpluses, since the pool of recipients of social security payments was small compared to the number of contributors (Table 3.5). These surpluses funded the expansion of other government expenditures without incurring large deficits. Subsequently, however, the aging of the population caught up with the system; the number of retirees as a proportion of active population rose, and so did the potential liabilities of the system. The ratio of the economically active population to the number of retirees fell by one-third--from 3 in 1970 to 2.6 in 1985--and is expected to fall even further. This reduces the amount of resources that can be collected to finance the system on a pay-as-you-go basis. At the same time, the pension scheme could not count on virtually any capital revenues, since the counterpart of the surpluses of the early years of operation of the system had been wiped out by the various inflationary episodes of the 1960s and 1970s.

3.31 The reaction to the crisis initially consisted of a combination of arbitrary reduction of benefits and increasing recourse to transfers of the central administration. From their legal, generous level of between 70 and 85 percent of salary, payments have fallen to under 40 percent of salary in recent years. After several court cases successfully challenged the reduction in benefits, the Government undertook a number of revenue measures,

TABLE 3.5: ARGENTINA - PROVINCIAL GOVERNMENT FINANCES, 1970-86
(Percent of GDP)

	Average 1970-74	Average 1975-79	1980	1981	1982	1983	1984	1985	1986
Current Revenues	5.03	6.89	9.51	7.99	7.29	6.51	6.48	5.07	4.95
Tax Revenues	4.82	5.94	8.30	6.91	6.24	5.18	5.27	3.80	3.81
Own	2.24	2.55	4.36	3.70	3.32	2.68	3.20	3.40	3.83
Coparticipated	2.58	3.39	3.94	3.21	2.92	2.50	2.07	0.40	-0.02
Non-tax Revenues	0.88	0.95	1.21	1.08	1.05	1.33	1.21	1.27	1.14
Transfers from Central Administration a/	1.51	1.93	0.36	1.30	0.79	3.74	2.74	4.42	4.80
Other Transfers	0.24	0.94	1.36	1.38	1.23	1.36	1.08	1.24	1.46
Current Expenditures	5.84	6.36	8.02	8.53	7.04	8.51	9.20	8.77	8.45
Surplus Before Investment	2.46	3.59	3.74	4.63	2.65	2.61	4.13	3.15	2.61
Capital Expenditures (Net)	2.04	3.46	3.53	3.42	2.48	2.86	2.62	2.56	2.68
Financing Requirements	0.42	0.13	0.21	1.21	0.17	-0.25	1.50	0.60	-0.07

Source: Ministry of Economy.
a/ For 1985 and 1986 coparticipated revenues are included in transfers from the Central Administration due to the lack of a coparticipation law.

Table 3.6: ARGENTINA - SOCIAL SECURITY FINANCES a/
(Percent of GDP)

	1950-59	1960-69	1970-79	1980-84	1985	1986	1987	1988	1989b/
Revenues	...	...	4.4	4.9	4.5	4.2	4.1	4.8	3.3
Expenditure	...	...	4.2	5.7	5.6	5.5	5.1	5.2	4.0
Surpluses	2.75	0.02	0.2	-0.8	-1.1	-1.3	-1.0	-0.4	0.7

Source: Tramite Parlamentario No. 108: Secretaria de Hacienda

a/ Period average.
b/ Projected.

with the dual objective of gradually restoring benefits to their legal level and insulating the central administration from transfers to the social security system. The measures comprised increases in wage-related contributions to the social security system and the earmarking of taxes on certain goods provided by public enterprises (gasoline, telephone, electricity). The additional revenues allowed the gradual closing of the financial gap in the social security accounts in the near term.

3.32 However, this success may be only ephemeral. Projections over the next few years indicate a continued increase in the ratio of retired people to active population. In absence of changes in the level of benefits, a continued balance in the social security accounts must rely in a continued real increase in revenues. This, in turn, must imply that either employers' and employees' contributions must keep rising, or that the revenues from the earmarked taxes on goods must rise in real terms, or both. However, the scope for either of these measures is limited because social security contributions are already very high (amounting to 42 percent of net wage). A continued increase in the earmarked revenues would imply a continued increase in relative prices of the goods to whose prices the taxes are linked. This, however, appears both improbable and undesirable.

F. Public Sector Enterprises

3.33 Public sector enterprises have contributed substantially to the overall deficit of the public sector--about a half on average during the 1980s (Figure 3.2). The continued deficits incurred in by the enterprises have been the result of several factors. The Government has pursued pricing policies that have oscillated between providing sufficient resources to cover costs and combating inflation through lagging real prices. Managerial criteria have seldom been used in formulating important business decisions. Legal constraints have been posed to public procurement, through the Compre Argentino law, resulting in inflated costs for the acquisition of inputs from the private sector. Noneconomic objectives, such as employment maintenance, shackle the sector, particularly in the railways. Finally, pricing distortions have been accentuated by the need to utilize the energy sector, in particular, as tax collector to cross-subsidize losses elsewhere.

3.34 To deal with these problems, the Argentine Government has recently pursued a dual strategy of, on the one hand, privatization as a strategy for some enterprises, and on the other hand, of strict predetermination of the rules regarding the transfers between the sector and the central administration. The most important efforts of privatization are the proposed partial sale of ENTEL (the state telephone company) and of Aerolineas Argentinas. While these have not yet been carried out because of congressional opposition, it is likely that they will bear fruit in the

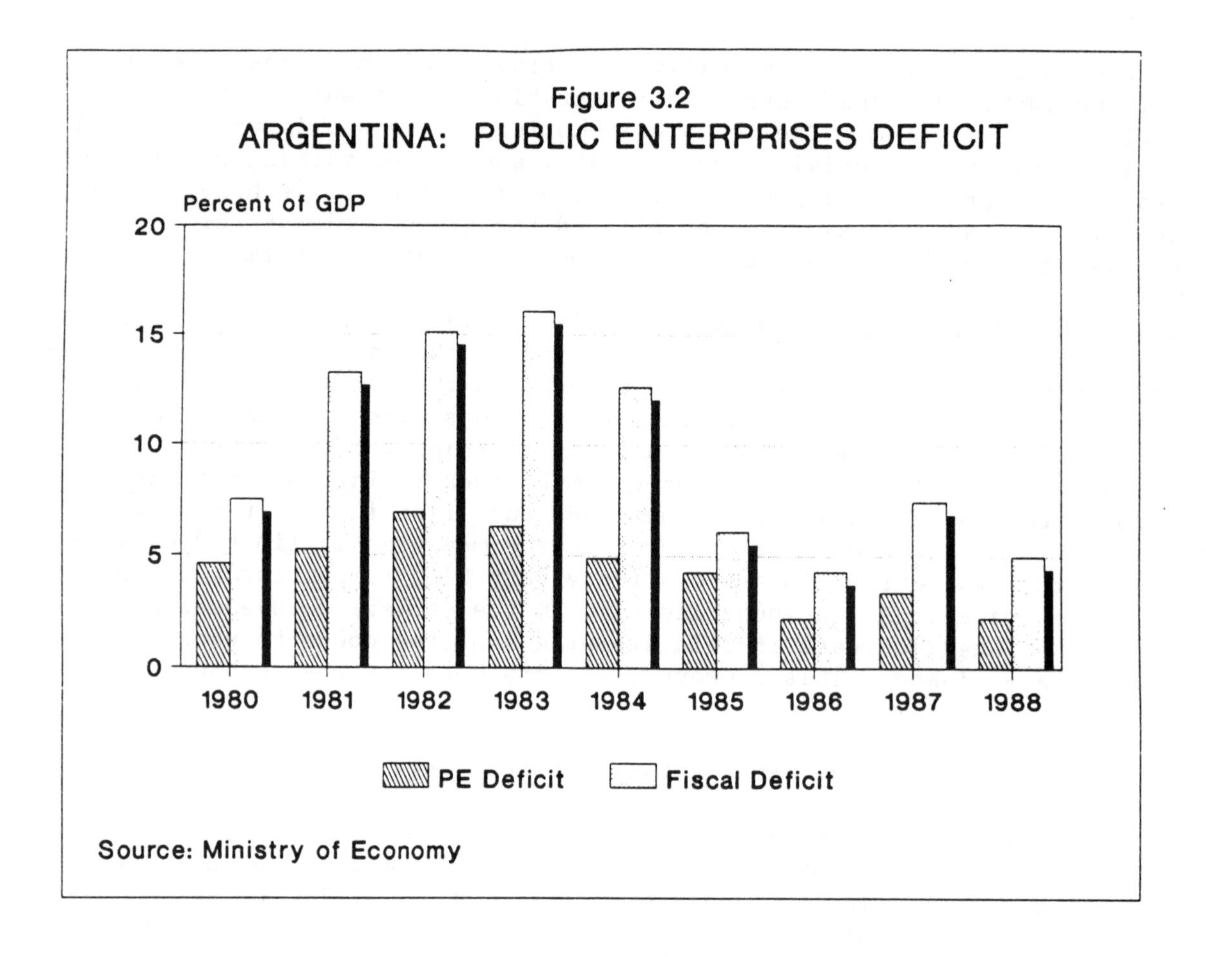

near future.5/ Through the Plan Houston and the Petroplan, private sector participation is being sought for oil exploration. On the other hand, the Government has sought to strengthen institutional and budgetary control through the creation of the Public Enterprise Board (DEP), which has been given broad authority on setting financial targets for individual enterprises. This has been accompanied by the setting of a clear rule for the transfer policy from the Central Government, namely limiting Treasury contributions to cover the servicing of financial external debt and what is due on account of special funds.

3.35 This principle has permitted a substantial reduction of the amount of transfers from the Central Government in 1988, limiting them to 1.1 percent of GDP. However, insulating the public enterprise sector from the central administration has been accomplished by an internal cross-subsidy system, whereby the money-making enterprises (essentially the state oil company) have had to provide funds to finance the loss-making ones

---

5/ Privatizations have been concluded for small government participations in firms previously handled by Fabricaciones Militares.

(predominantly the railways and social security). This system is dependent on reducing the deficit of loss-making enterprises, and guaranteeing sufficiently high real prices. Other unresolved issues (such as the royalty payments from the oil company that involve a strong subsidy element) also threaten the ability to generate sufficient resources. But the most fundamental problem with this strategy is the distortions arising from the cross-subsidies, since they lead to greater distortions in prices, discourage appropriate investment levels, and affect final demand arbitrarily.

G. Recommendations

3.36 The separation strategy pursued by the Government has been instrumental in containing the deficit in 1988, and it should remain in effect for the future. However, the fundamental problems outlined in this chapter and in more detail in Part II of this report must be resolved if a viable public sector is to emerge from the present crisis. A separation strategy without fundamental structural reform can only lead to increasing cross-subsidization, insufficient provision of public goods, and deterioration of the quality in the public administration. The Government's structural reform agenda, therefore, must include action on reducing expenditures, tax policy, public enterprises, social security, and provincial finances.

Expenditure Reductions

3.37 Expenditures of the nonfinancial public sector have fallen substantially since 1983. Current expenditure levels are still high relative to capacity for mobilizing tax resources as well as relative to other countries at similar levels of per capita income. The process since 1983 has been not part of conscious government policy, but the result of successive marginal contraction imposed by the threat of inflation. Consequently, the pace of expenditure reduction has always been too slow to achieve stabilization goals and the process itself has been inefficient. Too often the Government has contracted investment in the hopes that the fiscal crisis would pass rather than make the difficult cuts in employment and inefficient programs.

3.38 A major component of a structural reform program would involve expenditure reductions. This would have to focus on employment reduction throughout the public sector. Employment increased by 20 percent in 1983-88 in the central administration and 28 percent in the provinces--despite acute budgetary constraints. Since average public sector wages have fallen to very low levels, the employment reductions should be deep enough to permit some increase in average wages even with some gains in reducing the overall wage bill.

3.39 Similarly, the Government must mount a comprehensive effort to identify programs that could be cut to achieve savings and improve the efficiency of the public sector. In the central administration, the health, education and housing budgets should be carefully scrutinized since there is a need for greater efficiency in delivering these services and for additional savings at the same time. In education, for example, expenditures for primary and secondary education have been reduced far more than expenditures on higher education, even though the latter entail subsidies to the relatively wealthy and could be supported through a combination of increased direct charges with more scholarships for poor youth who would otherwise qualify for entrance. In transportation, the Government may wish to consider partial divestiture of the port facilities, which are inefficient and costly; greater reliance on toll roads and privately constructed toll roads could increase badly needed investment in this sector as well as reduce the cost to the budget. In agriculture, reforming the National Grain Board (Junta Nacional de Granos) along the lines suggested in Annex Chapter IX would provide some savings. The greatest savings are to be found in the public enterprise sector where divestitures and other recommendations (see below) could achieve substantial savings, and the social security administration.

Tax Policy

3.40 A reversal of the deterioration of performance must reduce tax expenditures, one of the main factors that has led to the progressive erosion of the tax base. The industrial promotion system, in turn, is the main source of these problems. Recommendations for future action would include:

(a) A drastic reduction in the scope of the industrial promotion system. The Government should take advantage of the current suspension of the system, and avoid reinstituting it;

(b) The special regimes for Tierra del Fuego and those favoring related industrial sectors should be abolished in a manner comparable to the promotion in the four provinces;

(c) The exemptions from import tariffs on capital goods and other imports should be abolished, leading to a reduction in the dispersion of effective protection rates and an increase in revenues; and

(d) Promotion schemes for nontraditional exports should be abolished and replaced by a more realistic exchange rate level.

3.41 Measures that would at the same time promote a more efficient utilization of resources, foster public enterprises reform and increase public sector revenues in the area of taxation of the energy sector, would include:

(a) A simplification of excise taxes could be based on the following guidelines. All oil products should pay the VAT. Other excise taxes should be merged in only one ad valorem tax, fixed on the supply plant price. This tax could be set at a rate such as to generate a revenue equivalent to present levels;

(b) Fuels used within the transportation sector (gasolines, diesel, and gas oil could have a higher tax rate than other derivatives to incorporate a concept of "road user charges." Equity and cost considerations might justify a higher tax rate on extra gasoline and a lower tax rate on kerosene; and

(c) Finally, YPF should be subject to income tax; if the finances of the public sector require additional transfers, the Government should use its authority to set dividends to do so.

3.42 Trade taxes have potentially distortive effects and, therefore, the move away from them should be pursued with determination. Regarding import duties, the Government should consider the removal of the exemption on capital goods imports, while maintaining the momentum towards a reduction in dispersion and lower average rates. The recent policy shift towards higher export taxes should be reversed and, if a greater contribution from the agricultural sector is desired, an increased role for income taxation or a national land tax should be considered. During a transition phase out of export taxes, an across-the-board uniform export tax could be used in lieu of product-specific export taxes; also, export tax payments be credited against value-added tax payments if and when made. A land tax already exists in most provinces, but is inefficiently administered; economies of scale could be obtained by combining the administration of a national tax with a revamped provincial tax. This would involve a substantial overhaul of existing cadastres.

3.43 Solving the problems of the tax administration will involve a determined medium-term effort on the part of the Government, as well as the implementation of measures in the tax system that will make its administration more simple. DGI needs to proceed in its effort of modernization and productivity increase. This will involve: (a) a continued high priority given to the development of the computerized taxpayer current account; (b) a review of the working of the new organizational structure by giving more emphasis to the operational structures; (c) the upgrading of the computing and data processing function and reversal of the excessive centralization of information processing; (d) a change in work practices aimed at generating exclusive dedication to DGI and increased workload; and (e) further development of the internal auditing function and of management control.

3.44 Additionally, the strengthening of administration would involve political efforts to: (a) reduce the complexity of the tax system, through the abolition of low-yield nationally administered taxes; (b) exploit, as far as possible, the figure of the substitute taxpayers, thus reducing the administrative workload for DGI; (c) consider the possibility of the complete abolition of the simplified VAT system; and (d) approve the law pending in Congress that would substantially increase DGI's powers.

Public Enterprise

3.45 **Prices.** The Government soon needs to correct the level of real prices of output. In the near term, these should be raised in real terms at least up to the average level of the second quarter of 1988. By 1990, tariffs in the sector should be sufficiently high relative to costs so as to generate a surplus on their noninterest current account equal to about 3 percent of GDP, thereby covering about half of their investment after interest expenses. Improvements over 1988 levels could be made up of a combination of either expenditure reductions or real tariff increases.

3.46 As it adjusts prices, the Government should formulate the new pricing policy to reflect an efficient resource allocation at least in terms of producer prices. In the energy sector, prices should reflect, as automatically as possible, the opportunity cost of production. Prices received by YPF and GdE should be brought up to international levels. However, consumer prices of YPF production should not rise whereas end-user prices of gas need to be increased substantially. Prices of AyEE and Hidronor's output should increase to reach the level of the long-run marginal cost of producing electricity and to lower consumption which has been unusually high in Argentina compared to similar economies. Prices of train fares need to be increased somewhat and should be kept at competitive levels afterwards. Prices of telephone services should recoup the real erosion experienced during the last semester of 1988.

3.47 **Royalties.** The issue of royalty payments to the provinces on petroleum and gas entail a heavy implicit tax on the sector to support provincial finances. The current reference price on which these royalties are paid dates from the early 1980s and is too high; it should be reduced to international levels as soon as possible.

3.48 **Budgetary Control.** The Government needs to enhance budgetary control over the public enterprises. The respective roles of the DEP (or its successor, if any), the Ministry of Economy, the SIGEP and the various Secretariates in the Ministry of Public Works should be clearly delineated. The collective budget for the public enterprises, the Ministry of Economy's cash and commitment budgets and the financial statements should all be drawn up within the framework of a uniform information system.

3.49 **Institutional Framework.** The market structure and regulatory framework of each sector requires definition so that enterprises are subjected either to the discipline of the market or well defined regulations regarding pricing. For those enterprises designated to function as regulated monopolies, the regulatory framework should be clearly established in the law so as to permit maximum managerial discretion within a carefully elaborated legal framework, thus minimizing the imposition of political and noncommercial objectives.

3.50 **Enterprise Restructuring.** Without enacting internal reforms within each enterprise to induce managerial responsibility, accountability, and autonomy, changes in the regulatory and competitive environment will not have their full desired effects. The Government should therefore continue and accelerate restructuring programs currently under discussion to strengthen management, improve personnel policies, and enhance financial controls. Management could be strengthened by restricting political appointments to boards of directors and/or a few senior-level posts as well as providing improved salaries, attention to qualifications in appointments and promotions, and sound training. These efforts could permit considerable reduction in expenditures and improvements in long-term efficiency.

3.51 **Procurement.** The Compre Argentino law should be modified to allow foreign competition in the bidding for contracts of the public enterprises. This would effectively abolish the implicit subsidy system to the private sector and reduce the operating costs of the public enterprises.

3.52 **Employment Rationalization**. Redefining the role of public enterprises surfaces the need to examine employment levels and ways to reduce public employment without causing affected workers unnecessary hardships. The Government should examine programs of early retirement; in the railways alone, one-third of the workforce is over the age of 55. It should also consider programs of severance pay and/or programs of income maintenance for laid-off workers in accordance to their tenure with the enterprise.

3.53 **Cross Subsidies**. The Government needs to redefine the mechanisms and degree of cross-subsidization among the public enterprises and needs to establish a transparent scheme for resource transfers. If the Government wants to subsidize selected loss-making enterprises, it should transfer these to the budget so that these expenditures are annually subjected to the budgetary review process. Highest priority should be given to reducing the demand for expenditures in Ferrocarriles, which absorbs most of the cross-subsidies and accumulates losses of nearly one percent of GDP annually.

3.54 **Expenditure Reduction through Divestiture and Closure**. For those enterprises where national interests permits, the Government should continue its privatization efforts. Maximum efforts should be made to accelerate those planned divestitures and review for possible inclusion remaining enterprises and activities that could be privatized. Other activities --such as some passenger lines--could be closed with considerable public savings. For divestitures, it should establish clearly defined legal procedures to ensure an unbiased selection of private investors. Aerolineas and Ferrocargo should be subjected to public tender as soon as possible. Care should be taken to define the regulatory framework for the joint-venture between Telefónica de España and the state telephone company (ENTEL) to ensure transparent tariff setting mechanism based on clearly defined criteria. For enterprises deemed necessary to remain in the public domain, enterprises should be subject to the discipline of competitive market forces and be allowed to price independently of central government dictates.

Provincial Finance

3.55 Although the Central Government managed to maintain the integrity of the new revenue sharing system, in the face of concerted demands for substantial new resources at the end of 1988, it is unclear how long it will be able to do so unless urgent reforms are undertaken to reduce the provincial public sector deficit, as well as to provide the provinces with longer term financing for their capital investments. The goal should be not just to reduce the deficit, but to transform the provinces into agents of development that generate savings surpluses which can be invested wisely to increase total provincial product, thereby augmenting their future tax revenues. The Central Government should work with the provinces to take several measures that would improve their finances. The Central Government should design incentives into future revenue sharing laws for the provinces to reduce their deficits. Provincial initiatives might include measures in several areas.

3.56 **Increase Own-Source Revenues.** Improved billing and collection procedures (e.g., better assessment, information systems and management techniques) could substantially increase provincial tax revenues without raising nominal rates (see Annex Chapter II). Improved fiscal performance would require not only better billing and more aggressive collection, but also improvements in the quality of services that should help increase the willingness to pay taxes and fees.

3.57 Nontax revenues could also be increased. To the extent possible, the cost of providing local services should be recovered from charges on the beneficiaries. Such charges should be related to individual consumption or, where this is not possible, to a measure of individual benefit received. This is clearly not the case in most of Argentina's provinces, where the "decentralized agencies" such as the public utility companies (e.g., Water and Sewerage, Electricity, etc.) often represent an important drain on public finance. Water is rarely metered, and delinquency in payment is high due to inefficient collection procedures, and also to the perception that the amounts charged are unfair, given that they are based on constructed area rather than actual consumption.

3.58 **Reduce Expenditures and Increase Their Efficiency.** Personnel policy is of highest priority in the provincial government in any effort to increase efficiency and lower expenditures. Provinces must make efforts to shed excess labor hired over the last half-decade; this could be done with systematic programs of transitional income maintenance and vouchers for vocational or other training. If the number of unproductive workers were reduced, remaining civil servants might be granted pay increases to regain some of the lost competitiveness with the private sector, while still reducing the overall wage bill. Each provincial government should receive technical assistance to do detailed studies on how to improve its efficiency in these and other ways.

3.59 Improved procurement procedures could reduce expenditures for goods and services by at least 10 percent, and would also be helpful to the modernization of industry (see Annex Chapter II). Buy-Provincial legislation should be eliminated.

3.60 Elementary improvements in organization and management could save much time and money for both the Government and those needing services. None of the six governments studied were investing more than token effort in these kinds of improvements. The use of computers could dramatically increase productivity in many areas.

3.61 **Improve the Information Systems on Provincial Public Finance.** It is impossible to do macroeconomic planning without information on the provinces when they are spending well over 12 percent of GDP and running up deficits of over US$1 billion per year. A first step in improving provincial financial management and the consolidated public accounts for macroeconomic planning should be to establish an efficient information system for the collection and analysis of the provincial budget data.

Social Security

3.62 To address the problems of Argentina's social security system, the Government might be attracted to adjustments aimed at mobilizing more resources for the current system: Increasing formal-sector employment, increasing the wage tax on the self-employed, and reducing the current transfer of 10 percent of social security income to the health insurance fund for retired persons. None of these options, nor any combination of them, offer much hope for increasing system revenues (let alone redressing the inefficiency and inequity of the present system).

3.63 A different pattern of benefits, based perhaps on an alternative theory of the function of social security, with sharply reduced pension obligations at higher income levels, may have to be introduced, to substitute for entitlements that will be unsustainable in the future. Benefits could be reduced in a way that could actually increase both the efficiency and equity of the benefits package:

(a) Present low retirement ages--age 55 for women and 60 for men--drive the system towards deficit, even at high quota rates. There is no alternative to increasing retirement age by 5 or 10 years so that years of contribution would rise relative to years of receiving benefits;

(b) Reducing the rate of salary replacement from 82 percent to about 40 percent; alternatively, the government could permit a low, basic rate of salary replacement with voluntary purchase of additional coverage; and

(c) Using an extended modest salary base for calculating pension rights (not the last salary but, for example, a 10-year real average).

A gradual phasing in of benefit changes for future beneficiaries, as was done in the United States' social security reform of the early 1980s, might offer an acceptable way of reducing otherwise unsustainable obligations.

CHAPTER IV: MONETARY POLICY: DEALING WITH FISCAL AND QUASI-FISCAL DEFICITS

4.01 The present hyperinflation is, in large measure, a manifestation of the erosion of the Central Bank's capacity to run monetary policy. The Central Bank has been tightly constrained by the need to secure financing for the nonfinancial public sector deficit and by the narrowness of austral-denominated financial markets. Yet monetary policy has often had to shoulder a disproportionate share of the burden of stabilization, leading to unsustainably high real interest rates. In addition, the Central Bank has become a loss-making financial intermediary. The losses have complicated the conduct of policy by adding a source of expansion of the monetary base. The resources the Central Bank has drawn away from the commercial banks have gone to finance relatively inefficient, if not loss-making, public-sector activities, leaving the private sector starved of resources essential for working-capital and commercial finance. The scarcity of such finance has contributed to Argentina's disappointing real economic performance over the past fifteen years.

A. The Central Bank in the Financial System

4.02 The Argentine Central Bank carries out monetary policy under an unusually strong set of constraints. The public has reduced holdings of domestic currency to a minimum, and the capital account remains virtually completely open. The balance sheet of the bank itself has been severely undermined by its utilization as a conduit for subsidies and for the absorption of the internal and external debts by the public sector. The virtual lack of access of the public sector to voluntary term lending has also thrown the burden of the financing of the deficit on the shoulders of the Central Bank.

4.03 While use of the Central Bank to pursue other objectives has a long history in Argentina, it is useful to recall the main developments in the past two decades, since they have shaped the current structure of the Central Bank balance sheet, contributed to the emergence of the problem of the quasi-fiscal deficit, and constrained the operations of monetary policy. The main events to be recalled are: the 1977 reform and the creation of the Monetary Regulation Account (CRM); the nationalization of the private external debt; and the nationalization of the private internal debt.

Reform: 1977-1980

4.04 The second Perón government effectively nationalized the banking system between 1973-76 through the imposition of 100 percent reserve ratios. In contrast, the military government that took power in 1976 started to move in a more market-oriented direction. In a compromise between liberalization (which called for low reserve ratios) and the need to establish monetary control to deal with continuing high inflation (which implied a need to restrain monetary growth), the Authorities settled on the required-reserve ratio of 45 percent contained in the 1977 reform. To minimize the effect of the reserve requirements on borrowing-lending spreads (now freely determined in financial markets), the Central Bank was directed to pay interest on commercial banks' required reserves. In turn, the interest would be financed through a fee collected on demand deposits. The Central Bank fund through which these operations were carried out was called the Monetary Regulation Account (CRM).

4.05 Early operations of the CRM involved losses for the Central Bank. Continuing inflation discouraged the public from holding demand deposits, thus limiting CRM revenues; CRM expenses remained high because the reserve accounts on which it paid interest were high. This was an instance of Central Bank decapitalization--a problem that has persisted. However, the reform eventually gained temporary success. Decreasing inflation brought real bank-deposit growth, increased commercial bank lending and intensified bank competition. The M2/GDP ratio rose from 13 percent in 1976 to a 1978-1981 average of 21 percent (Table 4.1). By 1979 the CRM losses disappeared, and the Authorities gradually reduced commercial-bank reserve requirements.

4.06 Important weaknesses emerged in the liberalization process, however, many of which have persisted. Regulation and supervision remained weak following the reform. The 1978-1981 macroeconomic policy, centered on fighting inflation through an increasingly overvalued exchange rate, effectively forced domestic interest rates to rise sharply in order to attract financial resources from overseas. As a consequence, banks began to compete for funds, and interest rate levels began to rise sharply. Sound banks were increasingly forced to match the higher interest rates offered by banks facing liquidity problems--in part because time deposits were insured. The liberalization process began to falter as credit operations became over-extended and the Government's macroeconomic policies subjected the financial system to excessive strain. Interest rates rose so high in private financial markets that the explosive growth of real private debt became a significant problem. This problem was compounded by heavy public and private external borrowing. The stage was set for an eventual "rescue operation" by the Central Bank, that would further undermine its balance-sheet.

Counter-Reform: 1980-85

4.07 A financial crisis erupted in early 1980, as several overextended banks and corporations either went bankrupt or required relief to remain in business. Meanwhile, the massive devaluation that ended the Tablita experiment also resulted in considerable difficulties for external debt service. By mid-1982 the external debt and the mushrooming corporate debt forced the Authorities effectively to bail out the entire system by ordering financial institutions to carry out a wholesale corporate-debt rescheduling for sixty months at negative real interest rates. The Central Bank effectively funded this by rediscounting it through a "consolidated loan" carrying sharply negative rates of interest in a period of near-hyperinflation. In addition, the Central Bank assumed virtually all of the private sector's external debt through an exchange-rate guarantee scheme, since it could no longer make foreign exchange available for private external-debt service. Controls on interest rates were reintroduced; marginal required-reserve ratios on regulated deposit accounts were increased virtually to 100 percent to allow the Central Bank to fund the purchase of the domestic debt. This "counter-reform"--carried out against the Authorities' longer-term objectives as an emergency measure to cope with an unprecedented crisis --returned the system to something much like its 1973-1977 structure. It

Table 4.1: ARGENTINA - MONETARY AGGREGATES, ANNUAL AVERAGES 1970-1988
(Thousands of 1970 Australes; stocks at end of year)

	M1 a/			M2 b/		
	1970 Prices	Annual Rate of Change	Percentage of GDP	1970 Prices	Annual Rate of Change	Percentage of GDP
1970	5.36	..	15.4	8.9	..	25.7
1971	5.13	-4.3	14.0	8.7	-2.4	23.8
1972	4.26	-16.8	11.6	7.4	-15.7	20.1
1973	5.89	38.2	14.3	10.0	36.0	24.3
1974	7.01	18.9	15.5	11.7	16.5	25.7
1975	4.90	-30.1	10.7	6.4	-45.4	14.0
1976	3.62	-26.0	8.5	5.7	-10.4	13.4
1977	3.14	-13.4	7.2	7.5	32.0	17.2
1978	3.35	6.8	7.7	8.5	13.1	19.5
1979	3.37	0.8	7.2	10.5	23.3	22.5
1980	3.84	13.8	7.6	11.3	7.4	22.3
1981	2.70	-29.7	6.1	9.4	-16.9	21.0
1982	2.46	-8.9	6.0	6.3	-32.2	15.5
1983	2.34	-4.8	5.5	6.5	2.2	15.2
1984	1.94	-17.1	4.4	5.3	-17.8	12.2
1985	2.44	25.9	5.6	6.2	16.0	14.2
1986	2.73	11.8	5.9	7.9	28.4	17.1
1987	3.68	34.8	5.0	11.7	47.5	15.9
1988 c/	1.66	-55.0	5.8	7.7	-34.5	23.8

Source: Central Bank of Argentina (BCRA).

a/ Currency plus private sector demand deposits.

b/ M1 plus private sector holdings of time deposits and certificates of deposit.

c/ 1988 figures are for end of third quarter.

represented another instance of the extreme difficulty of carrying out structural reforms in the financial system under unfavorable macroeconomic conditions, and put a heavy mortgage on the Central Bank's future operating ability.

4.08 Several measures accompanying the counter-reform undermined monetary control. The increased public expenditure occasioned by the South Atlantic conflict in April 1982 and the external financing constraint that resulted from the eruption of the international debt crisis led to unprecedented nonfinancial public sector deficits, and to heavy domestic credit creation through Central Bank advances. The Government determined that public enterprises were henceforth to be financed exclusively by the Treasury; accordingly, a rediscount mechanism was established at the Central Bank, which was used to cancel outstanding public enterprise liabilities with domestic commercial banks. After July 1982, the CRM fell into sharp deficit, because it had to pay interest on the higher reserve requirements at the same time that the private sector's willingness to hold money and long-term bank obligations was decreasing. This was also accompanied by monetary-base creation, because the CRM developed the practice of capitalizing interest and other adjustments on the loans rescheduled by the commercial banks, while it paid interest on the reserve requirements on short-term time deposits received by the financial system.

4.09 The banking system that emerged from the 1980-1982 crisis was thus a peculiarly distorted one, in which the Central Bank, having absorbed the private domestic and external debts, operated as a loss-making intermediary between the commercial banks' devalued assets and their resource accumulation. The operation of the system tended to ensure, moreover, that the Central Bank's losses would be financed preponderantly through money creation.

B. Dealing with the Quasi-Fiscal Deficit: 1985-1989

4.10 The progressive undermining of the Central Bank's balance sheet has resulted in an increasing claim on domestic savings to cover losses on this account. In the Argentine debate this has been referred to as the problem of the quasi-fiscal deficit.1/ Depending on economic developments, the deficit can be substantial, as was the case in 1987 and early 1989. It is defined as comprising the operating loss (or profit) of the Central Bank on account of intermediation operations, plus any adjustments necessary to take into account the fact that some rediscounts issued to the financial system and other entities may never be recoverable, i.e. they represent expenses that should be part of the NFPS budgetary allocation. The deficit can be further decomposed into an external and an

---

1/ Defining the quasi-fiscal deficit itself, however, has proven a controversial exercise. As discussed below, the annual submission of the rediscount budget to the Congress details the program of credit from the Central Bank to various entities that policy makers believe ought to be processed through appropriation by legislature. This definition --though important as it has been in limiting the recourse to Central Bank direct financing--is unduly narrow if the allocative effects of deficits of the public sector are at issue. In particular, there is no compelling reason to exclude from the deficit losses that arise from "pure" Central Bank operations.

internal component. The first represents the loss that the Central Bank incurs when servicing external debt or providing an exchange rate guarantee (net of the interest earned on gross international reserves). The second represents instead the loss (or profit) on account of domestic operations.

4.11 **Interest Rate Rules**. The vulnerability of the Central Bank to its own deficit, revealed dramatically in 1987 and in early 1989, is partly the result of different interest rate rules applicable to individual categories of assets and liabilities. These were determined by the reforms initiated after 1985. The new economic team, acting to revitalize the banking system and to limit the contribution of the CRM to the worsening inflation rate, announced a banking-sector reform package. The required-reserve ratios on different kinds of bank deposits were reduced and made more uniform. The fees on demand deposits that funded the CRM were eliminated. Commercial banks were authorized to set up unregulated time deposits with minimum seven-day maturities and subject to a series of other Central Bank regulations, which constituted an important liberalizing step. These measures enabled the banking system to capture and lend a significant quantity of resources that might otherwise have remained in the informal market. However, much of the lending was to be sequestered to finance public sector deficits.

4.12 One of the most important aspects of the reform was the institution of "forced investments" which the commercial banks were required to hold with the Central Bank to replace the high reserve requirements. The crucial difference between forced investments and the previous high reserve requirements is that the forced investments' balances would generally be unavailable to the banks even when their deposits declined; in addition, a large proportion of the interest on forced investments would be capitalized in each commercial bank's account, rather than paid out. Recourse to forced investments was increased in late 1985 through the issue of the BONOR. The interest rate on the forced investments was generally linked to the average cost of funds for the banks, plus a generous intermediation spread. Interest on the BONOR was paid in cash, but in practice the forced investment requirement was used to offset the resulting money creation. Several modifications were introduced to the system over the next two years, virtually guaranteeing the capitalization of most of the interest due.

4.13 Both the interest rate rules for Central Bank assets and the inherent quality of those assets have contributed to compounding the problem. Most Central Bank assets, as discussed, are in the form of rediscounts to failed financial institutions or to public banks such as BHN and BANADE, whose ability to service the debt is minimal. Essentially in consideration of this, the average interest rate charged on assets has tended to be de-linked from market rates, and, albeit marginally positive in real terms, has exposed the Central Bank to severe losses whenever monetary policy or other influences have resulted in high real interest rates. The most eloquent example is the December 1988 - May 1989 period.

4.14 Overall, three interest rate rules have been in use during the past two years: interest rates linked to LIBOR and the exchange rate (in practice charged only on assets, particularly export financing); interest rates fixed in real terms (charged on most rediscounts to the financial system) and interest rates linked to financial market rates. Table 4.2 shows the relative shares of different interest rules at end-1988, together with the average nominal interest rate on assets and liabilities. Due to the high market real interest rates prevailing in December 1988, the average rate charged on assets was considerably lower than that on liabilities.2/

4.15 Besides the losses that can be accumulated on account of interest rate rules, some Central Bank assets may turn out unrecoverable. Prudent practice suggests that provisions be made against these contingencies. Indeed, as discussed above, credit to the financial system has traditionally been an important and not easily controllable source of money expansion. Rediscounts to the National Mortgage Bank (BHN) and to the Development Bank (BANADE), as well as to provincial banks have been a major vehicle for granting subsidies outside the budgetary process or for financing provincial government deficits outside the co-participation mechanisms. This process became particularly marked during 1987, when the monetary base gave way to powerful expansionary pressures. Important legislative and gubernatorial elections in September made it difficult to carry out restrictive policies during the second and third quarters of the year. Central Bank "rediscount" credit to financial institutions increased sharply. The Authorities carried out some costly interventions in troubled financial institutions. In addition, the Central Bank provided rediscount credit to the BHN, which had lost access to an important block of pension funds during 1985 in a reform associated with the Austral Plan but had continued nevertheless to lend heavily. The Central Bank also provided rediscount credit to the BANADE. Over the twelve months from October 1986 to October 1987, Central Bank rediscounts increased by a monthly average of US $183 million.

4.16 Strongly voiced doubts on the soundness of the assets underlying the expansion of rediscounts led to the setting of severe rules regarding some of the rediscount lines. Beginning in late 1987, provincial banks were disallowed use of their formerly automatic window at the Central Bank; rediscounts for BHN were limited to cover only the obligations with suppliers that could not otherwise be rescinded, and the BHN ceased issuing new housing contracts; and adequate loan-loss provisioning was begun for advances to banks in liquidation. In addition, the annual rediscount program is now annexed to the national budget presented to Congress.

2/ As discussed elsewhere, the interest accrued on both assets and liabilities is mostly automatically capitalized (less than 20 percent of interest on liabilities is paid out, and virtually no interest on assets is paid in). This, in turn, has raised the question of the economic significance of the normal loss/profit, since the ability of some of the debtors ever to repay the Central Bank has been considered doubtful at best.

Table 4.2: ARGENTINA: SELECTED ITEMS OF THE BALANCE SHEET OF THE CENTRAL BANK - DECEMBER 1988

	Percent of GDP	Percent	Average Interest Rate
Assets			
Total	10.3	100.0	8.2
Market Interest Rates	2.2	21.1	
Fixed Real Rates	5.4	52.4	
Linked to Exchange Rates	0.8	8.0	
Other	1.9	18.5	
Liabilities			
Interest Bearing	8.6	100.0	11.4
Market Interest Rates	8.4	97.5	
Fixed real Rates	0.0	0.4	
Linked to Exchange Rates	0.2	2.1	
Monetary Base	5.3		

Source: Central Bank

4.17 Nonetheless, the burden on domestic resources of the Central Bank remains high (Table 4.3). Ironically, through its losses the Central Bank is preventing a full utilization of the inflation tax revenues by the non-financial public sector.

C. Constraints on the Operation of Monetary Policy

Instability Among the Sources of Monetary Expansion

4.18 Keeping the growth of monetary aggregates under control has often been difficult in view of the erratic nature of some of the sources of money base creation and of the high and variable interest rates. An important provision of the Austral Plan was the elimination of direct Central Bank financing of the nonfinancial public sector deficit, one of the main sources of expansion of the money base for a number of years. However, the public sector still represents a substantial source of direct and indirect money creation, via two main channels. First, the social security system utilizes the Central Bank as a "buffer" between the receipt of contributions and the payment of pensions. Any shortfall in revenues can thus be automatically financed through the use of the Central Bank window, leading to an increase in the base. Second, and most important in recent times, the Central Bank is also the lender of last resort in case

Table 4.3: Quasi-Fiscal Surplus of the Central Bank, 1988
(Percent of GDP)

	QI	QII	QIII	QIV	Annual
Domestic Result					
Interest received	10.0	15.8	16.8	11.5	13.5
Interest paid	11.5	14.1	12.3	10.6	12.2
Net interest	-1.5	1.7	4.5	0.9	1.4
Exchange operations	0.0	0.0	0.3	1.6	0.5
Nominal result	-1.5	1.7	4.8	2.5	1.9
External Result	-0.9	-0.2	-0.9	-0.3	-0.6
Provisions for BHN (45% nominal increase)	-1.9	-1.8	-2.8	-2.3	-2.2
Quasi-Fiscal Surplus	-4.4	-0.3	1.0	-0.1	-1.0
Adjustments for Inflation					
Net interest-bearing assets a/	1.8	2.0	3.2	0.5	1.9
Monetary Base	4.8	6.0	5.6	1.4	4.5
Total	6.5	8.1	8.8	1.9	6.4

Source: Central Bank of the Republic of Argentina (BCRA) and IBRD staff.
a/ Interest-bearing liabilities minus assets net of 45% of BHN rediscounts.

the Government is unable to roll over any amount of maturing debt. In the context of sharply reduced private sector confidence during 1988, this channel became one of the most important sources of monetary creation, since the Government was unable to roll over significant quantities of maturing debt without forcing interest rates up intolerably.

4.19 Another important source of excessive money creation has been the monetary financing of the quasi-fiscal deficit of the Central Bank itself. As discussed, the Central Bank has been simultaneously increasing the proportion of market-related-interest bearing liabilities in its balance sheet, and has allowed an almost complete liberalization of those rates. As a consequence, the interest paid on its liabilities that is not automatically capitalized has also become an important source of money creation, and one that grows worse when a restrictive monetary policy is applied. Thus, ironically, a restrictive monetary policy may lead to greater money creation in a vicious and unstable circle.

Narrowing Austral-Denominated Markets

4.20 The variability and high level of real interest rates observed in Argentina in recent times are to a certain extent the result of the application of restrictive monetary policies to shrinking financial markets. Table 4.1 shows the dramatic decrease in monetization observed in Argentina in the past few years. Narrowly defined money (M1, currency and unremunerated demand deposits) fell from 15.4 percent of GDP in 1970 to a low of of 2.4 percent at the end of the second quarter of 1985; the remonetization associated with the early success of the Austral Plan was quickly reversed when inflation picked up in 1987 and continued through the third quarter of 1988. A similar pattern is observable for other aggregates as well, albeit to a more limited extent: M4 (M1 plus time deposits and certain other assets), for instance, fell from a high of 27 percent of GDP in 1970 to about 15 percent in mid-1988. The stock of interest-bearing time deposits now amounts to about three times M1.

4.21 Inflation has caused the sharp fall in the demand for money. As a result, the monetary Authorities have found themselves with an increasingly tighter scope for maneuver. Unexpected increases in the supply of money may in fact very rapidly lead to loss of international reserves or fuel inflationary expectations.

Monetary Policy Practices: The Costs of Liberalization

4.22 The oscillation of policies vis-à-vis the financial sector between the complete nationalization of deposits and a relatively free interplay of market forces has produced considerable variation in the strategy and instruments of monetary policy over the years. Currently, Argentina's monetary programming, given a target for external reserve buildup and assumptions on inflation and growth of GDP, sets available credit to the private sector as a residual after net credit to the public sector has been determined.

4.23 To carry out the reconciliation of the demand for monetary base with the supply, the Central Bank has utilized a variety of instruments; more recently, given the difficulties involved in the use of government paper, the sterilization of excess money supply has been carried out mainly through the use of liabilities issued by the Central Bank, or through increases in forced investments. Market instruments consisted of short-term Central Bank bills (CEDEPs), whose yield was determined by the market.

4.24 The main considerations in the choice of these instruments have been cost, stability, flexibility and repercussions on the financial system. On the cost side, forced investments have not been much less expensive than Central Bank bills, although it can be presumed that, given the narrowness and proneness to nervousness of markets, an increased use of this instrument might lead to a higher risk premium being demanded. Additionally, the use of Bank bills complicates money management, since, as seen, only a portion of the interest due on forced investments is actually paid out to the banks, but all of the interest on CEDEPs is; if restrictive policies result in increased real rates, a greater burden is placed on monetary policy.

4.25 The problems associated with the use of the CEDEPs led the Central Bank progressively to reduce their use, and to rely more aggressively on forced deposits and increased reserve requirements as an instrument of monetary control. While this represents a setback in the intent to give market forces a greater role in determining the allocation of credit, it is of relatively secondary consequence. The real issue, in fact, is the continued need for recourse to austral-denominated credit on the part of the public sector, which leads to a sequestration of resources away from the private sector, whether this be in the form of market-driven bills or compulsory loans from the financial system.

D. Recommendations

4.26 Argentina's Central Bank needs to be reformed and recapitalized. The objective would be to create an institution that would carry out monetary and exchange rate policy governed by well-defined, transparent rules. Such reform and recapitalization would require a law approved by Congress, with appropriate changes in the present organic law governing the Central Bank. One way or another, the law would effectively separate the present Central Bank into an "old" and a "new" Central Bank. The old Central Bank would incorporate at least the loans to bankrupt public and private sector banks, as well as those external obligations not normally part of a central bank's external obligations--essentially those arising from the Central Bank's assumption of private external debt in the early 1980s. The Government could purchase these assets at their nominal value from the present Central Bank in order to set up the old Central Bank; in this way, the new Central Bank could be recapitalized. The Government could pay with (for instance) a ten-year bond, which it would service by turning over marketable short-term bonds that the new Central Bank could use in open market operations. The Government might also provide some additional capitalization, perhaps through the proceeds of a foreign loan, or perhaps a special bond issue. The old Central Bank would be audited and then liquidated.

4.27 The present forced investments and rediscounts could be dealt with in various ways, depending on their relative magnitudes at the moment of the stabilization. (The magnitude might be relatively small if the hyperinflation went on for relatively long.) The dynamics of the hyperinflation are such that Central Bank rediscounts to the private and public banks are rising rapidly while forced investments are being released. At the moment of stabilization, a commercial bank that still had a net liability position with the Central Bank could receive the balance in long-term government bonds (paying a relatively low real interest rate) or as new bank reserves (on which it would earn little or no interest).

4.28 If the hyperinflation continues, the Central Bank would have little choice but to release the entire stock of forced investments and/or increase the flow of liquidity rediscounts to prevent the commercial banks from becoming illiquid in the face of withdrawal demands. Eventually, the commercial banks' net liability position (i.e., forced investment less rediscounts) with the Central Bank would fall to zero--at which point the problem of the forced investments stocks would dissipate or even become negative. The commercial banks would disappear in the process, however, since the private sector will have drained their deposits from them. In the end, the commercial banks would have virtually no remaining deposit liabilities against a portfolio of uncollectable loans. The private sector would convert the currency withdrawn into dollars, so that the entire monetary system would disappear. This would amount to a complete "meltdown" of the monetary system.

4.29 The new Central Bank law would make the Central Bank directors politically independent, in the sense that, once appointed by the president and confirmed by the Congress, they might be removed only for closely defined impeachable offenses. Their terms ought to be relatively lengthy, and staggered. They would have full power to deny loan requests to any private or public agency. Rules regarding rediscounts and lending to the Government should be clear and well-defined; private credit recipients should be required to repay them in cash before subsequent credit can be granted. Credit to the Government should be prohibited.

4.30 Activities outside the normal purview of a monetary authority should be transferred outside the Central Bank. For example, in order to make public finances more transparent, and to make the Central Bank's profit-and-loss account (quasi-fiscal deficit) more accurately reflect the costs of monetary authority activity, all obligations arising from the 1983-1987 new-money and rescheduling agreements should be transferred to the Central Government. This would leave the Central Bank holding obligations only to other central banks, and to the IMF. Second, deposit insurance and superintendency activities should be carried out by separate institutions. Establishing separate Authorities for these activities would enable the Central Bank's directors to concentrate on monetary and exchange-rate policy. Finally, Central Bank rediscounts should not be used for any subsidization. For example, at present rediscount operations for export activity incorporate a subsidy element that should be eliminated.

4.31 For the financial system, the following measures are suggested: The net flow of all Central Bank credits to the BHN, BANADE, and other public banks should cease immediately. This, together with the reforms of the Central Bank, would sever the intermediation relationship between the Central Bank and the public banks. It would also imply that restructuring plans already contemplated for these banks would have to be accelerated or the institutions would have to be closed.

CHAPTER V: MEDIUM-TERM PRICE STABILITY AND EXTERNAL FINANCE

5.01 Putting Argentina on the path of sustained recovery and high growth requires steady improvement in the now stagnant domestic savings. This improvement has to be rapid enough to allow for increased investment and servicing of some portion of the external debt. A strategy to revive domestic savings would have to center on restoring price stability and therefore emphasize deficit reduction. The objective would be to reduce the nominal fiscal deficit immediately to levels of available foreign finance so that the Government would not have to borrow from the financial system or print money. However, the imbalance of the combined public sector is now so large--perhaps as much as 13 percent of GDP in 1989--that drastic measures--actions hitherto considered politically impossible--will be required rapidly to eliminate the structural fiscal deficit.

5.02 For the foreseeable future, voluntary borrowing by the Government in domestic capital markets appears impractical. A necessary requirement for price stability, consequently, will be the availability of sufficient external borrowing to finance the overall public sector deficit so as to eliminate recourse to domestic borrowing. In other words, if expenditures should rise (either for reasons of domestic policy or because of exogenous shocks, such as a rise in foreign interest rates or a real devaluation), inflation will be the ineluctable result--unless increased foreign finance is available. To achieve stability over the medium term, adequate foreign finance is therefore as essential as tight control of the fiscal deficit.

5.03 In the near term, conventional new money packages to support price stability seem out of reach. Argentina has accumulated about US$4.0 billion in arrears to external--mainly private--creditors, and ceased regular payment to commercial creditors in April 1988. Continued arrears seem unavoidable in the short term. However, the funding of a program for price stability and growth cannot long rely on a strategy of accumulating arrears. Failure to pay creditors creates uncertainty among private investors, undermines exchange rate policy, and threatens trade credit lines that are the lifeblood of foreign commerce.

5.04 Therefore, once a medium-term economic program is in place, discussions with external creditors should seek to regularize foreign financial ties, and reestablish the conditions for restored creditworthiness. This would aim: (i) to put the Government in a position to service its external debt without the need to refinance interest payments in new concerted packages; (ii) to fund adequately the Government's stabilization efforts under reasonable assumptions of economic management; and (iii) allow the private sector to attract suppliers' credits, trade credits and term finance. Since private term finance is unlikely to materialize without some price stability, controlling inflation is virtually inseparable from restored creditworthiness.

5.05 The projections in this chapter indicate that these conditions can be satisfied only through substantial reductions in the external debt. There appears to be no scenario--even prolonged recession--that would permit an adjustment in external accounts sufficient to reduce external financing requirements to the levels proposed by the commercial banks in

September 1988. The banks proposed at that time financing of US$2 billion, equal to 32 percent of interest payments for 1988-89--and this only with credit enhancement from the World Bank. Even assuming capitalization of arrears through end-1989, the economy would have to contract massively to allow the current account to move toward equilibrium at a rate consistent with these low levels of external finance. This would in effect replay the short-term pattern of adjustment in 1982-83, and would require a substantial further fall in living standards through a further real devaluation of the austral that would have to be sustained on a permanent basis. Moreover, recession would reduce savings, and increasing the trade balance would imply lower investment.

5.06 But this is only part of the problem. Since the public sector does not own the country's exports, the Government would still have to raise the resources to service the debt. During a recession, revenues would fall without autonomous policy action, yet increasing taxes would depress private consumption even further and would be politically unsustainable. Resorting to the inflation tax--the principal vehicle used in 1982-83--is no longer possible because a stable moderate inflation scenario no longer exists.

5.07 This chapter examines the magnitudes of the needed efforts--domestic and foreign--necessary to provide price stability and growth in Argentina. It presents a base-case scenario of stabilization based on improved public finances and then examines the foreign resource requirements necessary to support these efforts.1/ It should be underscored that this projection is a scenario--not a forecast--and is predicated upon a strong program of stabilization and structural reform of the public sector; until the design and implementation of the program are clear, the projections should be seen as illustrating the general magnitudes of the required internal and external adjustments; more precise projections can be prepared only after a well-defined policy course is charted. The first section focuses on realistic targets for the public sector, outlines growth objectives and draws implications for the current account of the balance of payments. The second section examines the magnitudes of external financing that would support a strong domestic program.

A. Domestic Macroeconomic Objectives

Public Finance

5.08 A successful stabilization effort can be defined as the substitution of tax-based resources to finance expenditures for inflation-based financing. A prerequisite is a reduction of the overall demands on financial markets through a decreased budget deficit. However, it is not sufficient that the deficit be brought down--the deficit must be reduced in a manner that assures sustainable progress and through instruments that do not introduce further distortions. The sustainability of the progress is as important as the size of the reduction of the deficit. The public will

---

1/ The general form of the model and detailed assumptions are presented in Annex 1.

be willing to hold money and eventually government debt only if it perceives the process to be sustainable. Consequently, the reduction of the deficit must be achieved through institutional reforms that change rules and practices that in the past have been the major sources of what might be called the structural deficit.

5.09 The objective of fiscal policy should therefore be to bring down the combined public sector deficit to levels consistent with expected net foreign lending; in the short term, foreign resources would seemingly have to come primarily from the international financial institutions and arrears to the commercial creditors. Otherwise borrowing in domestic credit markets will maintain pressure on domestic interest rates, and borrowing from the Central Bank will perpetuate inflation.

5.10 This task may not be feasible for 1989 as a whole. The loss of control over inflation this year has decimated revenue collection in real terms, increasing the deficit of the nonfinancial public sector. Moreover, the unusually high interest rates have widened the quasi-fiscal deficit of the Central Bank beyond levels experienced in the past. Increases in external interest rates in 1988, together with real devaluation implicit in exchange rate movements in the second quarter, have added an additional 2.2 percentage points of GDP to fiscal expenditures. These factors are pushing up the combined deficit of the public sector, probably to its highest level in the past five years. Finally, foreign finance is constrained, and so the deficit financeable with foreign resources obtained through orderly processes is considerably lower than in the past.

5.11 **Fiscal Adjustments.** The scenario discussed below presumes that a stabilization program will be enacted during the second semester of 1989. Moreover, it assumes that efforts will continue so that the deficit of the nonfinancial public sector will be eliminated by 1991. Assuming that the domestic component of the quasi-fiscal deficit of the Central Bank is zero,2/ the quasi-fiscal balance would then be limited to the external component of the Central Bank's operating results. The combined deficit would be consistent with a declining growth of the nominal money stock, and long-term inflation rates falling sharply from over 30 percent per month in 1989 to 4 percent annually in 1991.

2/ It is difficult to project realistically the domestic component since it is the result of the spread between the asset side income rates (based upon an average real lending rate that is marginally positive) and costs of funds on the liability side (based upon market rates). In conditions of financial stability, it is reasonable to expect that the domestic component would be slightly positive. Therefore, an assumption of zero domestic deficit is not unwarranted.

Table 5.1: Argentina - Public Sector
(Commitment Basis as % of Current GDP) a/

	1987	1988	1989	1990	1991	1992	1993	1994	1995	1996	1997	1998
Total Current Revenues	19.6	17.1	15.8	19.1	20.1	21.1	22.4	23.1	23.1	23.1	23.1	23.1
Total Current Expenditures	24.5	21.0	20.9	20.7	20.0	20.5	21.7	22.1	22.1	22.1	22.1	22.1
Wages & Salaries of General Government	4.2	3.7	3.4	3.4	3.4	3.7	3.7	3.7	3.7	3.7	3.7	3.7
Goods & Services of General Government	2.2	1.9	1.8	1.8	1.8	2.0	2.0	2.0	2.0	2.0	2.0	2.0
Transfers & Other CE	13.8	11.9	10.0	9.2	9.3	9.8	11.1	12.0	12.7	13.1	13.6	14.0
Interest	4.3	3.5	5.7	6.3	5.5	5.0	4.9	4.4	3.7	3.3	2.8	2.4
Domestic b/	0.8	0.5	0.9	1.0	0.7	0.6	0.6	0.5	0.5	0.4	0.4	0.3
Foreign	3.5	3.0	4.8	5.3	4.8	4.4	4.3	3.8	3.3	2.8	2.4	2.0
Public Savings	(3.0)	(2.5)	(5.2)	1.4	3.6	4.6	4.7	6.0	6.0	6.0	6.0	6.0
Public Enterprise Non-interest Savings	1.9	1.4	(0.1)	3.0	3.5	4.0	4.0	5.0	5.0	5.0	5.0	5.0
Capital Expenditures	5.1	4.4	3.9	3.9	4.3	4.5	4.5	4.9	5.0	5.0	5.0	5.0
General Government	1.8	1.1	0.9	0.9	1.3	1.5	1.5	1.9	2.0	2.0	2.0	2.0
Public Enterprise	3.4	3.3	3.0	3.0	3.0	3.0	3.0	3.0	3.0	3.0	3.0	3.0
Nonfinancial Public Sector	(7.3)	(5.9)	(8.0)	(1.4)	0.3	1.1	1.2	2.1	2.0	2.0	2.0	2.0
Quasi-Fiscal Surplus of Central Bank	(1.0)	(1.1)	(5.5)	(2.0)	(1.8)	(1.6)	(1.5)	(1.4)	(1.1)	(1.0)	(0.8)	(0.7)
Overall Balance Financed by:	(8.3)	(7.0)	(13.5)	(3.4)	(1.5)	(0.5)	(0.3)	0.7	0.9	1.0	1.2	1.3
External Borrowing (net)	2.0	0.2	(0.2)	(1.7)	(1.3)	(0.8)	(1.1)	(1.4)	(1.6)	(1.8)	(1.9)	(1.8)
Net Arrears and Unidentified	0.0	1.9	6.0	4.3	2.7	1.5	1.6	1.2	1.0	0.9	1.0	0.8
Net Domestic Financing	6.3	4.9	7.6	0.7	0.0	(0.1)	(0.1)	(0.6)	(0.3)	(0.1)	(0.3)	(0.3)

Source: Annex 1.
a/ Excludes provinces.
b/ Includes only the real component.

5.12 A first priority in reducing the nonfinancial public sector deficit would be to reverse the deterioration of tax revenues. This could be accomplished through: (i) an immediate reduction in tax expenditures (for example, a suspension of export subsidies, emergency suspension of the industrial promotion program and vigorous implementation of the new industrial bond scheme); (ii) increased efforts at strengthening the General Tax Administration (DGI); (iii) a rapid passage through the legislature of measures overhauling taxation of the energy sector, broadening the base of the value-added tax to replace export taxes, increasing the role of direct taxation (through an increase in the scope of personal income taxation and of land taxation); and (iv) a better integration of provincial and federal taxation on income and wealth. If these efforts were put into place, it should be possible to restore over 5 percentage points of GDP to the revenue base by 1992, including an additional increase from a positive Olivera-Tanzi effect (Table 5.1).

5.13 Current expenditures, already compressed, would have to remain under tight control. Substantial expenditure reductions would have to be achieved in the public enterprises, provinces and social security system. In the initial years (1990-91), increases in **wages** and purchases of **goods and services** would be held to the rate of growth of GDP, and rise only slightly after that time; a reform of the civil service rules which would allow for reductions in the overall workforce would permit some scope for the much needed increase in salary levels. **Transfers** to the provinces would have to be limited to amounts stipulated in the co-participation law, and, to offset automatic increases going to the provinces associated with revenue sharing in the tax reform; transfers to social security would have to fall by about 1 percent of GDP over the 1990-91 period, while transfers to the public enterprise sector (other than for agreed debt service) would have to remain at zero.3/ Meeting these targets would require additional fiscal discipline in the provinces and continued institutionalization of financial relationships between the provinces and national government as well as a major reform of the social security system. **Interest** expenditures in the short run are expected to increase because of the near-term rise in LIBOR and increased debt stock.4/ The overall objective, then, would be to constrain expenditure levels of the public sector to their current share of GDP, implying an absorption of the near-term increase in interest payments.

3/ In the projection, interest payments of the public enterprises are consolidated with the central administration, and since transfers are limited to those interest payments, the assumption of no transfers is warranted.

4/ The interest rate on commercial bank debt in Argentina is set roughly six months before due date, so interest rate changes have effect with a one semester lag. The annual rate for 1990 is 9.8 percent and 9.1 percent for 1991. Thus, when the text refers to interest rates applicable to 1990, it means the average prevailing in July 1, 1989 to June 30, 1990.

5.14 The **public enterprises** in 1987 ran a surplus on their consolidated noninterest current account, but that has eroded in the year since the Plan Primavera. Reforming the management in this sector would require institutional changes in the public enterprises designed to subject them to the discipline of the market, make price setting more autonomous from macroeconomic policy, reduce expenditures, and make spending more efficient. Adequate pricing, liberating the sector from the shackles of the "buy national" legislation (which costs as much as 1 percent of GDP), and more efficient management could readily produce a 4 percent turn around in their noninterest current account. An aggressive program of whole and partial divestiture would begin to pay dividends in the form of lower current expenditures over the medium term. These changes, together with tax reform and expenditure reductions, could increase public savings to over 4 percent of GDP by 1993.

5.15 This savings performance would permit some increase in **public investment** in 1991. At the same time, the austerity of this program would require that the public sector use its limited resources more efficiently. Since three-quarters of investment in the nonfinancial public sector (excluding the provinces) takes place in the public enterprises, they will have to be a central focus of efficiency gains. Highest on the reform agenda are elimination of buy-national procurement practices, improved competitive bidding, selective divestitures and new management arrangements, and better project selection. In addition, the Government should reallocate resources from low- (or negative-) return sectors--such as the railways--to high-return sectors, such as gas and oil production.

5.16 These measures could be expected to reduce the nonfinancial public sector deficit by more than three-fourths in 1990 relative to 1988 and eliminate it by 1991. This pattern of adjustment, after taking into account the quasi-fiscal loss of the Central Bank, virtually ends reliance on domestic borrowing from 1990 under the assumption that 2.6 percent of GDP in net foreign financing (including unidentified) is available in 1990 and 1.4 percent in 1991. External financing requirements will necessarily be met in the very short run through the accumulation of arrears to nonpreferred creditors until an internationally supported medium-term program can be worked out.

Output and Growth

5.17 Stabilizing the economy will permit the gradual recuperation of domestic savings and investment. In this scenario, savings would have to increase from their present extremely low rates of less than 7 percent of GDP to nearly 14 percent in 1994 (Table 5.2). Increased domestic savings rates are necessary to finance continued high levels of external interest payments and increased domestic investment in productive activities. The public sector, which has relied excessively on the inflation tax and sequestration of resources from the financial system, must play a leading role in the recovery of savings during the period of 1990-92, during which time private savings would be expected to respond timidly; after that, private savings could well be a driving force for growth. Stabilization has to proceed so that savers choose to save in domestic financial markets

Table 5.2: Argentina - Key Macroeconomic Indicators: Sustained Price Stability Scenario
(In percent unless otherwise specified)

	1987	1988	1989	1990	1991	1992	1993	1994	1995	1996	1997	1998
Annual Real Growth Rates:												
GDP (constant market prices)	(0.2)	(3.1)	(2.2)	2.3	2.8	2.8	3.8	3.8	3.8	3.8	4.2	4.2
GDP per capita (constant market prices)	(1.7)	(4.6)	(3.7)	0.8	1.3	1.3	2.3	2.3	2.3	2.3	2.7	2.7
Private Consumption per capita	-	(5.8)	(2.3)	(1.4)	0.1	(0.3)	1.7	1.7	2.0	2.0	2.5	2.5
National Accounts (share of current GDP):												
Total Investment	13.2	12.5	11.3	12.6	13.5	14.1	14.7	15.3	15.7	16.1	16.5	16.9
Private	8.2	7.5	7.4	8.7	9.2	9.6	10.2	10.4	10.7	11.1	11.5	11.9
Public	5.1	4.4	3.9	3.9	4.3	4.5	4.5	4.9	5.0	5.0	5.0	5.0
National Savings	8.1	10.8	6.3	8.4	10.4	11.9	12.6	13.9	14.9	15.7	16.2	16.8
Private	11.0	13.3	11.5	7.0	6.8	7.3	7.9	7.9	8.9	9.7	10.2	10.8
Public	(3.0)	(2.5)	(5.2)	1.4	3.6	4.6	4.7	6.0	6.0	6.0	6.0	6.0
Foreign Savings	5.1	1.7	5.0	4.2	3.1	2.2	2.1	1.4	0.8	0.4	0.3	0.1
Public Sector (commitment basis as % of current GDP):												
Public Savings	(3.0)	(2.5)	(5.2)	1.4	3.6	4.6	4.7	6.0	6.0	6.0	6.0	6.0
Capital Expenditures	5.1	4.4	3.9	3.9	4.3	4.5	4.5	4.9	5.0	5.0	5.0	5.0
Nonfinancial Public Sector	(7.3)	(5.9)	(8.0)	(1.4)	0.3	1.1	1.2	2.1	2.0	2.0	2.0	2.0
Quasi-Fiscal Surplus of Central Bank	(1.0)	(1.1)	(5.5)	(2.0)	(1.8)	(1.6)	(1.5)	(1.4)	(1.1)	(1.0)	(0.8)	(0.7)
Overall Balance	(8.3)	(7.0)	(13.5)	(3.4)	(1.5)	(0.5)	(0.3)	0.7	0.9	1.0	1.2	1.3
Balance of Payments:												
Exports GNFS (real growth rate)	---	12.2	(10.3)	8.2	5.4	5.3	4.9	5.1	5.0	5.0	5.1	5.1
Livestock & Cereals	---	(5.8)	(30.3)	3.7	1.6	1.6	1.6	1.6	1.6	1.6	1.6	1.6
Manufacturing	---	29.3	(5.1)	9.3	7.3	7.3	6.8	6.8	6.5	6.5	6.5	6.5
Exports of GNFS/Current GDP	10.2	11.8	16.0	16.6	17.1	17.6	17.8	18.1	17.5	16.9	16.0	16.0
Imports GNFS (real growth rate)	---	(11.0)	(17.1)	5.2	6.1	5.6	7.6	7.6	7.4	7.5	7.8	7.9
Imports of GNFS/Current GDP	9.9	8.1	11.1	10.9	11.2	11.5	11.8	12.2	12.0	11.7	11.6	12.0
Import Elasticity (including NFS)	---	3.6	7.6	2.3	2.2	2.0	2.0	2.0	2.0	2.0	1.9	1.9
Resource Balance/Current GDP	0.3	3.7	4.8	5.7	5.9	6.1	6.0	5.9	5.5	5.1	4.4	4.0
Current Account Balance/Current GDP	(5.1)	(1.7)	(5.0)	(4.2)	(3.1)	(2.2)	(2.1)	(1.4)	(0.8)	(0.4)	(0.3)	(0.1)
Terms of Trade Index (1987 = 100)	100.0	115.0	111.7	107.5	108.5	109.6	110.6	111.8	113.0	114.3	112.4	110.5
Current Account Balance (million US$)	(4,238)	(1,615)	(3,263)	(2,899)	(2,276)	(1,761)	(1,768)	(1,290)	(841)	(458)	(396)	(224)
Net Financing Requirement (million US$) a	39	2,108	4,511	3,436	2,296	1,327	1,547	1,329	1,162	1,268	1,553	1,325
Coverage Ratio b/	5.8	67.6	56.6	61.8	71.6	80.9	82.2	90.8	99.5	107.3	110.2	116.4
Debt Indicators:												
Total DOD (million US$) c/	58,299	58,810	60,771	62,616	64,129	65,122	66,064	66,467	66,354	65,787	65,082	64,124
Total DOD/Current GDP	70.4	61.3	93.1	90.9	86.8	82.2	77.1	71.6	62.8	54.6	47.1	42.7
Debt Service Total/Current GDP	8.3	8.0	14.6	13.5	12.5	11.7	12.0	12.3	11.2	9.9	9.0	8.2
Interest Total (million US$)	4,362	5,198	5,592	6,314	6,046	5,968	6,215	6,010	5,851	5,784	5,565	5,150
Interest Total/Current GDP	5.3	5.4	8.6	9.2	8.2	7.5	7.2	6.5	5.5	4.8	4.0	3.4
Preferential Debt Service/Current GDP d/	6.4	7.2	11.1	11.1	9.6	8.7	8.0	7.1	6.0	5.3	4.5	4.0
Prices and Exchange Rate (1987 = 100):												
Domestic Inflation Index (annual <>)	---	424	3904	27	4	4	4	4	4	4	5	4
Real Exchange Rate Index	100.0	90.3	133.6	144.3	144.3	144.3	144.3	144.3	137.1	130.3	123.7	123.7
GDP (billion US$)	82.8	96.0	65.2	68.9	73.9	79.2	85.7	92.8	105.7	120.6	138.1	150.1

Source: Annex 1.

a/ Net credit from sources other than IBRD, IDB, IMF and Bilaterals.

b/ The resource balance as a percentage of total interest.

c/ Includes use of IMF credit, short term debt and debt service arrears.

d/ Amortizations to preferred creditors plus total interest as a percentage of GDP.

and investors choose to invest in Argentina. It also implies that reforms in the financial system have to be consolidated so as to permit financial markets to mobilize and mediate savings efficiently, that inflation must continue to decelerate and stabilize at a new low level to make saving attractive, and that domestic income must grow to provide a basis from which to generate savings.

5.18 Higher domestic savings rates, together with foreign savings, are needed to finance new investment--investment which is a *sine qua non* for expanded export potential. In the projection, investment is seen to grow from 11 percent of GDP in 1989 to 15 percent in 1994. Much of this would go to replace badly deteriorated capital stock. The private sector would be a leading force during this period, increasing investment in response to new growth opportunities stemming from the export sector. Such improvements in both saving and investment depend critically upon private sector confidence in the macro and sectoral policy framework and on the stability of these policies over time. Exchange rates, trade and financial market policies are among the most important in this regard.

5.19 In these circumstances, it is entirely possible to foresee a moderate recovery from the long recession of 1988-89. Output could expand at rates in excess of 3.0 percent for the 1990-94 period and over 4.0 percent after that time. Exports would become a leading sector, based on industrial goods and nontraditional and processed agricultural goods. Investment too would provide a strong impetus to growth. Increases in output at these rates assume incremental capital-output ratios (ICOR) that are slightly lower than those prevailing in the positive-growth years of the 1970-87 period.5/ If the Government were to implement structural reforms in trade, finance, and the real sectors (energy, agriculture and industry) and external finance were to be available, output might well expand more rapidly and the ICOR could be expected to improve.

5.20 This scenario is austere. It would permit only a marginal recovery of per capita consumption beginning after 1992. Marginal savings rates are assumed to be high in 1990-94 (averaging .51) to support increases in investment. Only if investment were substantially more efficient than in previous periods and the economy were to grow more rapidly than in the last two decades would the austerity be eased. More rapid growth is not out of the question since reforms in the trade and industrial policy regime, agriculture, energy and public investment could readily reverse declining productivity trends, in which case the austerity in this scenario would be eased.

External Balance: Current Account

5.21 The medium-term adjustment in public finance and availability of external finance will influence the equilibrium in exchange markets, which is anticipated to make Argentine exports quite competitive. A potential

---

5/ The simple average of the ICORs for the 14 years with non-negative growth was 6.0; the assumption of the projection is 4.3 for 1990-94 and 4.0 thereafter.

near-term deterrent to exports has been the introduction of an emergency export tax on all exports; this is expected to be replaced with a more efficient tax system in any medium-term scenario.

5.22 The near-term external environment, while remaining open to Argentine exports, would be decidedly less favorable to growth than in previous years because of slower OECD growth and rising interest rates in the near term. Over the medium term, the external environment should be rather propitious, especially if the imbalances in the US fiscal accounts are reduced and protectionist sentiments do not gain ascendance.

5.23 In this base-case medium-term scenario, exports stimulated by reforms of the trade regime and the prevailing competitive exchange rate would play a leading role in economic growth, expanding by more than 5 percent annually for the period 1990-94. Although the drought in 1989 will probably have sharply negative effects, especially on soya production, rapid growth in nontraditional exports would be the dynamic sector. Slower OECD growth will mean continued slow expansion for nontraditional exports of cereals, grains, and livestock,6/ but there is also some room for the growth of agricultural exports beyond the growth of consumption in industrialized countries, as Argentina's loss of market shares in the recent past has mainly been due to supply factors. Nontraditional exports, led by new capacity in chemicals, plastics, machinery and transport products, could grow at over 7 percent annually for the period. Exports would rise from about 12 percent of GDP in 1988 to 18 percent by 1994. This policy scenario implies a competitive exchange rate for all exports, at least as high as the average level prevailing in 1987.

5.24 Imports, meanwhile, would be subject to the twin offsetting forces of trade liberalization and a higher exchange rate; after contracting in 1988-89, real imports are projected to rise rapidly at first--about 6 percent over 1990-94--as past severe contractions are overcome.

5.25 This would produce a trade balance that would rise from US$3.5 billion in 1989 to US$5.1 billion by 1994 and US$5.9 billion by 1997 (Table 5.3). The trade and nonfactor service balance would be sufficient to cover a rising share of the interest burden. In 1989, the ratio of the resource balance to total interest payments is anticipated to be about 56 percent; under the scenario assumption, this interest coverage would rise to nearly 80 percent by 1992, and surpass 90 percent in 1994 (Table 5.3).

6/ The IDB estimated that a sustained one percent fall (rise) in OECD growth produces a cumulative 0.8 percent fall (rise) in Argentina commodity export earnings over a six year period; the effect in the same year as the initial one percent growth change is 0.4 percent. See IDB, External Debt and Economic Development in Latin America, 1984: 131-137.

5.26 This trade performance, together with a LIBOR rate of 9.8 percent in 1990 falling to 8.4 percent in 1991 and roughly 8-9 percent thereafter, would produce some improvement in the current account balance over the coming years, albeit slowly. Fiscal adjustment will have to be a driving force in improving the net foreign asset position of the country over the medium term. In the short term, the rise in LIBOR will negatively affect the current account in 1989 and 1990, estimated to be US$3.3 billion and US$2.9 billion respectively. The deficit would decline to US$1.3 billion by 1994 as improvements in the trade balance would be partially offset by interest payments on increased indebtedness. Nonetheless, because the economy is growing at a reasonable and sustainable rate, the ratio of the current account deficit to GDP is projected to fall from about 5 percent in 1989 to under 2 percent by 1994.

B. External Financing

Financial Requirements under Austerity

5.27 Even with the strong efforts at public adjustment and austerity described above, Argentina would require large amounts of external finance in the form of new money or interest capitalization. The base scenario assumes that public external creditors continue to lend to Argentina while bondholders and financial markets do not.7/ For 1990-94, net borrowing requirements would average about US$2.0 billion per year (Table 5.4), roughly 50 percent of interest owed to commercial banks and suppliers of unidentified finance in 1990-94. It is important to note, however, that the trend over the period would be toward a lower current account deficit.

5.28 The upside potential for additional finance should be underscored. Non-debt-creating flows in the form of foreign direct investment could well play a more important role if the policy environment were conducive. Official resources--from the Paris Club, IDB and World Bank--could be substantially larger if the policy framework were especially strong. Moreover, two to three years of financial stability could evoke potentially large reflows of flight capital. Argentine foreign assets are about equal to its foreign debt and some presumably large portion could be attracted back to Argentina if the macroeconomic environment were to provide evidence of sustained stability and open up potentially high-return private investments. Nonetheless, these flows cannot be relied upon and would be justified only in the most optimistic scenarios.

7/ Specifically, the scenario assumes that the World Bank and the IDB will each continue to contribute net disbursements in accordance with their lending programs of US$200-300 million per year over 1990-94 (Table 5.4). Bilateral creditors are assumed to provide US$400 million per year in new commitments; this level would be insufficient to prevent net negative disbursements of about US$200-300 million per year during the same period. Suppliers are expected to contribute US$50 million in 1990, rising to US$150 million per year in 1992 and thereafter; this would offset existing payment obligations and leave their net disbursements marginally positive. Commercial banks and bond holders would be repaid together with roughly US$1 billion per year of debt that is not refinanceable.

Table 5.3: Argentina - Balance of Payments, 1987-1998
(In Millions of Current US Dollars)

	1987	1988	1989	1990	1991	1992	1993	1994	1995	1996	1997	1998
Trade Balance	540	3,810	3,459	3,585	3,986	4,461	4,752	5,111	5,506	5,934	5,936	5,889
Exports of Goods	6,360	9,134	8,159	8,668	9,628	10,678	11,762	13,013	14,375	15,889	17,257	18,762
Imports of Goods	5,820	5,324	4,700	5,083	5,642	6,217	7,009	7,903	8,869	9,955	11,321	12,874
NFS Balance	(285)	(298)	(295)	320	342	365	358	346	315	273	199	106
Exports of NFS	2,112	2,167	2,279	2,744	2,978	3,229	3,503	3,800	4,121	4,473	4,860	5,274
Imports of NFS	2,397	2,465	2,574	2,424	2,636	2,864	3,145	3,454	3,806	4,200	4,661	5,168
Resource Balance	255	3,512	3,164	3,904	4,328	4,826	5,111	5,457	5,821	6,207	6,135	5,994
Net Factor Income	(4,485)	(5,127)	(6,427)	(6,804)	(6,604)	(6,587)	(6,878)	(6,747)	(6,662)	(6,665)	(6,531)	(6,219)
Factor Receipts	218	211	254	240	229	232	255	255	261	276	283	280
Factor Payments of which:	(4,703)	(5,338)	(6,681)	(7,043)	(6,833)	(6,818)	(7,133)	(7,002)	(6,922)	(6,941)	(6,814)	(6,499)
Dividend Repatriation	(558)	(660)	(675)	(729)	(787)	(850)	(918)	(992)	(1,071)	(1,157)	(1,249)	(1,349)
MLT interest (exc. IMF)	(4,241)	(4,605)	(5,029)	(5,776)	(5,565)	(5,516)	(5,756)	(5,578)	(5,437)	(5,376)	(5,172)	(4,782)
Other	96	(73)	(977)	(539)	(480)	(453)	(459)	(431)	(414)	(408)	(392)	(367)
Net Current Transfers	(8)	0	0	0	0	0	0	0	0	0	0	0
Current Account Balance	(4,238)	(1,615)	(3,263)	(2,899)	(2,276)	(1,761)	(1,768)	(1,290)	(841)	(458)	(396)	(224)
Net Direct Investment	(19)	1,147	734	793	888	959	1,036	1,118	1,208	1,305	1,409	1,522
Debt Conversion	0	382	365	300	300	300	300	300	300	300	300	300
Other	(19)	765	369	493	588	659	736	818	908	1,005	1,109	1,222
Net MLT Flows	1,729	(252)	(640)	(949)	(594)	(237)	(528)	(879)	(1,229)	(1,800)	(2,238)	(2,265)
IBRD	662	299	227	8	231	373	347	361	298	(93)	(310)	(325)
IDB	29	78	325	163	289	387	392	459	385	258	187	124
Bilaterals	384	477	79	(170)	(238)	(162)	(397)	(313)	(355)	(244)	(279)	23
Financial Markets	1,244	454	0	0	(263)	(396)	(793)	(1,311)	(1,573)	(1,706)	(1,832)	(2,094)
Bonds	(96)	(656)	(666)	(753)	(562)	(452)	(118)	(119)	0	0	0	0
Supplier Credits	(492)	(685)	(536)	(31)	39	78	99	102	73	50	24	6
Private Non-guaranteed	(2)	(219)	(69)	(166)	(90)	(65)	(58)	(57)	(58)	(65)	(28)	0
Net ST Flows	113	(39)	(1,530)	(242)	(140)	(97)	(76)	(47)	(45)	(35)	(20)	(18)
Unidentified Sources (Net)	39	2,108	4,511	3,436	2,296	1,327	1,547	1,329	1,162	1,268	1,553	1,325
Overall BOP Surplus	(2,334)	1,358	(120)	139	174	191	210	231	254	280	308	339
Financing:												
<> in Net Reserves (-=Inc.)	2,334	(1,358)	120	(139)	(174)	(191)	(210)	(231)	(254)	(280)	(308)	(339)
<> in Gross Reserves (-=Inc.)	1,111	(1,785)	1,858	(139)	(174)	(191)	(210)	(231)	(254)	(280)	(308)	(339)
Net IMF	614	18	38	0	0	0	0	0	0	0	0	0
Purchases	1,253	541	750	689	970	855	590	968	932	723	779	950
Repurchases	(639)	(523)	(712)	(689)	(970)	(855)	(590)	(968)	(932)	(722)	(779)	(950)
Other BCRA Reserves	609	409	(1,776)	0	0	0	0	0	0	0	0	0
Memo:												
Gross Reserve Level	3,075	4,860	3,002	3,141	3,315	3,506	3,716	3,947	4,202	4,482	4,790	5,128
Liquid Reserves a/	1,617	3,363	1,505	1,644	1,818	2,009	2,219	2,450	2,705	2,985	3,293	3,631
Gross Unidentified	39	2,108	4,511	3,436	2,303	1,685	2,656	3,011	3,228	3,615	4,337	4,259

Source: Annex 1.
a/ Foreign exchange plus commercial bank deposits in dollars.

New Money

5.29 It seems unlikely that amounts of this magnitude will be forthcoming in the conventional framework of new money arrangements of the 1980s. The discussions between Argentina and commercial creditors in the wake of the Plan Primavera resulted in the commercial banks' offer of only US$2 billion for the 1988-89 financing period, about half the amount that the Argentine Government proposed as necessary to close the gap; moreover, this offer was contingent upon guarantees from official resources.

5.30 The fact that the Government ceased regular servicing of its external commercial long-term debt in April 1988 complicates any debt rescheduling. It paid two installments totalling US$170 million in interest in 1988 to facilitate on-going discussions with commercial banks. By end year 1988, arrears on commercial banks debt totalled US$2.0 billion. Other arrears brought the total to US$2.5 billion.8/ During the first semester of 1989, the Government accumulated additional arrears to the commercial banks for an estimated US$800 million.

5.31 No less important, indicators of creditworthiness, while showing some improvement over the ten-year period, are not sufficiently better to remove the shadow of uncertainty over private sector borrowing for term finance, let alone facilitate eventual future borrowing for the public sector. The debt-to-GDP ratio under the scenario of price stability would fall from over 90 in 1990 to under 50 by 1997; the interest burden, under the assumption of a LIBOR of 8-9 percent in 1992-98, is improved relative to the peak year of 1990, but would not fall below 5 percent of GDP until 1996 (Table 5.2).

5.32 Macroeconomic management appears to have no alternative but to seek substantial debt service reduction. The projections show the need to lower the net external transfer in the short and medium term--and in a way consistent with sustained and substantial improvement in debt indicators. A strong macroeconomic program supported by an agreement to reduce debt service would position the Government to service its remaining external debt without the need to resort to new concerted lending packages; it would provide adequate external finance for the medium term, and thereby reduce one source of uncertainty hampering private investment; and it would facilitate the private sector's efforts to obtain regular trade credits, suppliers' credits and term finance. The Government should, therefore, work as closely as possible with official institutions to design a strong macroeconomic program that would provide an incentive for all external creditors to participate in resolving Argentina's external financing problem in a medium-term framework.

8/ Arrears to the Paris Club amounted to US$230 million; the remaining US$270 million was distributed among arrears to Bolivia (US$80 million), suppliers (US$80 million), bondholders and multilateral organizations.

Table 5.4: Argentina - Financing Requirements, 1990-94
(In Millions of Current US Dollars)

	1990	1991	1992	1993	1994	Total	Average Annual
Net Borrowing Requirement: Uses	3,391	2,759	2,350	2,364	1,782	12,646	2,529
Current Account Deficit	3,722	3,094	2,735	2,743	2,202	14,496	2,899
Change in Gross Reserves	139	174	164	214	221	912	182
Other Capital Flows n.e.i.	0	0	0	0	0	0	0
Less: Direct Foreign Investment	(471)	(509)	(549)	(593)	(641)	(2,762)	(552)
Net Borrowing Requirement: Sources:	3,391	2,759	2,350	2,364	1,782	12,646	2,529
Identified Net Flows:	(1,191)	(817)	(463)	(949)	(1,279)	(4,700)	(940)
World Bank	8	221	376	301	307	1,213	243
IDB	163	289	387	392	459	1,690	338
Bilaterals	(170)	(311)	(294)	(529)	(446)	(1,751)	(350)
Financial Markets	0	(263)	(396)	(793)	(1,311)	(2,763)	(553)
Bonds	(753)	(562)	(452)	(118)	(119)	(2,004)	(401)
Change in Central Bank Reserve Liab. a/	0	0	0	0	0	0	0
Others b/	(439)	(191)	(84)	(202)	(169)	(1,085)	(217)
Unidentified Net Flows:	4,582	3,577	2,813	3,313	3,061	17,346	3,469
Memo:							
90% of Paris Club Amortizations	440	563	550	736	670	2,959	592
Financial Market Amortizations	0	263	396	793	1,311	2,763	553
Net Financing Requirement c/	4,142	2,750	1,867	1,784	1,080	11,624	2,325

Source: Annex 1.

a/ Includes net IMF.

b/ Supplier credits plus private non-guaranteed.

c/ Unidentified net flows less 90% of Paris Club amortizations and already rescheduled amortizations to commercial banks.

ANNEX CHAPTER I: PUBLIC ENTERPRISES

A. Main Issues

1.01 The public enterprises in Argentina have, in the aggregate, constituted the most persistent and important source of deficit spending in the public sector. About half of the fiscal deficit can be attributed to the largest public enterprises (PEs). Repeated attempts to improve their performance have failed due to a combination of adverse political and economical circumstances, partially within the state enterprise sector but also related to the broader macroeconomic environment. The Government has been trying to reduce the financing needs of the PEs but has continued to impose a number of noncommercial objectives and to use the price of public production in the pursuit of its anti-inflationary policy. At the same time special interests have created a regulatory environment that paralyses the PEs and that gives rise to a complex system of inefficient transfers.

The Deficit of the PEs

1.02 In the last eight years, the deficit of the PEs before transfers ranged between 2.2 and 6.9 percent of GDP. These deficits represented between 40 and 60 percent of the overall nonfinancial public sector (NFPS) deficit (Annex Figure 1.1 and Annex Table 1.1). The magnitude of these deficits should be interpreted against the size of the major PEs. Annex Table 6.2 shows that the total value of public production has averaged around 11.5 percent of GDP. Its share in GDP has been declining during the last two years. The output of YPF (the State Petroleum Company) equals about 5 percent of GDP while the electricity companies represent 2 percent and the communications and transportation sector each around 1.5 percent of GDP.

1.03 In the recent past, the deficit of the PEs has fallen in absolute value due to some cost reductions but, more importantly due to cuts in capital spending. In the period 1986-88, the deficit averaged at 2.6 percent of GDP while in the period 1980-85, its average was 5.4 percent.

1.04 The fuels sector has been responsible for the observed changes in the total deficit. Since 1986, both YPF and the gas company (GdE) have reduced their deficits to below 0.4 percent of GDP from more than 2 percent in 1984 (Annex Table 1.3). The three electricity companies (AyEE, SEGBA, and HIDRONOR) have had a constant deficit of 1 percent of GDP. In the transport sector, Ferrocarriles (the railroad company) is responsible for a deficit of 1 percent of GDP as well. A striking observation is the fact that the deficit of Ferrocarriles is twice as high as its operational income. Though most of the other companies' financial borrowing requirements are due to investment Ferrocarriles' total operational expenditure has been twice as high as its revenue.

Institutional Environment

1.05 Public enterprises in Argentina are present in almost all areas of the domestic economy. In addition to the 13 most important enterprises under the supervision of the Public Enterprise Board's (DEP) and the

PART II

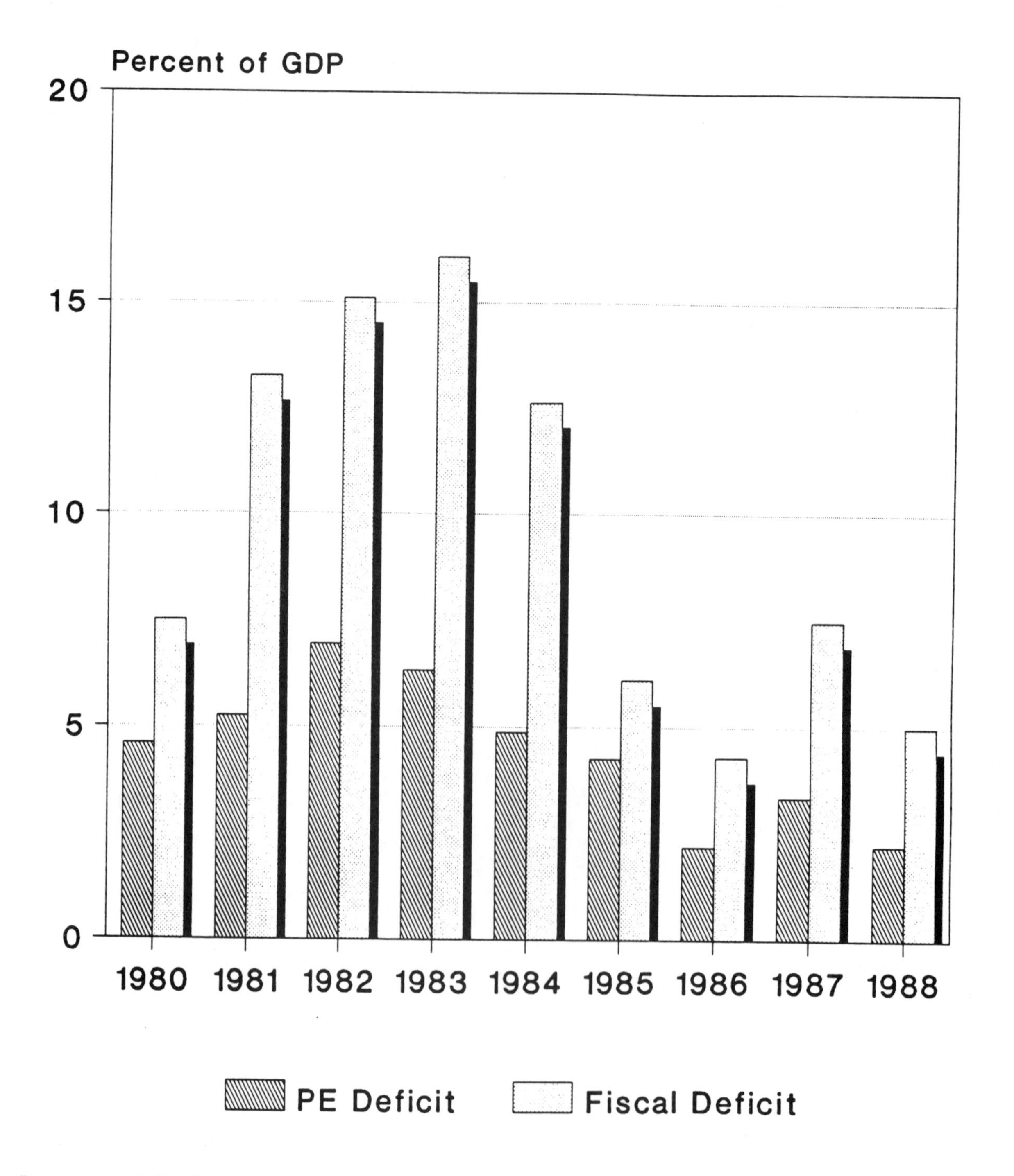
Annex Figure 1.1
ARGENTINA: PUBLIC ENTERPRISES DEFICIT
Percent of GDP
20
15
10
5
0
1980
1981
1982
1983
1984
1985
1986
1987
1988
PE Deficit
Fiscal Deficit
Source: Ministry of Economy

Annex Table 1.1: ARGENTINA - PUBLIC ENTERPRISE ACCOUNTS AND THE FISCAL DEFICIT a/
(Percent of GDP)

	1980	1981	1982	1983	1984	1985	1986	1987	1988 b/
Current Account	-1.24	-1.91	-3.87	-2.57	-1.44	-1.51	0.45	0.00	0.87
Revenue	8.64	10.27	9.72	11.03	10.53	13.57	12.06	11.45	11.45
Expenditure	9.87	12.18	13.59	13.60	11.98	15.08	11.61	11.45	10.58
Personnel	3.35	2.96	2.20	3.03	3.21	3.08	2.91	2.71	2.82
Good and Services	3.77	4.55	6.11	7.22	6.36	8.91	6.78	6.43	5.86
Interest	1.58	3.52	4.09	2.46	2.03	2.52	1.54	1.20	1.40
Other	1.19	1.15	1.19	1.07	0.38	0.58	0.73	0.66	0.50
Capital Account	-3.35	-3.33	-3.04	-3.73	-3.43	-2.75	-2.66	-3.35	-3.19
Revenue	0.21	0.16	0.36	0.15	0.14	0.19	0.11	0.10	0.06
Expenditure	3.57	3.49	3.41	3.88	3.57	2.94	2.77	3.45	3.25
Financing Requirement (1)	4.59	5.24	6.91	6.30	4.87	4.26	2.20	3.35	2.22
Fiscal Deficit (2)	7.48	13.26	15.11	16.08	12.64	6.09	4.30	7.46	5.00
(1)/(2)	0.61	0.40	0.46	0.39	0.39	0.70	0.51	0.45	0.44
Transfers from:									
Treasury	0.72	0.77	1.24	6.20	2.37	1.86	1.53	1.59	1.11
Special funds	0.78	0.78	0.67	0.84	0.67	0.81	0.66	0.80	0.90

Source: Ministry of Economy and DEP.

a/ Includes 13 DEP enterprises and binationals
b/ Preliminary

03-Mar-89

Annex Table 1.2: ARGENTINA - TOTAL OUTPUT OF DEP ENTERPRISES
(Percent of GDP)

	1984	1985	1986	1987	1988
Combustibles	5.44	7.80	6.23	6.29	6.12
YPF	4.26	6.10	4.68	4.97	4.76
GdE	1.16	1.62	1.48	1.27	1.29
YCF	0.02	0.08	0.07	0.05	0.07
Electricity	1.91	2.17	1.95	1.92	1.92
AyEE	0.73	0.87	0.78	0.73	0.68
SEGBA	1.06	1.14	1.03	1.03	1.07
HIDRONOR	0.12	0.16	0.14	0.16	0.17
Communications	0.94	1.27	1.75	1.41	1.48
ENTEL	0.70	0.96	1.40	1.07	1.23
ENCOTEL	0.24	0.31	0.35	0.34	0.25
Transportation	1.65	1.78	1.58	1.59	1.64
FA	0.50	0.49	0.43	0.34	0.40
AA	0.72	0.85	0.81	0.89	0.84
ELMA	0.32	0.31	0.27	0.29	0.32
AGP	0.11	0.13	0.07	0.07	0.08
TOTAL	9.94	13.02	11.51	11.21	11.16

Source: DEP and company accounts.

Annex Table 1.3: ARGENTINA - SELECTED PUBLIC ENTERPRISE DEFICITS
(Deficit=+)

	1984	1985	1986	1987	1988
		Percent of GDP			
YPF	1.25	1.57	0.17	0.08	0.15
GdE	0.76	0.44	0.24	0.28	0.09
AyEE	0.37	0.50	0.44	0.46	0.33
SEGBA	0.26	0.26	0.16	0.40	0.20
HIDRONOR	0.24	0.23	0.18	0.15	0.18
Ferrocariles	1.42	1.02	0.98	0.96	0.81
ENTEL	0.14	-0.07	-0.19	0.33	0.32
		Percent of enterprise revenue			
YPF	29	26	4	2	3
GdE	66	27	16	22	7
AyEE	51	57	56	63	48
SEGBA	25	23	16	39	18
HIDRONOR	200	144	129	94	105
Ferrocariles	284	208	228	282	200
ENTEL	22	8	14	31	26

Source: DEP and company accounts.

03-Mar-89

bi-national entities, the Central Government owns nearly 100 smaller enterprises including radio and television, hotels and airlines. Provincial and municipal governments own about 80 enterprises, and national and local governments share the ownership of other 30 enterprises. Some of these enterprises were acquired to prevent their liquidation, while others were acquired for social and political reasons which may have long since lost relevance.

1.06 It has been difficult for the Government to improve control over the public enterprises. Most of the changes in the institutional environment have been short-lived and did not lead to a durable improvement in the performance of the public enterprises. Since the beginning of 1988, DEP operates as a holding for the 13 largest public enterprises and controls the execution of the budgets of these 13 enterprises. This institution revives the idea of the Corporation of National Enterprises (CEN) that functioned between 1974 and 1979. A political struggle with the different Secretariats in the Ministry of Public Works ultimately led to the abolishing of the CEN. The DEP faces the same problem, but has the additional disadvantage that it was established by executive decree rather than by law. The DEP has functioned independently in the first half of 1988, but thereafter became subordinate to the macroeconomic policy of the Central Government.

1.07 Although the creation of the DEP in its present form has helped to improve the efficiency of the PEs, it has not led to financial autonomy because of the continuing application of inefficient rules and regulations governing operations. Many of the public enterprises must keep a large amount of workers on their payroll, disregarding changes in business conditions under which they operate. They all must comply with the Compre Argentino law that regulates procurement and amounts to a subsidy of the private domestic industry. Social considerations keep prices of some public production deliberately low to subsidize particular consumer groups and there is a complex scheme of inter-enterprise subsidies and transfers within the public sector generally leading to inefficient investment decisions.

Links with Macroeconomic Policy

1.08 On top of the intrinsic inefficiencies created by the over-regulated environment and the problems with budgetary control, the Government has been using the public enterprises for two macroeconomic objectives: (i) balance of payments financing, and (ii) lowering inflation. Public enterprises incurred large amounts of foreign debt during the period 1978-82 in order to satisfy the Government need for foreign exchange. The burden of the ensuing debt service became so overwhelming that the Government decided to take over the debt service in 1983. Unfortunately, this foreign debt is still present in the enterprises' accounts, preventing them from obtaining independent financing and distorting their financial picture. Repeatedly the Government used the price of public production to mitigate accelerating inflation. As a result real prices have been very volatile and their manipulation is linked to the size of the operational deficits of the public enterprises. Consequently, sources of deficit can be found on both revenue and expenditure side of the PE accounts.

B. Revenues: Pricing Policy

1.09 The Central Government has often overruled the price setting power of other institutions within the framework of its attempts to lower inflation. Under normal circumstances, the Secretariat of Energy sets energy prices while the DEP makes recommendations for all other companies under its purview. The Ministry of Public Works subsequently endorses the recommendations of the DEP. This implies that the Secretariats of transport and communication have lost their role in the pricing policy concerning the enterprises in their sector.

1.10 Even if one abstracts from periods in which prices of public production are used as an instrument of anti-inflationary policy, prices do not adequately reflect the cost of efficient resource allocation in the Argentine economy. They are often set to obtain specific targets related to the PE deficits or to a particular subsidy policy of the Government rather than in relation to costs. As such, these prices are not adequate to guide investment decisions in the domestic economy.

1.11 The anti-inflationary policy of the Government has induced a cyclical pattern in the real price evolution over the period 1980-88 (Annex Figure 1.2). After a period of falling real prices of public production, the need to mitigate the PE deficits or the failure of the anti-inflationary program led to a recuperation of these prices. The period 1980-1988 is marked by three such episodes (Annex Figure 1.2): (i) in 1982, real prices fell by almost 20 percent below their average 1981 level and took more than two years to recover; (ii) in the second half of 1986 and throughout the first three quarters of 1987, prices of the public enterprises were used once again to try to reduce inflation. In the first half of 1988, real prices were allowed to recoup their past loss and in the second quarter of 1988 reached again their average 1985 level; and (iii) since the implementation of the Plan Primavera in August-September of 1988, real prices have deteriorated again during the Government's renewed attempt to reduce inflation. The Government anticipates a substantial increase in the real output price in mid-1989 to correct the current situation. In terms of the enterprises' own cost, real prices fell in 1987 by about 5 percent and started to fall again in the third quarter of 1988 (after recuperating by 10 percent) as a result of the Plan Primavera (Annex Table 1.4).1/

1.12 These fluctuations in real prices translate almost by definition into similar fluctuations in the PE current account. The real price reduction of 1982 is associated with a fall in PE savings by 2 percent of GDP,

---

1/ The real price index computed in Annex Table 1.3 overstates the volatility of the actual real output prices because it uses the nonagricultural wholesale price index as a proxy for the cost evolution in the public enterprises. A large part of these costs, inter-company deliveries, interest payments and salaries, do not evolve according to that index. Unfortunately, comparable data using this weighted cost index are available only for 1987 and 1988.

Annex Figure 1.2

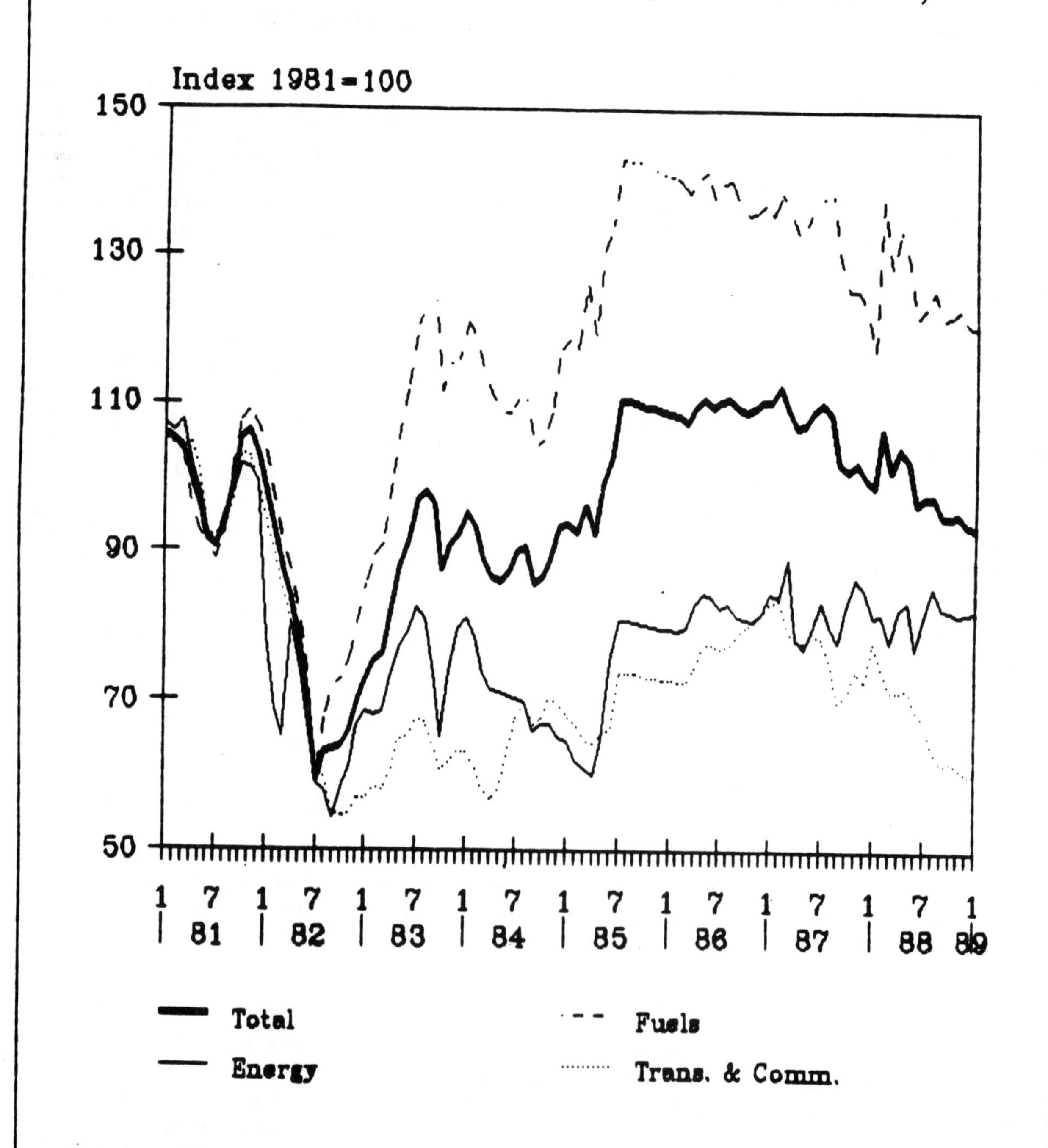

Annex Table 1.4: ARGENTINA - REAL PRICE OF PUBLIC PRODUCTION [a]/

Index 1980=100

	1981	1982	1983	1984	1985	1986	1987[b]/				1988[b]/			
							I	II	III	IV	I	II	III	IV
General level	104.5	84.9	100.5	101.6	120.2	117.5	115.39	109.62	106.48	110.97	114.29	120.15	117.46	110.33
Fuels and energy	104.4	88.4	116.3	117.6	141.7	134.2	127.48	120.90	120.54	123.83	128.11	136.00	134.67	127.30
Fuels	105.9	91.4	124.8	127.5	156.4	146.5	139.15	132.54	132.18	135.37	141.83	150.54	147.97	139.99
YPF	105.9	91.1	129.6	143.1	166.2	154.9	147.36	140.27	139.71	143.21	153.47	163.54	161.84	153.47
GdE	106.0	93.3	99.8	86.1	105.0	103.7	104.68	100.05	100.54	102.44	93.93	96.32	90.51	84.33
YCF	92.7	81.7	108.3	105.1	135.5	139.9	143.24	136.51	136.28	139.90	126.98	128.20	138.25	126.46
Energy	98.9	75.4	80.3	75.7	79.6	80.6	81.78	75.36	74.98	78.67	74.42	79.06	82.63	77.61
SEGBA	98.5	74.0	79.4	74.6	79.4	80.7	75.20	69.32	68.59	71.70	66.65	73.94	83.89	80.16
AyEE	99.8	19.0	82.4	78.1	79.9	82.5	91.33	84.12	84.24	88.78	84.70	86.49	80.80	73.91
Trans. & commun.	105.4	79.0	71.8	73.4	81.5	81.2	82.50	79.15	75.39	76.10	76.94	77.12	70.37	63.88
Transport	99.9	77.2	86.0	89.4	91.9	85.4	101.76	99.16	94.40	99.87	100.76	99.01	89.80	80.35
FA	98.5	71.3	77.5	80.9	78.8	74.2	69.67	64.61	64.31	60.66	63.40	74.46	78.22	73.54
AA	101.5	83.8	108.2	106.7	117.4	112.0	118.52	116.92	110.44	120.80	120.09	110.59	94.61	82.88
AGP	102.6	98.6	88.8	97.0	104.4	92.7	93.55	88.08	83.08	84.88	93.13	106.12	100.25	89.86
Communication	109.4	78.6	61.0	61.2	73.7	77.6	64.38	60.33	57.51	53.74	54.54	56.54	52.10	48.39
ENTEL	109.4	81.2	61.3	59.2	57.0	51.6	46.58	43.98	41.57	38.42	37.47	37.89	35.32	32.75
ENCOTEL	109.5	75.4	60.7	63.7	95.0	110.3	134.21	124.49	120.02	113.85	121.49	129.68	117.91	109.75
Industry and Services														
OSN	90.0	58.3	45.1	40.2	37.0	51.8	66.80	60.43	63.72	56.90	54.88	55.02	51.47	45.80
Price/own cost							102.52	99.57	97.49	100.19	102.09	107.24	106.93	105.02

Source: Ministry of Economy and SIGEP.

[a]/ Prices Perceived by Enterprises. Nominal price deflated by non-agricultural Wholesale Prices

[b]/ New weights: share in 1988 budget

and that of 1987 with a reduction in savings by 0.5 percent of GDP. In 1988, the impact of the Plan Primavera price reduction was offset by the rise in the price during the first half of the year and by cuts in real wages. Since the investment program is determined independently, these manipulations of the real price modify the overall deficit of the public enterprises in a straightforward manner.

1.13 These strong fluctuations in the real price of public production induced by policy considerations unrelated to the objectives of the PEs provide for a highly unstable environment and prevent the PEs from pursuing strategies aimed at improvements in efficiency. Together with other inefficiencies in the regulatory environment this policy induces the need for transfers from the Treasury to cover operational losses.

C. Expenditures

Interest Burden

1.14 **External Debt**. The public enterprises were used as magnets to attract foreign capital needed to equilibrate the overall balance of payments and to support the exchange rate. From 1978 to 1983 the public enterprises were strongly encouraged to borrow in foreign currency even as the peso was becoming increasingly overvalued. Initially, this was done to reduce the burden on the Treasury of financing enterprise deficits. Later it was used to defend the peso against a wave of capital flight induced by fears of a major devaluation. When the peso devalued in 1981-82, the service of the external debt became a threat to the solvency of the PEs. This policy had ultimately an adverse effect on the financing needs of the PEs as the total deficit before contributions increased from 3.3 percent of GDP in 78 to 6.9 percent in 82 while at the same time investment fell. In September 1983, the Treasury assumed full responsibility for servicing all of the then outstanding stock of external debt of the public enterprises.

1.15 The total stock of foreign debt rose from US$3 billion at the end of 1977 to US$11.7 billion at the end of 1983. Since 1983, the external debt has increased at a lower rate, the increase being due to investment projects in gas, oil and hydropower, and recently, also in telephone services (the MEGATEL plan). In June 1988, the total foreign debt amounted to US$14.9 billion (Annex Table 1.5). In absolute terms, YPF and GdE hold about half of the total stock, while Hidronor and AyEE together hold one fourth of it. The public enterprises, in the aggregate, would need about 23 months of sales to pay off the outstanding external debt.

1.16 Recently, rising international interest rates have increased the burden of this debt service to both the enterprises and the Treasury. The total stock of outstanding debt represented an interest obligation of US$1 billion in 1988. The Treasury assumes US$890 million while the enterprises themselves owe US$110 million on suppliers' credits. No amortization of this stock took place in 1988.

Annex Table 1.5: ARGENTINA - EXTERNAL DEBT OF DEP ENTERPRISES [a]
(US$ millions, end of period)

	1979	1980	1983	1986	1987	1988[b]
TOTAL	5,389	7,870	11,715	13,641	14,875	14,884
Financial	4,035	6,417	9,929	11,383	12,968	13,182
Commercial	1,354	1,452	1,786	2,258	1,907	1,702
Fuels and Energy	4,097	5,916	9,202	10,376	11,653	11,677
Fuels	2,426	3,476	5,810	6,605	7,504	7,627
YPF	1.937	2,906	4,366	4,554	4,840	4,927
GdE	423	452	1,264	1,878	2,470	2,522
YCF	66	118	180	173	185	178
Energy	1,671	2,440	3,392	3,771	4,149	4,050
SEGBA	532	661	781	717	732	678
AyEE	873	1,497	2,221	2,201	2,366	2,392
Hidronor	266	282	390	853	1,051	980
Transp.& commun.	1,292	1,955	2,514	3,265	3,222	3,207
Transport	1,143	1,653	1,930	2,600	2,529	2,521
FA	257	440	712	1,197	1,161	1,075
AA	316	516	713	840	867	786
AGP	26	33	20	10	8	7
ELMA	544	664	485	552	543	653
Communication	149	302	584	665	693	686
ENTEL	149	293	576	656	688	679
ENCOTEL	0	9	8	9	5	7

Source: SIGEP

a/ Reported numbers follow accounting practices of individual enterprises, and are not entirely comparable.

b/ June 30.

1.17 In addition, the foreign debt assumed by the Government still appears in the financial statements of the public enterprises. This makes it virtually impossible for any of the enterprises to obtain credits in international markets. In the case of YPF the transfer mechanism for the interest payments is particularly inefficient. The amount of transfers is denominated in australes and YPF is allowed to deduct this amount from its tax liabilities to the Government. However, the tax deduction is budgeted whereas the actual payments correspond to the austral equivalent of the dollar obligation. In 1987, the austral depreciated faster than was budgeted and YPF has a corresponding gap in its financial statements resulting in their non-ratification by the audit company. Recently, a decree has been drafted that would give YPF a government bond paying the exact equivalent of the foreign interest payments. This would avoid recurrence of the present situation, but it would not significantly improve YPF's chances to obtain independent financing.

1.18 **Domestic Debt.** Although public enterprises are no longer allowed to borrow from the domestic financial market, they still have a large amount of domestic debt outstanding. On March 31, 1988, total domestic debt amounted to US$3 billion representing 161 days of sales of the PEs. Recent preliminary data on the evolution of arrears of the PEs show a dramatic increase from US$85 million at the end of 1987 to US$1.1 billion at the end of 1988 (Annex Table 1.6). Among the reasons for this evolution are the fall in the real price of public output, the cutoff from foreign and domestic financial markets and the Government's decision not to finance operational deficits any longer.

1.19 The PEs owe more than US$1 billion to the private sector and US$356 million to the national Government. On the other hand, the provincial Governments owe the PEs US$302 million. Since September 1988 total arrears have increased by US$632 million. All of these claims are adjusted for inflation and most of them bear interest as well. The recent increase in interest rates in Argentina aggravates the arrears problem. All enterprises are net debtors to the private sector and the National Government except Aerolineas, ELMA (the international shipping company) and OSN (water and sewerage). These three are net creditors to the National Government and OSN and ELMA are also net creditors to the private sector. All companies except SEGBA are net creditors to the provincial governments.

1.20 The picture becomes even more complicated if one includes the inter-PE arrears. On December 31, 1988, these amounted to US$371 million, an increase by US$220 million over December 1987. These claims gave rise to litigations between the public enterprises as the result of different inflation adjustment and interest imputations between different enterprises. YPF, Gas del Estado and AyEE have been the largest creditors among the public enterprises, while SEGBA, YCF and Ferrocarriles have been the largest debtors.

Wages and Employment

1.21 The public enterprises determine neither the quantity of labor nor the real wage rate paid to their employees. Both are largely set by political factors.

Annex Table 1.6: ARGENTINA - DEP ENTERPRISES DOMESTIC ARREARS a/
(US$ Millions)

	1987	----- 1988	-----	1988/87
	12/31	9/30	12/31	% change
Credits				
Intra-Co Debt b/	152.9	266.9	371.4	142.9
National Adm.	168.3	107.9	177.4	5.4
Provincial Adm.	122.3	265.3	318.5	160.3
Private Sector	166.2	41.3	42.6	-74.4
Total	609.7	681.4	909.8	49.2
Debts				
Intra-Co Debt b/	152.9	266.9	371.4	142.9
National Adm. c/	215.5	286.4	533.0	147.4
Provincial Adm.	0.1	1.5	16.0	18197.5
Private Sector	326.5	616.0	1111.6	240.5
Total	694.9	1170.7	2032.0	192.4
Net Indebtedness d/				
National Adm. c/	47.2	178.4	355.6	653.5
Provincial Adm.	-122.3	-263.8	-302.5	147.4
Private Sector	160.3	574.6	1069.0	566.9
Total	85.2	489.3	1122.1	1216.7

Source: Public Enterprise Board (DEP)

a/ Preliminary estimates.
b/ Debt owed to other DEP public enterprises; figure taken from debtor company.
c/ Includes debt with specific investment funds, disbursed primarily to DEP enterprises.
d/ Debts - Credits.

03-Mar-89

1.22 The absolute level of employment reflects the social orientation of the ruling political party or the type of government. Under the Peronist Government the PE labor force reached a peak of 424,000 in 1975. The military Government reduced the amount of workers drastically, but under the new democratic Government total employment increased again since 1983 and now stands at 293,000 (Annex Table 1.7). The Government recognized the need to reduce the labor force, but the effort stagnated in the light of the 1989 elections. Especially the railroad company, which employs one third of the total, has an excessive amount of workers. Its wage bill is almost twice as high as its total operational revenue.

1.23 Efforts to improve productivity through layoffs have been hampered by the application of a law that guarantees lifetime employment in civil service to a substantial number of PEs. Given the size of this labor force and its importance to the labor unions it will take a lot of political courage to modify the existing law.

1.24 Wages in the PEs are often negotiated in the same macroeconomic package that reduces real output prices in order to mitigate inflation. This is a necessary condition to keep a handle on the PE deficit during periods of anti-inflationary policy. However, labor unions are quite sensitive to cuts in the real wage. As a result this part of the austerity package is often the first to give way due to social tension and strikes. Real wages rebound faster than real prices and prevent a positive response of PE saving to a recovery of their real output price (Annex Figure 1.3). The evolution of the PE's current account in 1983 and 1987 clearly reveals this tendency.

1.25 As a consequence, total labor cost fluctuates beyond control of the PEs. It ranges between 2.2 and 3.2 percent of GDP and its share in total cost is largely determined by the level of employment (Annex Table 1.1).

Goods and Services: The Compre Argentino Law

1.26 Almost 25 percent of the purchases of goods and services relate to transactions among the PEs themselves at regulated prices. These prices are often set by political authorities disregarding the economic cost of production. This system gives rise to an intricate network of cross subsidies and large outstanding claims among PEs. At this stage no comprehensive overview of the cross-subsidy system and its implications exists, but some typical examples indicate the nature of the problem. YPF is required to sell gas to Gas del Estado below its own production cost and it buys coal from YCF at a price three times the international price of coal.

1.27 The claims resulting from these transactions often give rise to litigations among public enterprises. The DEP reports many non-matching claims between debtor and creditor companies resulting from different inflation adjustment and imputation of interest rates. This may be one of the reasons why half of the outstanding debt among public enterprises is more than six months overdue.

1.28 Purchases of the remaining inputs are regulated by the Compre Argentino law which favors the domestic private industry. Since most of the private industry is of an oligopolistic nature, sales to public enterprises are often executed at a premium and imply subsidies to private

Annex Table 1.7: ARGENTINA - DEP ENTERPRISES: LEVEL OF EMPLOYMENT
(Number of Workers)

	1970	1975	1980	1983	1984	1985	1986	1987	1988[a]/
YPF	33,615	50,555	33,602	32,772	33,725	32,455	32,488	34,780	34,800
GdE	8,844	10,906	10,469	9,912	10,238	9,928	9,464	9,251	9,660
YCF	2,735	5,249	3,949	3,722	3,858	3,826	3,676	3,592	3,401
AyEE	13,834	26,044	19,468	10,567	11,249	11,445	11,255	11,074	11,250
SEGBA	23,809	26,334	21,774	20,130	21,336	21,657	20,999	21,535	21,764
HIDRONOR	84	759	911	1,057	1,250	1,426	1,562	1,680	1,705
ENTEL	44,395	50,543	45,280	47,833	48,200	47,100	46,300	46,900	46,657
ENCOTEL	50,426	54,506	44,506	42,968	42,843	41,484	40,292	38,815	37,874
FA	147,475	153,308	96,935	103.102	107,837	102,941	99,897	97,218	97,300
AA	6,586	8,233	10,096	9,822	10,303	10,598	10,323	10,283	10,400
AGP	5,504	6,127	5,219	5,045	5,024	4,925	4,820	4,565	4,130
ELMA	4,854	5,268	5,294	5,387	5,234	5,003	4,808	4,726	4,842
OSN	21,593	26,296	13,296	9,434	9,202	9,603	9,374	9,346	9,360
TOTAL	363,754	424,127	311,248	301,751	310,299	302,391	295,258	293,765	293,143
% of Active	4.2	4.58	3.10	2.90	2.90	2.78	2.67	2.62	2.58

Source: Company response to the questionnaires and company financial statements. For 1988: SIGEP.

a/ Data on September 31. Preliminary for SEGBA, HIDRO, NOR, ENCOTEL.

Annex Figure 1.3

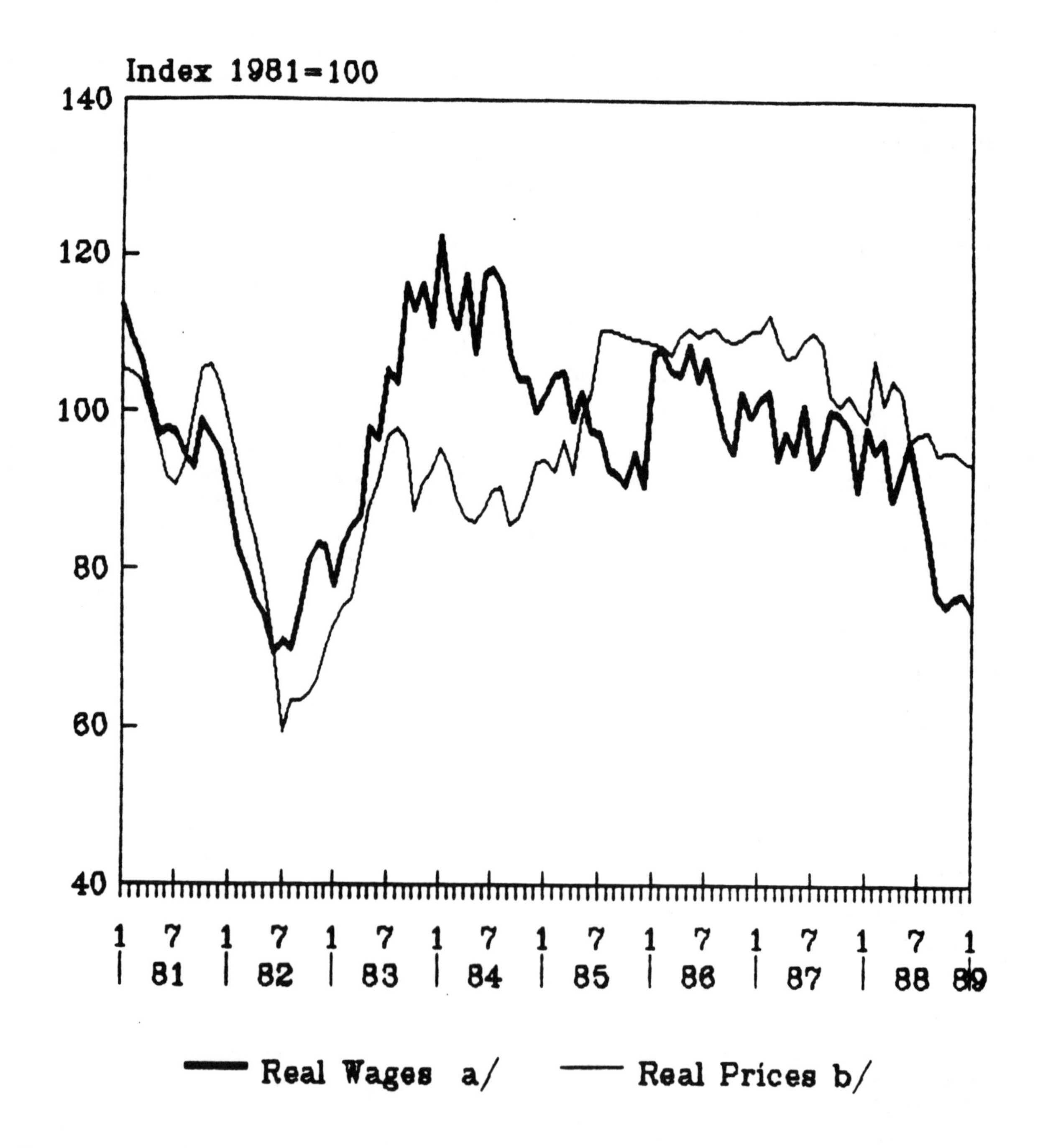

industries. Although it is difficult to quantify this phenomenon the following observations may indicate the nature of the problem. In 1983, political authorities decided to contract out a large fraction of the maintenance work that was traditionally done by the PEs and decided to reduce the own production of intermediate inputs by the enterprises themselves. The share of goods and services in total recurrent expenditure immediately increased from 45 percent in 1983 to 52 percent in 1984 and reached 68 percent in 1985 (Annex Figure 1.4). Since then, the share has declined to 53 percent, but still remains higher than any period before 1983. A measure of the intermediate input per unit of production rose from 136 in 1982 to 165 in 1987. Under normal circumstances, this measure should fall indicating the presence of economies of scale in highly capital intensive industries, a characteristic of most of the PEs.

Investment

1.29 The political decisions on investment are made in the respective Secretariats of Energy, Transportation, Communication and Industry. The often ambitious programs are financed through a system of special funds and by means of foreign and domestic borrowing. In recent years both sources of foreign and domestic financing have almost dried up. Public enterprises are no longer allowed to borrow in the domestic market and foreign private banks are unwilling to increase their exposure to the debt-burdened Argentine economy.

1.30 Special taxes on combustibles and electricity replenish special accounts from which funds are drawn for various investment projects. Unfortunately, only a fraction of the revenue is reinvested in the sector of the public enterprises. In addition, there is a substantial earmarking of funds so that efficiency concerns are not the primary criteria to decide on various investment projects. The revenue of the special funds fluctuates with the price of output, and is, therefore, heavily influenced by the episodes of anti-inflationary policy of the Government. On average, the special funds generate an amount of investment between 0.75 and 1 percent of GDP, about a third of total investment.

1.31 Because of the reduction in available financing, real fixed investment fell from an average of 3.2 percent of GDP in 1981-84 to 2.5 percent in 1985-86. In 1987, some international institutions financed large projects for gas, hydropower and telephone services. Real investment picked up to 3.1 percent of GDP in 1987, but fell again to 2.8 percent of GDP in 1988 (Annex Table 1.8). The sectoral distribution of investment is shown in Annex Figure 1.5.

D. Transfers

1.32 The public enterprises are linked with other sectors of the private economy and of the public administration through a complex system of transfers. Some of these transfers are implicitly reflected in low prices to consumers of public output or high prices to suppliers of goods and services, and consequently, difficult to quantify.

Annex Figure 1.4

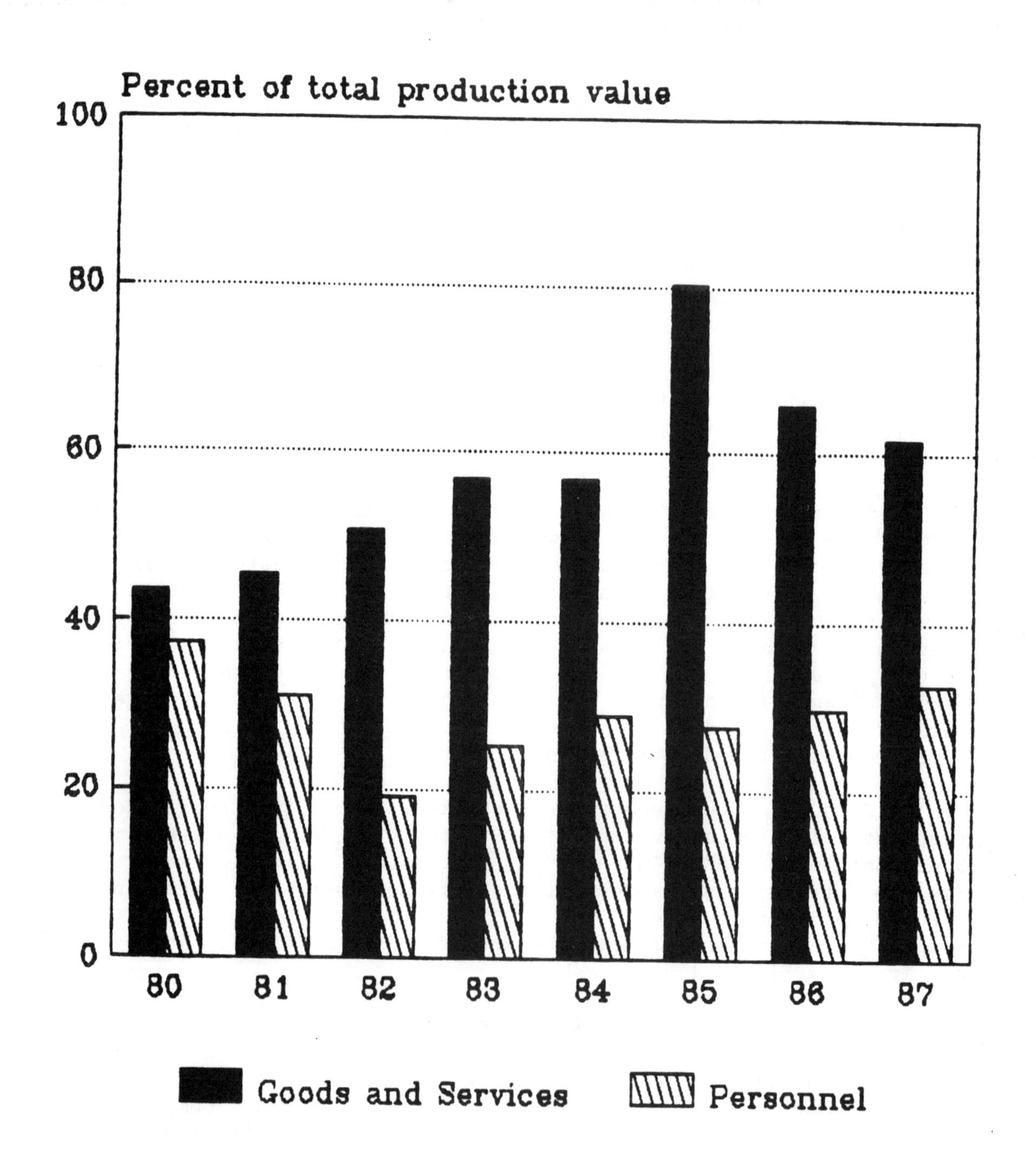

Source: SIGEP

a/ Excluding HIDRONOR and ELMA

Annex Table 1.8: ARGENTINA - REAL FIXED INVESTMENT IN DEP ENTERPRISES
(Percent of GDP)

	1981	1982	1983	1984	1985	1986	1987	1988
TOTAL	2.99	3.17	3.32	3.18	2.75	2.25	3.11	2.76
Fuels and Energy	1.99	2.29	2.38	2.09	2.13	1.60	2.13	1.72
Fuels	1.03	1.34	1.33	1.32	1.53	1.14	1.59	1.19
Energy	0.96	0.95	1.05	0.77	0.60	0.46	0.53	0.53
Transp. and Commun.	0.94	0.83	0.89	1.05	0.58	0.62	0.95	1.02
Transport	0.40	0.46	0.51	0.69	0.31	0.27	0.25	0.24
Communication	0.54	0.37	0.38	0.36	0.27	0.36	0.70	0.78
Industry and Services OSN	0.06	0.05	0.05	0.04	0.04	0.03	0.04	0.02

Source: SIGEP

Annex Figure 1.5

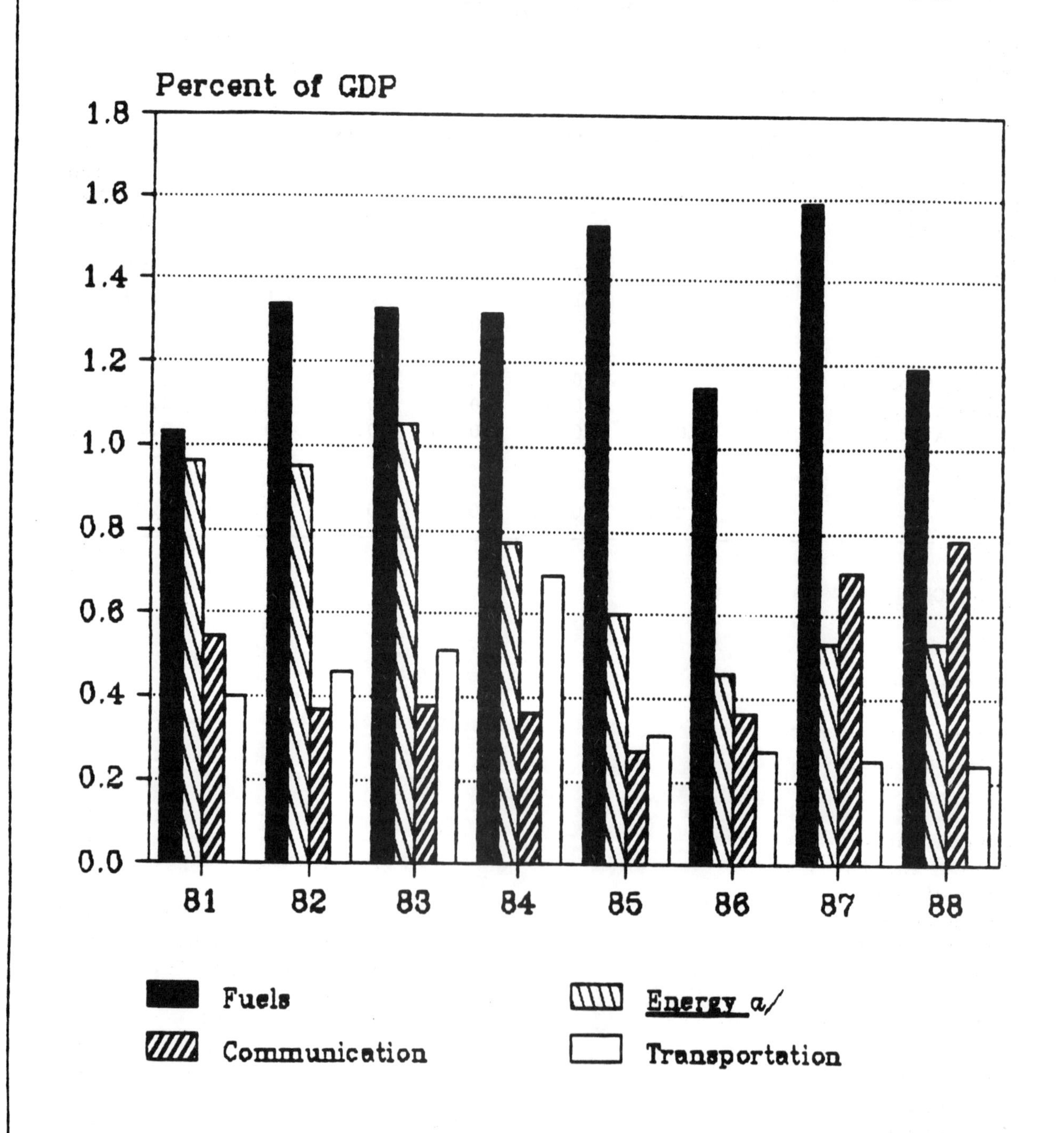

Source: SIGEP and DEP.

a/ Does not include Binationals.

1.33 The tax administration, the social security, and the investment funds are major recipients of taxes that weigh most heavily on the fuel sector. Electricity and telephone services also provide revenue for the rest of the economy. Nevertheless, market prices are biased towards the use of electricity rather than combustibles as a source of energy. As a result of these differences between consumer and producer prices, it is difficult to base investment decisions on efficiency criteria.

1.34 On the other hand, the Treasury used to finance operational deficits and has been servicing a large part of the outstanding stock of foreign debt of the public enterprises.

1.35 Due to the lack of accurate data on the various transfer mechanisms, it is very difficult to accurately describe the overall net position of the different entities that are part of the system. Nevertheless, the description of some major aspects of the transfer system is sufficient to indicate the nature and magnitude of the problem.

1.36 First of all, a large number of earmarked investment funds are financed by means of taxes on fuels and electricity. Revenue from these two categories amounted to 1.5 and 0.2 percent of GDP, respectively in 1987 (Annex Table 1.9). However, only 0.8 percent of GDP is invested in the public enterprises themselves and only 0.6 percent in the sectors that generate this revenue. Large amounts of these funds are channeled through to the bi-national enterprises, the provinces and the rest of the public sector. In the last two sectors, they are primarily used for road construction. The sectoral bias of these transfers is even worse; the fuel sector, contributing 1.5 percent of GDP to the funds receives only 0.6 percent. The reverse is true for the electricity sector that receives twice as much financing as it contributes.

1.37 The State Petroleum Company also pays royalties on the extraction of oil and gas to the provinces. Although this is a common feature of extraction industries in many countries, in Argentina the price on which the royalties are computed is artificially determined and does not reflect international prices. Since this artificial price has been much higher than the current international price in the recent past, the Government decided to compensate YPF for the excess payments. However, in 1988, the Government decided to change its policy and YPF now has to pay all royalties. The total payments implied a revenue for the provinces of 0.57 percent of GDP in 1988. Of this amount 0.33 percentage points is due to the difference between the artificial price and the actual well-head price of the products. YPF has refused to pay this excess amount and deducts this transfer to the provinces from the taxes it owes to the Central Government.

1.38 In January 1988, the Government issued a decree in order to raise revenue for the social security system by imposing a tax on fuels, 24 percent on naphtha and kerosene and 17 percent on natural gas and gasoil and telephone services (24 percent). This surcharge generated an estimated 1.1 percent of GDP in 1988 and its contribution is likely to increase in 1989 to a budgeted 1.6 percent.

Annex Table 1.9: ARGENTINA - PUBLIC SECTOR INVESTMENT FUNDS SOURCES AND USES
(Percent of GDP)

	1985	1986	1987
Sources			
Fuels a/	1.98	1.61	1.48
Electricity b/	0.19	0.24	0.24
Total	2.17	1.85	1.73
Uses			
Public enterprises	0.81	0.66	0.80
Fuels c/	0.14	0.02	0.06
Electricity d/	0.52	0.50	0.52
Binationals	0.16	0.14	0.22
National public sector e/	1.21	1.04	0.80
Provinces	0.12	0.12	0.09
Private Sector	0.01	0.02	0.01
Other f/	0.01	0.02	0.03
Total	2.17	1.85	1.73
Net use			
Fuels	-1.84	-1.59	-1.42
Electricity	0.33	0.26	0.28
Rest of Economy	1.52	1.32	1.14

Source: DEP and Ministry of Public Works

a/ 10 percent tax on crude oil, 50 percent on nafta, 10 percent on kerosene, fuel and diesel oil, 30 percent on gas oil and 10 percent on fraction of natural gas.
b/ 21 percent tax on total sales of electricity.
c/ YPF, YCF, Gas del Estado
d/ AyEE, SEGBA, HIDRONOR
e/ Includes transfers to other national investment funds for projects mainly contracted out to the private sector.
f/ Includes net use of loans, net transfers to/from previous/future periods.

1.39 In June 1980 the Government decided to limit its transfers to the public enterprises to the foreign debt service and it gave the DEP the power to organize an internal transfer scheme among the public enterprises to finance the wage bill of Ferrocarriles. This enterprise received 0.15 percent of GDP from the rest of the 13 DEP enterprises that were obliged to deposit a percentage of their variable cost into a forced savings account (Decree 683/80). Not all enterprises were able to deposit the required amount so that this scheme could lead to a net transfer of resources out of the more efficient companies where money could have been invested in profitable projects.

1.40 In the other direction, the Treasury has transferred on average 1 percent of GDP to the public enterprises in the period 1984-88 primarily for foreign debt service and to cover operational deficits (Annex Table 1.10). The Government recognized that these transfers were due to some noncommercial objectives imposed on the enterprises. One example is the payment for gas to Bolivia at a price three times above the domestic end-user price. However, in mid-88 the Government decided not to finance operational deficits any longer. This explains the fall in transfers for operational deficit from 1 percent of GDP in 1987 to 0.7 percent in 1988.

E. Budgetary Process and Control

1.41 The budgetary process for the public enterprises has involved the Ministry of Economy, the enterprises themselves and the different institutions that existed between those two levels at various points in time. Traditionally, individual Secretariats were responsible for the designated enterprises under their jurisdiction. An aggregate budget found its way down from the Ministry of Economy to the Budget and Planning Directorate of the Ministry of Public Works where it was divided among the Secretariats. The individual enterprise budgets were reconciled with the aggregate budget at the Secretariat level. The control on the execution of the budget proceeded along similar lines. Consequently the overall process of drafting and controlling the budget was a lengthy and inefficient one. At various points in time the Government attempted to shorten and depoliticize the process in order to gain control over the public enterprises. These attempts have been hampered by the absence of an integrated information system.

1.42 The Corporation of National Enterprises (CEN) was the first attempt to consolidate the budgetary process and control. However, the CEN soon became a bureaucracy of its own and the Secretariats criticized the concentration of power within the CEN. After a short life of five years the CEN was dismantled in 1979. Only the State auditing company (SIGEP) survived and became an independent entity. However, it only audited the financial statements of the public enterprises. SIGEP is not involved in the budgetary control process.

1.43 The Public Enterprise Board is the second attempt to simplify the budgetary process. Initially the DEP lacked decision making power and an inter-ministerial committee consisting of the Ministers of Public Works and

Annex Table 1.10: ARGENTINA - TRANSFERS FROM TREASURY TO PUBLIC ENTERPRISES a/
(Percent of GDP)

	1984	1985	1986	1987	1988
Operational Deficit b/	1.21	1.02	1.09	1.03	0.72
Debt Service	0.90	0.60	0.44	0.56	0.39
Other c/	0.26	0.22	0.0	0.0	0.0
TOTAL	2.37	1.86	1.53	1.59	1.11

Source: Ministry of Economy and DEP.

a/ 13 DEP enterprises and binational entities.

b/ Includes compensation to gas del Estado for purchases from Gas Bolivia.

c/ Includes repayment of domestic loans quaranteed by the government and not itemized transfers.

03-Mar-89

Economy made final decisions. Since the beginning of 1988 the DEP started to work as an independent holding company and was vested with the power to make decisions on management, procurement and personnel of the public enterprises. The DEP is also responsible for the reconciliation of the consolidated public sector budget with those of the individual enterprises. Although there has been an improvement in efficiency it still takes four to five months into the budget year before a final budget is agreed upon.

1.44 The DEP attempts to follow the budget execution on a monthly basis and has established its own independent information system. Unfortunately, this information system is not linked to the financial reporting of the public enterprises. The public enterprises merely respond to the information requests from the DEP.

1.45 The Ministry of Economy constructs a quarterly cash budget and monitors its execution independently from the DEP and other institutions. It has established its own cash reporting system since 1984. This information system is an extra-accounting process without links to the financial statements of the enterprises or the DEP system. As such it also suffers from the lack of independent control. The SIGEP was supposed to audit the cash reporting by the public enterprises, but has been unable to perform this task due to a lack of resources. The information on the cash execution is often reported with a 3-4 month lag notwithstanding the legal obligation to report within 15 days after the end of each quarter.

1.46 The SIGEP has been the only independent control agency for the public enterprises. Its task has been limited to the auditing of the financial statements of the enterprises. The audits are often performed with lags of more than a year due to late reporting by the public enterprises. For many years and for about half of the public enterprises under its supervision, the SIGEP reports an audit "with reservations" or simply refuses to affix its signature to the company accounts.

F. Performance Issues in 1988-89

1.47 **Financial Autonomy.** The performance of the public enterprises in 1988 was governed by the dilemma between the long-run goal of improvement in efficiency and financial autonomy on the one hand, and the short-run macroeconomic goals of reducing inflation combined with the continued use of the public enterprises to serve noncommercial objectives.

1.48 During the first half of the year the Government took energetic measures to achieve more financial autonomy for the PEs. It granted more power to the DEP and decided to clarify the financial relations between the public enterprises and the Treasury. The Treasury would no longer finance the operational deficits of the public enterprises and limit its transfers to interest payments on foreign debt. At the same time real prices of public production recovered their losses from the 1987 level. Although these measures were certainly dictated by the need to lower the overall fiscal deficit, they improved the efficiency of the public enterprises and constitute a first step on the way to deep restructuring of the public sector.

1.49 Unfortunately, the restructuring effort was stalled in the second half of 1988 when the Government was once again looking for a nominal anchor for the Argentine economy in order to mitigate accelerating inflation. Prices of public production have been falling in real terms since the Plan Primavera was enacted.

1.50 **Compre Argentino.** The real price reduction squeezed revenues of the public enterprises while at the same time very little effort was made to implement institutional changes that would lower the costs of the public enterprises. The amendment to the Compre Argentino law was not implemented so the public enterprises continued to subsidize the private domestic industry. Rules governing employment remained rigid and the wage guidelines of the Government led to unrealistically low real wages. Social tension and strikes during the last quarter provoked increases in real wages in the last months of 1988.

1.51 **Cross-Subsidies.** The Government continued to impose the pursuit of noncommercial objectives on the public enterprises and tried to reduce its own deficit by increasing the burden of royalty payments on YPF. It did not relax its stance on operational deficit financing so that the companies had to finance the wage bill of Ferrocarriles by means of a forced savings scheme. Other aspects of the macroeconomic policy, rising real interest rates and an overvaluation of the exchange rate increased the financial burden on the public enterprises. In addition, public enterprises were virtually cut off from domestic and foreign financing.

1.52 The logical outcome of this situation is a reduction in investment and a build up of arrears with other sectors of the economy. The crisis in the electricity sector in the last quarter of 1988 shows the contradiction between the need to improve the quality of the capital stock and the cuts in the investment program as a result of the precarious financial situation of the Central Government. Arrears with the rest of the domestic economy more than doubled in the last quarter of 1988. Annex Table 1.11 shows the steady deterioration of the accounts of the PEs since the second quarter of 1988: a fall in savings, capital revenue and fixed real investment.

G. Policy Recommendations

1.53 The Government has rightfully perceived the need to grant more financial autonomy to the public enterprises and to improve their efficiency. Its decision to severe the link with the Treasury for the financing of operational deficits and the establishment of the DEP in its current format seem to have been taken primarily to reduce the impact of the PEs' financing needs on the Treasury's deficit and are not accompanied by structural changes that address the underlying inefficiencies. The Government still uses the price of public production for short-run macro-stabilization attempts disregarding its distorting effects on the allocation of resources.

Annex Table 1.11: ARGENTINA - DEP PUBLIC ENTERPRISES
BUDGET EXECUTION 1988
(Millions of 1988 Australes)

	Quarter				Annual
	I	II	III	IV	
Current Account					
Revenue	22,457.1	27,483.5	24,296.9	22,748.3	96,985.9
Expenditure	21,058.4	24,806.4	21,723.0	21,920.2	89,508.0
Savings	1,398.7	2,677.1	2,573.9	828.1	7,477.8
Capital Account					
Revenue	473.3	340.8	278.9	182.9	1,375.9
Expenditure	5,991.0	7014.8	5,802.8	6,716.9	25,525.5
Real Fixed Invest.	5,918.5	6,970.0	5,769.9	5,453.2	24,111.6
Deficit[a]	4,019.0	3,996.9	2,950.0	5,705.9	16,671.8

Source: DEP

a/ After contributions from the rest of public sector.

03-Mar-89

1.54 In the short term, the Government needs to correct the level of real prices of output, address the arrears problem and take measures to enhance control and prepare institutional reform. In the medium term the Government should enact a structural and institutional reform that will allow the public enterprises to operate autonomously in a competitive environment.

1.55 **Prices.** The Government needs to correct the level of real prices of output. In the near term, these should be raised in real terms at least up to the average level of the second quarter of 1988. By 1990, tariffs in the sector should be sufficiently high relative to costs so as to generate a surplus on their noninterest current account equal to about 3 percent of GDP, thereby covering roughly half of their investment after interest expenses. Improvements over 1988 levels could be made up of a combination of either expenditure reductions or real tariff increases.

1.56 As it adjusts prices, the Government should formulate the new pricing policy to reflect an efficient resource allocation, at least in terms of producer prices.

1.57 **Royalties.** The issue of royalty payments to the provinces on petroleum and gas entails a heavy implicit tax on the sector to support provincial finances. The current reference price on which these royalties are paid dates from the early 1980s and is too high; it should be reduced to international levels as soon as possible.

1.58 **Budgetary Control.** The Government needs to enhance budgetary control over the public enterprises. The respective roles of the DEP (or its successor, if any), the Ministry of Economy, the SIGEP and the various Secretariats in the Ministry of Public Works should be clearly delineated. If the DEP is to perform its budget control task effectively, it should be equipped with the necessary means to verify the information received from the individual enterprises. The DEP budget, the Ministry of Economy's cash and commitment budgets and the financial statements should all be drawn up within the framework of a uniform information system.

1.59 A uniform information system that would integrate the budget control mechanism with the financial accounts of the companies needs to be designed and implemented. It will facilitate the control of the execution of the budget, increase the speed of reporting and make the various information sets consistent. At the same time, it could make the aggregate operational and investment budgeting compatible with the planning at the level of the individual enterprise.

1.60 **Enterprise Restructuring.** Without enacting internal reforms within each enterprise to induce managerial responsibility, accountability, and autonomy, changes in the regulatory and competitive environment will not have their full desired effects. The Government should therefore continue and accelerate restructuring programs currently under discussion to strengthen management, improve personnel policies, and enhance financial controls. Management could be strengthened by restricting political appointments to boards or a few senior level posts as well as improved salaries, attention to qualifications, and sound training. These efforts could permit considerable reduction in expenditures and improvements in long-term efficiency.

1.61 **Institutional Framework**. The market structure and regulatory framework of each sector requires definition so that enterprises are subjected either to the discipline of the market or well-defined regulations regarding pricing. For those enterprises designated to function as regulated monopolies, the regulatory framework should be clearly established in the law so as to permit maximum managerial autonomy within a carefully elaborated legal framework, thus minimizing the imposition of political and noncommercial objectives.

1.62 **Procurement**. The Compre Argentino law should be modified to allow foreign competition in the bidding for contracts of the public enterprises. This would effectively abolish the implicit subsidy system to the private sector and reduce the operating costs of the public enterprises.

1.63 **Cross-Subsidies**. The Government needs to redefine the mechanisms and degree of cross-subsidization among the public enterprises and needs to establish a transparent scheme for resource transfers. If the Government wants selected loss-making enterprises to maintain uneconomic activities and to keep a given number of workers on its payroll, it should set up a mechanism for this that would subject these expenditures to the budgetary review process. Highest priority should be given to reducing the demand for expenditures in Ferrocarriles, which absorbs most of the cross-subsidies and accumulates losses of nearly one percent of GDP annually.

1.64 **Divestiture**. The Government should announce its program of privatization. This should accelerate planned privatizations and review for possible inclusion remaining enterprises and activities that could be privatized. In the process, it should establish clearly defined legal procedures to ensure an unbiased selection of private investors and competitive bidding.

ANNEX CHAPTER II: PROVINCIAL GOVERNMENT FINANCE

A. Overview

2.01 The efforts to control Argentina's excessive fiscal deficit will not succeed unless they include the provincial governments. Provincial expenditures rose rapidly over the 1970-86 period, reaching 11.2 percent of GDP in 1986; meanwhile provincial tax revenues were only 5.0 percent of GDP in 1986, and had actually fallen from 5.6 percent of GDP in 1980 (Annex Figure 2.1). The fiscal deficit of the provinces in 1986 before transfers from the Central Government was 6.3 percent of GDP (about US$4 billion). After transfers, the provinces showed a slight surplus, but at the expense of helping transform the National Administration's before transfer surplus of 5.4 percent of GDP into a deficit of 4.6 percent of GDP after transfers (Annex Figure 2.2).

2.02 The provincial governments have resisted reforms and stopped sending data on their finances to the Ministry of Economy. Thus, national macroeconomic planning has been "flying blind" with respect to this important part of the public sector since 1986. Data from case studies of six provinces show that their combined fiscal deficit more than doubled in 1987 to US$1.1 billion (or 1.8 percent of GDP) after transfers, as own-source revenues declined and expenditures, especially for salaries, continued to rise.1/ For all 22 provinces and the Municipality of Buenos Aires, the total fiscal deficit after transfers was over US$2.8 billion or about 4 percent of GDP.2/ The decline in own source revenues stems partly from the rise of inflation, but also from the increasing inefficiency of the provinces in collecting taxes, as well as the reticence of taxpayers to pay taxes and fees for services of ever deteriorating quality. This reduced fiscal effort by the provinces has made them more dependent on transfers from the Central Government to cover their ever rising current expenditures.

2.03 Prior to 1988, incentives facing provincial officials favored deficit spending since transfers took the form of discretionary grants of the Central Administration. The heavy reliance on these grants had generated a perverse competition among the provinces to run up large deficits to get a larger share of total grants. The implementation of the

---

1/ These data were collected as part of the World Bank's forthcoming study on provincial government finance. Field studies in six sample provinces (Buenos Aires, Chubut, Cordoba, Salta, Santa Fe and Santiago del Estero) were done in late 1988 in order to obtain finance data for 1987. These six provinces accounted for 69 percent of the 1985 national populations, 63 percent of households with unsatisfied basic needs in 1980, half of the floating debt in 1986, and 71 percent of total gross provincial product in 1980.

2/ Estimated using the results of the six case studies and the proportions of total revenues and expenditures in these six provinces during the 1981-86 period.

Annex Figure 2.1

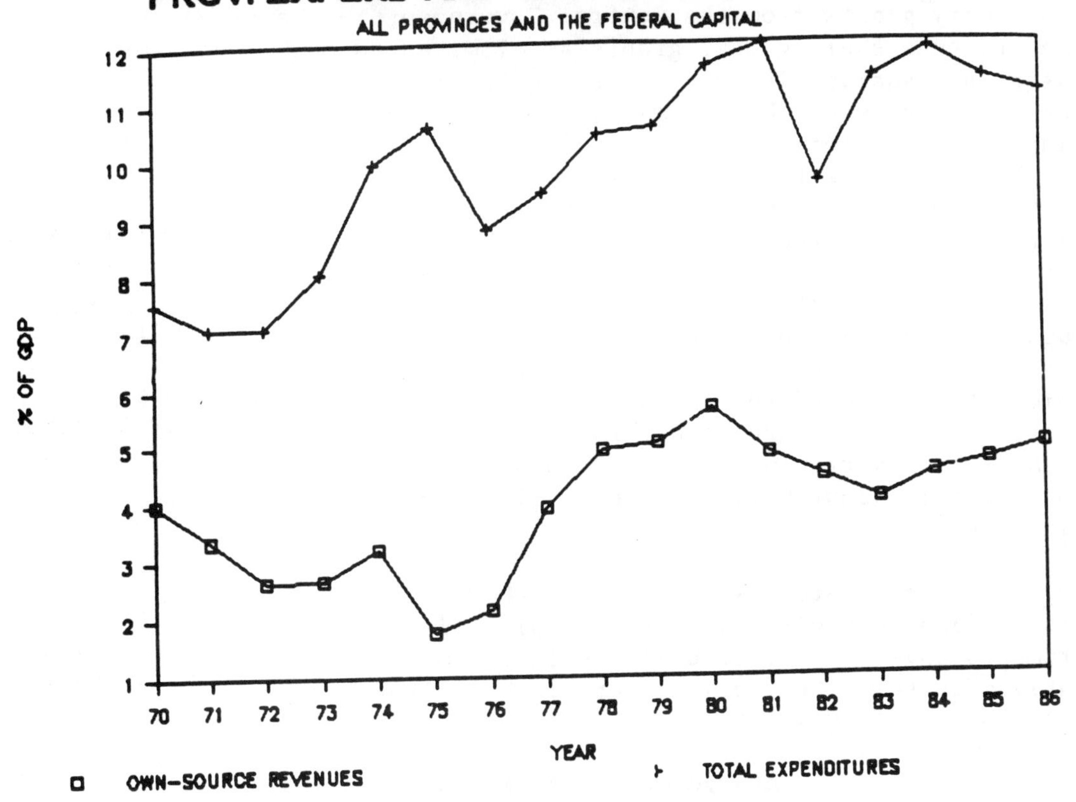

Annex Figure 2.2

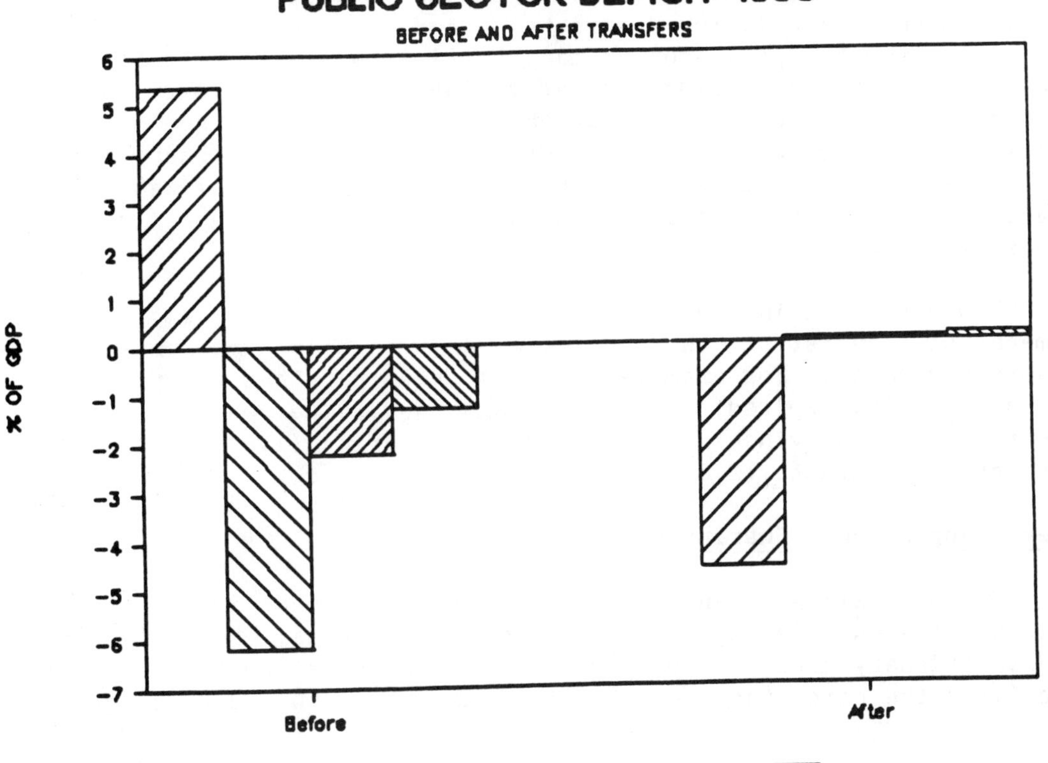

new revenue sharing law (Ley de Co-Participación) in 1988 restricts discretionary grants from the Central Government to the provinces. Under the new revenue sharing law, grants are limited to only 1 percent of total transfers and sharing is done according to criteria fixed in the law. It is not clear, however, how long the Central Government can hold the provinces to the distribution criteria of this law, unless efforts are undertaken to reduce the explosive rise of provincial fiscal deficits. In 1988, the central government was compelled to ask Congress to provide some extra resources for the provinces in March and then again in December.

2.04 The fact that provinces borrow on a very short-term basis (mostly through suppliers and overdrafts on the provincial banks) accentuates their need to go to the Central Government for transfers. Short-term debt in 1987 more than doubled for the six sample provinces to US$842 million, accounting for about three quarters of all borrowing. A rough estimate of this short-term borrowing for all 22 provinces and the Municipality of Buenos Aires is about US$2 billion in 1987, more than twice that of the previous year.

2.05 Provinces are using short-term basis through suppliers and overdrafts to finance their substantial capital investments. The provincial governments invested almost US$1.6 billion in 1986 (2.6 percent of GDP), slightly more than the public enterprises in 1986 and about 60 percent more than the National Administration.

B. Intergovernmental Fiscal Relations

2.06 The new revenue sharing law that was passed in 1987 and became effective in January 1988 evolved from the previous revenue sharing law of 1973 (20.221). Not only did it increase the amounts shared with provinces, it also fixed the criteria for distribution among the provinces by law. Responsibility for the provision of many services was also transferred to the provinces in the late 1970s and early 1980s, such as primary education, medium-size hospitals, water, irrigation and electricity. The formal revenue sharing system was interrupted from the end of 1984 when revenue sharing law expired, until January 1988, when the current law was implemented. From 1985-87, a series of ad hoc agreements substituted the revenue sharing law.

2.07 However, even in the period before the previous law expired, a chief means of transfers to the provinces was Discretionary Grants (Aportes del Tesoro Nacional). For example, in 1984 (the year that the previous revenue sharing law expired) an amount equal to 4.30 percent of GDP was transferred to the provinces via both revenue sharing (1.8 percent of GDP) and discretionary grants (2.5 percent of GDP).

The Prevailing Revenue Sharing Law

2.08 The prevailing revenue sharing law represents an important step towards more responsible fiscal federalism in Argentina: Not only does it limit discretionary grants to only one percent of total shared taxes, but it also fixes the percentages of total revenues from shared taxes going to the provinces as a whole (i.e., primary distribution), and to each individual province (i.e., secondary distribution). In primary distribution, the percentage of shared or ("coparticipated") taxes going to the provinces rose from 48.5 percent to 57.66 percent under the new law. Furthermore, taxes to be shared are deposited directly into the accounts of the provinces in the Banco de la Nacion Argentina in accord with the percentages used in secondary distribution within 72 hours of their collection.

2.09 Equally important, any additional transfers to the provinces must now be approved by Congress. This makes the revenue sharing system transparent, and raises the political cost to provinces of requesting financial assistance of the Central Government. The law is valid for two years, but will be extended automatically until a new one is adopted. As passage of revenue sharing laws has been difficult in Argentina, the current law could well stay in effect for some time to come.

2.10 The total amount transferred under the new law was projected to increase by about US$345 million over the 1984-86 average to US$2.5 billion in 1988. Total revenues of shared taxes dropped by about 11.5 percent in real terms for the first nine months of 1988 as compared to the same period of the previous year, thereby precipitating in part the crisis that will be discussed below.

2.11 The criteria used in dividing the total revenues of shared taxes among the provinces are very heavily weighted against Cordoba, Santa Fe and especially Buenos Aires whether the measure is population, output, or population with unsatisfied basic needs. Annex Table 2.1 shows that Buenos Aires received only 44 percent of its share relative to population and 51 percent relative to its share of the nation's poor.

The Recent Political Confrontation over Revenue Sharing

2.12 In December 1988, the provinces requested A$4 billion (US$260 million) to help them meet their payrolls and the second half of the annual bonus (aguinaldo) for public employees as well as other expenditures. Several governors organized demonstrations in Buenos Aires to underscore their claims. At stake in this confrontation was the integrity of the existing revenue sharing law. Finally compromising to additional transfers, the Central Government adeptly negotiated as a quid pro quo a major tax reform and measures to reduce and make more transparent the huge fiscal loss caused by the Industrial Promotion Program.

2.13 The final agreement involves two bonds: The Bono para El Saneamiento Financiero Provincial (hereafter, Bono Provincial), and Bono Federal:

(a) Bono Provincial. The Bono Provincial of A$3 billion (about US$200 million) was a non-negotiable bond of the Central Government transferred to the provinces using the same percentages established for secondary distribution in the revenue sharing law. The provinces will pay the loan back to the Central Government in two years with one year of grace at an interest of 2 percent per year plus a monetary correction.

Annex Table 2.1

Secondary Distribution in Revenue Sharing and Distribution of Population, Gross Provincial Product and Households with Unsatisfied Basic Needs by Province

Provinces	% for Secondary Distribution (a) 1988 (A)	Population 1985 (B)	A/B (C)	Gross Provincial Product 1980 (D)	A/D (E)	Households with Unsatisfied Basic Needs 1980 (F)	A/F (G)
TOTAL	100.00	100.0	1.00	100.0	1.00	100.0	1.00
Buenos Aires	19.93	45.4	0.44	44.0	0.45	39.4	0.51
Catamarca	2.86	0.9	3.34	0.6	4.89	1.1	2.56
Chaco	5.18	2.9	1.77	1.7	3.08	4.7	1.11
Chubut	1.38	1.2	1.19	2.4	0.59	1.3	1.07
Cordoba	9.22	9.8	0.94	10.5	0.88	8.2	1.13
Corrientes	3.86	2.7	1.43	2.0	1.95	3.9	0.98
Entre Rios	5.07	3.6	1.40	3.5	1.43	4.2	1.19
Formosa	3.78	1.2	3.03	0.6	5.99	2.0	1.90
Jujuy	2.95	1.8	1.65	1.8	1.66	2.7	1.10
La Pampa	1.95	0.9	2.27	1.2	1.69	0.7	2.64
La Rioja	2.15	0.7	3.17	0.4	5.51	0.8	2.82
Mendoza	4.33	5.0	0.87	6.0	0.72	3.9	1.12
Misiones	3.43	2.5	1.35	1.7	2.03	3.5	0.98
Neuquen	1.54	1.1	1.36	1.8	0.86	1.2	1.24
Rio Negro	2.62	1.7	1.51	1.7	1.51	2.1	1.25
Salta	3.98	2.8	1.41	2.1	1.90	4.0	0.99
San Juan	3.51	1.9	1.82	1.4	2.55	1.8	1.97
San Luiz	2.37	0.9	2.72	0.9	2.66	1.0	2.43
Santa Cruz	1.38	0.5	2.73	1.1	1.27	0.4	3.32
Santa Fe	9.28	10.0	0.93	13.3	0.70	9.2	1.01
Sgo. del Estero	4.29	2.5	1.75	1.4	3.13	3.9	1.10
Tucuman	4.94	4.1	1.20	3.9	1.28	5.0	0.98

SOURCES: Lei de Coparticipacion, Capitulo 1, Art. 4, Diciembre 22, 1987.
INDEC, ARGENTINA EN CIFRAS: TERRITORIO, Poblacion Estimada al 30/5/85.
Consejo Federal de Inversiones, Estimates of Gross Provincial Product, xerox.
INDEC, LA POBREZA EN LA ARGENTINA, Buenos Aires, 1985.

NOTES: (a) Percentages used in secondary distribution.

Payments will be automatically deducted from the shared revenues of the provinces. The provinces transferred the bonds at face value to the provincial banks in payment of their debts.

(b) The Bono Federal. This bond of A$2 billion (US$130 million) was to have been issued by the government, which was to receive 42.34 percent and the provinces 57.66 percent. The provincial share was then to be divided in accord with the percentages established in the revenue sharing law. However, due to the economic crisis, the bond was never issued by the Central Government.

The total face value of the two bonds (A$5 billion or US$330 million or about 0.5 percent of GDP) would represent only about 4 percent of total provincial expenditures, 11 percent of projected revenue sharing and 7 percent of all transfers.

2.14 The Central Government managed to maintain the integrity of the revenue sharing law through these measures. However, its inability to issue the Bono Federal is proof of its limited ability to transfer resources to the provinces, and, therefore, of the urgent need for reforms to make them more fiscally autonomous.

C. Provincial Revenues

2.15 Total provincial revenues from all sources tended to rise steadily during the 1970-86 period. The top line of Annex Figure 2.3 is total revenue from all sources, and the second, provincial own-source revenues. The gap between the two is the amount covered by transfers of all types from the Central Government, which increased greatly after 1980. Case study data shows that this positive trend in revenues reversed in 1987, when own-source revenues declined by 14.7 percent (discussed below). As this drop was partially compensated by a 2.7 percent rise in transfers from the Central Government, total revenues from all sources for the six provinces in 1987 suffered a real decline of 8.5 percent.

2.16 **Tax Revenues.** Total provincial tax revenue skyrocketed from only 1.0 percent of GDP in 1975, peaking at 4.4 percent of GDP in 1980, then dropping to 2.7 percent of GDP in 1983, before beginning a steady recovery to 3.8 percent in 1986 (Annex Figure 2.4). Total real tax revenues dropped by over 11 percent from the previous year, with all taxes showing losses in real terms.

2.17 The primary "motor" in this improvement was the turnover or gross sales tax, shooting from only 0.1 percent of GDP in 1975 to 2.1 percent of GDP in 1980, before dropping back down to 1.5 percent in 1983. However, the real estate tax and the automobile tax also showed positive trends until 1986, with only the transfer tax (i.e. stamp tax) losing ground in recent years (Annex Figure 2.5). This temporal instability in fiscal performance can be explained by the Olivera-Tanzi effect in times of rising inflation, but also by shifts in the efficiency of collection and billing procedures. There is a wide margin for improvement.

Annex Figure 2.3

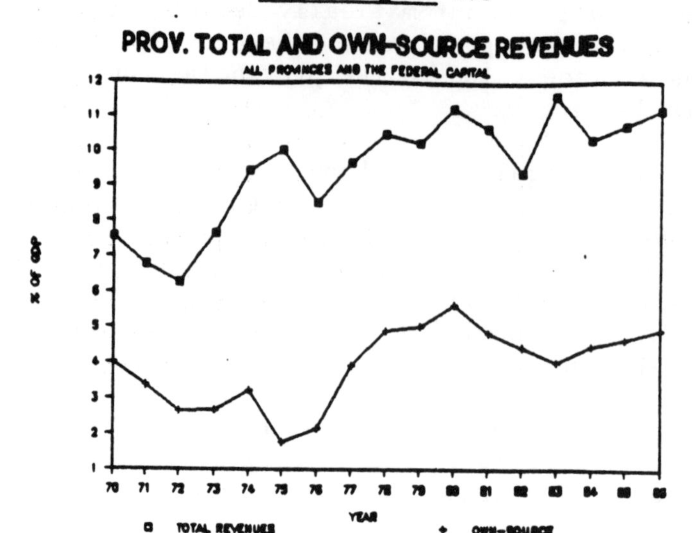

Annex Figure 2.4

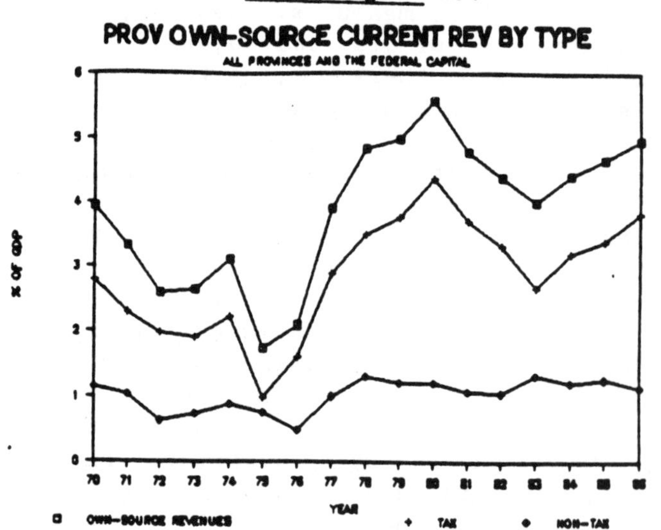

Annex Figure 2.5

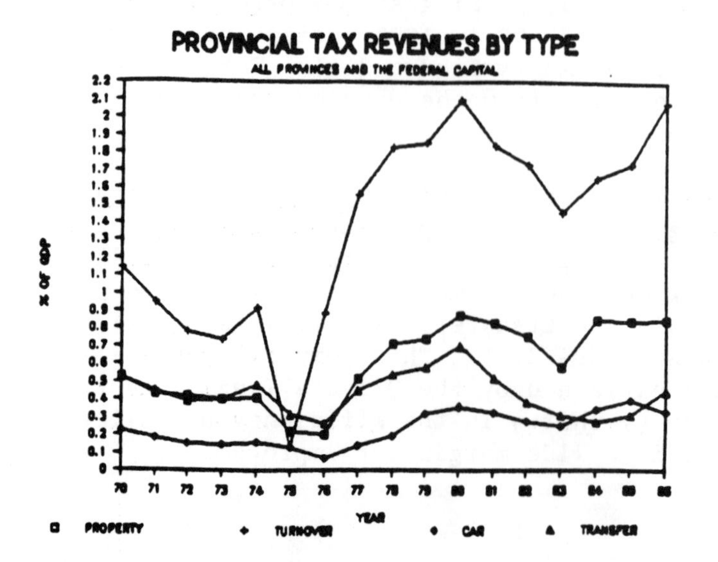

2.18 **Nontax Revenues.** The performance on nontax revenues from fees, tariffs, etc. has been quite weak, with their total more or less stable at around 1 percent of GDP in the 1980s. This low importance is partially due to the exclusion of the public utility companies from these Ministry of Economy data, as these are precisely the entities that charge user fees. For Buenos Aires, adding the utility companies boosts nontax revenues from 10.4 percent of total own-source provincial current revenues in 1987 to 22.7 percent. These nontax revenues including public utilities dropped by almost 38 percent in 1987 for the six provinces, adding to the longer-term negative slide. Clearly, there is a need for improvement to recuperate tariffs and user fees for public services. User fees are widely perceived as unfair because they are not closely related to services and many do not pay at all.

D. Provincial Expenditures

Total Expenditures

2.19 Expenditures for all provinces and the Federal Capital rose rapidly over the period from 7.5 percent of GDP in 1970 to 11.2 percent of GDP (or about US$8 billion) in 1986 (Annex Figure 2.6). The case study data shows that these expenditures rose in 1987 by 7.4 percent, even though total revenues declined. Current expenditures have risen much more rapidly in recent years, while capital expenditures have remained stable at about 3 percent of GDP.

Current Expenditures

2.20 The driving force in the increase of current expenditures was clearly personnel, but transfers to municipalities and other entities were also important.

2.21 **Personnel.** Payments for personnel varied widely over the period, ranging from 3.13 percent of GDP in 1977 to a high of 5.8 percent of GDP in 1984 (see Annex Figure 2.7). Provincial public employment in only the consolidated Central Administration increased by over 230,000 (34 percent) in all 22 provinces during the 1983/86 period to 913,000 in 1986 (Annex Figure 2.8). This trend continued in 1987 for most of the six provinces studied. For this reason, total real provincial expenditures on personnel for the General Administration alone rose by slightly over 33 percent during the 1983-86 period to over US$3.9 billion, excluding the semi-autonomous entities.

2.22 The provinces argue that this great increase in employment was necessary to meet social needs in health and education, which had been ignored during the military regime. Employment in health and education did in fact increase, but that employment in other areas of public administration grew at almost exactly the same rates. The total public employment in these six provinces increased by 123,000 over the 1983/87 period. The area of health and education increased at rates equal to those of the "other" category, including general administration.

Annex Figure 2.6

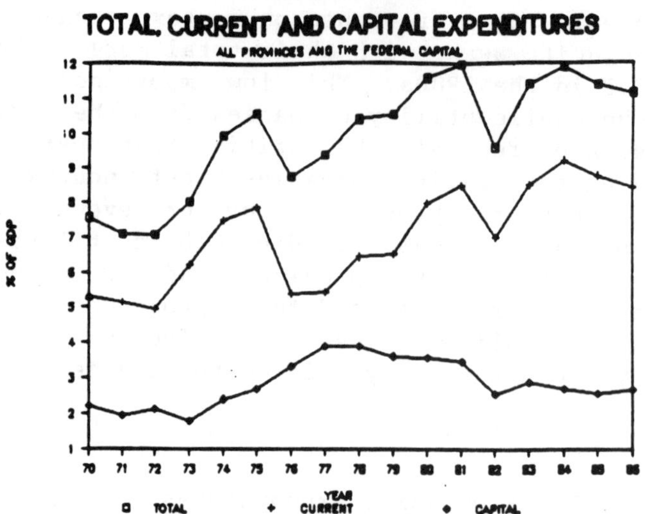

Annex Figure 2.7

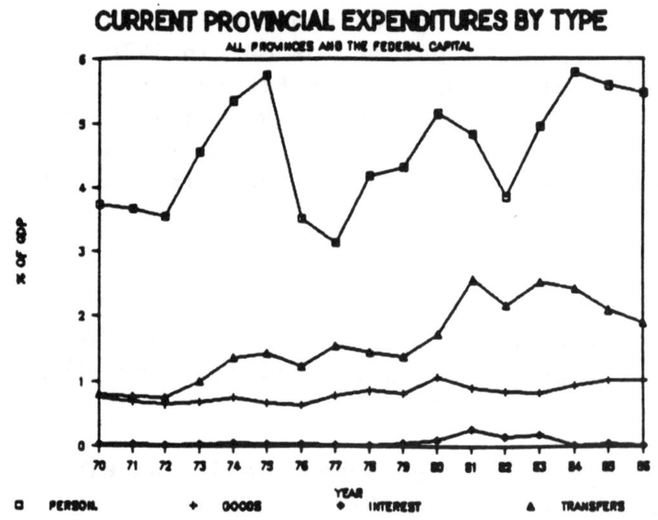

Annex Figure 2.8

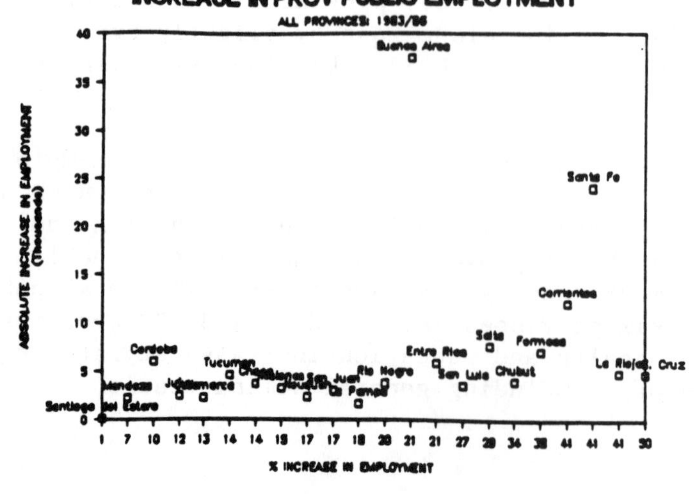

2.23 This growth of employment varied greatly among the six provinces. In Salta, total provincial public employment catapulted from 26,000 in 1983 to 55,000 in 1987, an increase of 29,000 or 111.5 percent in only four years. Of this, about 12,000 or 40 percent was in health and education with the other 17,000 in the general administration. The Province of Salta declared itself an "employer of last resort," and is now considering a number of policies to reduce the payroll, including assistance in opening small businesses, and "privatizing" services such as transport, copying, etc. In 1987, 90 percent of total expenditures in Salta went for the payroll, which exceeded total current revenues by over 30 percent.

2.24 Rapid increases in employment could mean that the efficiency in the provincial public sector has declined, if it has produced inefficient mixes of inputs in the production of public goods (e.g., the teacher without chalk and the doctor without any medical supplies).3/ The gross indicators available show the need for much more detailed and careful analysis of the efficiency of the social sectors in the provinces to determine how they might be more efficient producers of these important services.

2.25 **Goods and Services.** The expenditures on goods and services amounted to about 1 percent of GDP (about US$700 million) over most of the 1980s. Most provincial administrators interviewed agreed that improved purchasing procedures could cut at least 10 percent from expenditures with no loss in quality or efficiency. Implementing such procedures, however, is no mean task. Provincial governments not only adhere to the Buy Argentina program, but also often have their Buy Provincial programs.

2.26 **Interest Payments.** Interest payments account for less than 1 percent of total expenditures in all of the provinces, except Santa Fe where they reached almost 2 percent. The reason for this is that provinces "borrow" mainly through suppliers. The interest payments are, therefore, implicit and are included in the price of the goods and services received. Just how much the suppliers charge for implicit interest depends on their perception of risk for late and nonpayment.

2.27 **Transfers to Municipalities and Other Entities.** Provincial governments transfer resources to municipal governments via revenue sharing and discretionary grants, as well as to provincial enterprises and to social security entities. Of these, transfers to municipalities are normally by far the most important item.

2.28 Current transfers to municipalities, provincial enterprises and special organisms such as social security increased from less than 1 percent of GDP in the early 1970s to a peak of 2.6 percent of GDP in 1981, but tended to decline over the 1983-86 period, reaching 1.9 percent of GDP in 1986. In 1987, these transfers accounted for 22 percent of total expenditures in Santa Fe and about 16 percent in Buenos Aires.

3/ David L. Lindauer, Government Pay and Employment Policies and Government Performance in Developing Countries, Background Paper for the World Development Report of 1988, Washington, the World Bank, pp. 7-12.

E. Deficit/Surplus and Sources of Credit

2.29 The contribution of the provinces to the consolidated public sector deficit is high and probably rising. The fiscal deficit of the six provinces studied (excluding the non-consolidated entities and enterprises) jumped from US$554 million in 1986 to US$1.1 billion in 1987 (106 percent) due to both the decline of total revenues and grants of 8.5 percent and an increase in total expenditures of 7.4 percent. Including the non-consolidated entities would increase the total need for borrowing to US$1.5 billion (by about 35 percent) in 1987 for the six provinces studied.

The Importance of Short-term Credit

2.30 Almost all of the debt of these six provinces is financed through the variation of short-term assets and liabilities, which explains why the floating debt is so large and apparently growing. For example, of the total need for borrowing in 1987 of US$1.1 billion, US$842 million was financed through variations in short-term assets and liabilities.

2.31 This short-term credit can take a number of forms, including: (i) unpaid payment orders that are not overdue (voluntary credit); (ii) unpaid payment orders that are in arrears (involuntary credit); (iii) unified checking accounts; and (iv) bank overdrafts, usually on the provincial bank. The unified checking accounts (Fondo Unificado de Cuentas Oficiales) involve combining the balance of all the provincial checking accounts. The province can then use up to a stipulated limit of the balance of this unified account. Finding out the distribution of the floating debt among these different forms is not easy, as shifts often occur daily. However, in Salta the breakdown was estimated as follows: unpaid bills (both within the established payback period and in arrears), 83 percent; overdrafts on the provincial bank, 14 percent; and other sources, 3 percent.

2.32 The provinces are now "financing" their capital investments with short term credit. The estimates of the floating debt for each expenditure item for Cordoba showed it to be highly concentrated in the capital expenditures.

2.33 One form of "borrowing" currently being used by some provinces (Salta, Jujuy, Catamarca, Tucuman and La Rioja) is very close to the emission of money, in that it consists of issuing "bonds" that have all of the practical characteristics of money.

F. Recommendations

The Need for Provincial Financial Reform

2.34 Although the Central Government managed to maintain the integrity of the new revenue sharing system in the recent controversy over the floating of bonds to assist the provinces, it is unclear how long it will be able to do so unless urgent reforms are undertaken to reduce the provincial public sector deficits, as well as to provide the provinces with longer term financing for their capital investments. The goal should be not just to reduce the deficits, but to transform the provinces into agents of development which generate surpluses that they can invest wisely to increase total provincial product, thereby augmenting their future tax revenues.

2.35 The recommendations for developing such a strategy will involve: increasing own-source revenues, reducing expenditures and raising their efficiency, borrowing to finance capital investment, and improvements in the information system to allow adequate macroeconomic planning at the national and provincial levels.

2.36 Ways of implementing such a strategy in provinces that are extremely heterogeneous by almost any indicator must be developed. One way to deal with this diversity would be to provide technical assistance for the provinces to do their own financial and economic planning.

Increase Own-Source Revenues

2.37 **Improve the Efficiency of Tax Collection.** Improving the efficiency of tax billing and collection procedures should receive top priority, as this would produce high fiscal returns in a short period of time. All six provinces studied could substantially increase their own-source tax revenues without raising nominal tax rates. This could be achieved through improving cadastres, information systems and general tax administration. In many cases, own-source tax revenues could be increased by over 25 percent with just these basic management reforms. More efficient tax collection would also be more equitable, as it would reduce the very high existing levels of tax evasion and free-riding. Tax legislation, especially that regarding exemptions, should be reviewed. Tax administration should not be under-staffed, as is currently the case in some provinces. Qualified staff should be recruited as soon as possible either within the Government or by new hiring. Where feasible and cost effective, billing and collection should be privatized (e.g., the collection of delinquent property taxes in the province of Santa Fe). These measures to improve fiscal performance should be linked to those which improve the quality of services (see the section on improved efficiency below). This should, in turn, help increase the willingness to pay taxes and fees.

2.38 **Increase Nontax Revenues.** To the extent possible, the cost of providing local services should be recovered from charges on the beneficiaries. Such charges should be related to individual consumption or, where this is not possible, to a measure of individual benefit received. This is clearly not the case in most of Argentina's provinces, where the "decentralized agencies" such as the public utility companies (e.g., Water and Sewerage, Electricity, etc.) often represent a important drain on public finance. Water is rarely metered, and delinquency in payment is high due to inefficient collection procedures, and also to the perception that the amounts charged are unfair, given that they are based on constructed area, rather than actual consumption.

Reduce Expenditures and Increase Their Efficiency

2.39 **Personnel Policy.** Clearly, personnel policy is of highest priority in the provincial government in any effort to increase efficiency and lower expenditures. Provinces must make efforts to shed excess labor hired over the last half-decade; this could be done with systematic programs of transitional income maintenance and vouchers for vocational or other training. If the number of unproductive workers were reduced, remaining civil servants might be granted needed pay increases to regain some of the lost competitiveness with the private sector, while still reducing the overall wage bill. Each provincial government should receive technical assistance to do detailed studies on how to improve its efficiency in these and other ways.4/

2.40 **Procurement Procedures.** The case studied revealed that improved procurement procedures could reduce expenditures for goods and services by at least 10 percent, and would also be helpful to the modernization of industry. Buy-Provincial legislation should be eliminated, and the Buy-Argentina law modified in the same way as suggested for public enterprises in Chapter IV and Annex Chapter I.

2.41 **General Organization and Management.** Elementary improvements in organization and management could save a lot of time and money for both the government and those needing services. None of the six governments studied were investing more than token effort in these kinds of improvements. The use of computers could dramatically increase productivity in many areas.

Analyze the Impact of Access to Longer-Term Financing on Total Expenditures

2.42 Longer-term borrowing would certainly reduce the need for pressuring the Central Government to cover the borrowing needs for capital investments, but would have to be tied to the general structural adjustment already discussed. Provinces could be provided with technical assistance to not only do a financial action plan which would show how they could raise their revenues or lower their expenditures to create a surplus sufficient to amortize longer-term loans. They should also receive assistance for doing a medium-term investment plan which would identify areas of highest return for public sector investment in the province.

Improve the Information Systems on Provincial Public Finance

2.43 It is impossible to do macroeconomic planning without information on the provinces, when they are spending well over 12 percent of GDP and running up deficits of over US$1 billion per year. A first step in improving provincial financial management, and also the consolidated public sector accounts for general macroeconomic planning should be to establish an efficient information system for the collection and analysis of the provincial budget data.

2.44 **Provincial Governments.** The entire provincial public sector should be seen as a whole, including both the nonfinancial (that we have discussed here) and the financial subsectors (i.e. the provincial banks). The nonfinancial subsector must cover both the Central Administration and the decentralized entities and enterprises.

2.45 **National Level.** The provinces and the Central Government should be strongly urged to reach an agreement on how to set up an efficient information system for providing the consolidated public accounts as soon as possible after each exercise, as well as preliminary results during the exercise. This should be part of any new revenue sharing agreement.

4/ The Bank has been involved in numerous efforts to improve public sector employment policies. For a review, see: Barbara Nunberg, Public Sector Pay and Employment Reform, Policy Planning and Research Working Papers, WPS 113, World Bank, Washington, D.C., October 1988.

ANNEX CHAPTER III: SOCIAL SECURITY

A. Introduction

3.01 The solvency of pension and health care funds has been eroding in the 1980s, and prospects for sustaining benefits without recourse to general Treasury revenues are poor. The average pension has already fallen far below levels specified in the law, and the prospect is uncertain that even these levels can be sustained without recourse to general Treasury resources. Social security deficits in the medium-term could undermine efforts to put public finances into balance.

3.02 The social security system in Argentina is in reality a heterogeneous group of loosely coordinated accounts, some in reasonable balance (for example, the insurance fund for federal employees, where pensions are virtually a form of deferred compensation), others with histories of financial distress (the fund for autonomous workers, many provincial employee funds). Retirement programs (including survivors' and disability benefits are separate from the health programs, which are financed through social funds (*obras sociales*) which are discussed separately in this chapter. There are about 300 industry-associated funds that finance health-care services delivered in large part by private hospitals and physicians.

3.03 The social funds differ in clientele, financial soundness, the benefits they offer, and in the efficiency of their operations. However, they do share a lack of regulation and monitoring. The pension funds are overseen by the Secretariat for Social Security; the social funds by the Secretariat of Health. The major provincial retirement funds are linked to the Federal Government only by the latter's fiscal obligation to cover deficits they may incur.

B. Pension Program

Affiliates and Beneficiaries

3.04 About six million Argentine workers are affiliated to one or more of the pension and disability funds. Over three million inactive persons, including retirees, the disabled, and survivors of deceased contributors, receive payments from these funds (Annex Table 3.1).1/ The industry and trade fund has three million contributing affiliates; the fund for government employees 0.8 million, and the fund for the self-employed has 1.5 million. There are about 1.4 million persons receiving benefits. Uncounted additional thousands of affiliates are associated with provincial pension funds.

1/ None of the systems has personal accounts, nor does any individual have an account number that would reveal whether he or she pays into, or receives benefits from, more than one of the many funds. An employee might avoid payment into a fund by claiming that, through another job, he is paying into a different fund. Beneficiaries can simultaneously receive checks from, for example, a military pension, a civil service pension, and a self-employed worker pension, depending on past career service. There are currently no means available to the Social Security Secretariat to measure the frequency of such occurrences.

Annex Table 3.1: ARGENTINA - NUMBER OF AFFILIATES AND RETIREES, 1983 MAJOR RETIREMENT FUNDS
(Millions of Persons)

Fund	Number of Affiliates	Number of Retirees
Industry and Trade	3.0	0.67
Public Employees	0.8	0.19
Self-Employed	1.5	0.54
Total	5.3	1.40

Source: Golbert and Lo Vuolo. "Aportes para un debate sobre previsión social" Mimeo. Buenos Aires: UNDP and Ministerio de Salud y Acción Social, 1988: p.19.

Benefits

3.05 Argentine social security funds dispensed over 5 percent of GDP in 1989 through pension and disability benefit programs. Retirement ages in Argentina are generous by world standards, and even more so in light of the country's relatively long life expectancy: women can retire at age 55 and men at 60. Argentine legislation provides for pension benefits equal to 70 percent to 82 percent replacement of an employee's last working salary. Nonetheless, for most of the 1980s, actual benefits paid have been well below legally prescribed levels (Annex Table 3.2).2/ These data do show an unmistakable deterioration in the real value of pensions that government must be under considerable pressure to correct.

Annex Table 3.2: ARGENTINA - RATIO OF PENSIONS TO WAGES, 1980-87

Year	Average Pension/ Average Wage
1980	.65
1981	.64
1982	.62
1983	.58
1984	.46
1985	.50
1986 Est.	.20
1987 Est.	.38

Source: Golbert and Lo Vuolo, op cit, p.25.

2/ Data for the two most recent years must be regarded as provisional because underlying statistical data have not been published in recent years.

Contributions

3.06 Wage taxes are extraordinarily high and amount to about 32 percent of the wage bill in the formal sector of the economy; 23.5 percent is assigned to pension benefits, 7.5 percent or more assigned to health care financed through the social funds. However, there is a high rate of avoidance by rejection of formal-sector employment to work in the informal sector: Evasion is estimated to be one-third of the obligated enterprises and employees in 1986, the latest year for which government estimates are available. Some 2.6 million workers and their employers manage to avoid contributing.3/

Deficit Finance for Pension Benefits

3.07 Wage-bound revenues are not adequate to finance the legally-mandated benefits of retirees, their dependents, and disabled beneficiaries. Spending as a share of GDP rose through the 1970s from under eight percent to over ten percent. The crisis then brought on a precipitous decline to less than 5 percent of GDP for these services by 1984, this despite the fact that in the years 1981-84, Treasury payments to the retirements funds accounted for over 40 percent of the funds' total revenues (Annex Table 3.3).

Annex Table 3.3: ARGENTINA - SOURCES OF PENSION FUNDS' INCOME, 1979-87
(Percentage Distribution)

Year	Affiliate and Employer	Treasury	Special Laws	Other
1979	95.7	0.0		3.6
1980	86.4	8.0		4.2
1981	40.4	45.8		13.5
1982	40.9	47.9		12.0
1983	37.4	42.7		10.5
1984	45.2	43.0		4.1
1985	66.8	25.4	9.0	3.8
1986	63.3	23.3	6.4	8.0
1987 Est.	73.9	13.3	8.0	4.8

Source: Golbert and Lo Vuolo, op cit, p.24.

3.08 In the 1980s, the gap between system revenues from currently employed workers, and system expenses for retired and disabled beneficiaries, grew to several percentage points of GDP. That gap was filled by resort to general-purpose revenues from the Treasury and a serious erosion--about 55 percent--in the real value of payments to beneficiaries. The Treasury sought to extricate itself from the high level of obligations, during what was perceived to be a temporary situation, through special

---

3/ Golbert and Lo Vuolo, op cit, p.18.

budgetary legislation that paved the way for nonwage taxes on fuel to be used to pay for pension obligations. The funds have become more obviously dependent on general revenues since 1986, when 1.4 percent of GDP was transferred to the Social Security Secretariat to pay for pension benefits. Special taxes on fuels are now earmarked to pay for pension benefits. Increasing the wage tax to close the deficit is not feasible because higher quotas will probably lead to greater tax evidence and evasion.

3.09 The new tax on fuels and certain public services will imply additional resources equivalent to two percent of GDP for the pension system, and the new health insurance law, ready for passage by Congress and executive approval, raises employer contributions from 4.5 percent to 6 percent of the aggregate salary bill. These taxes will not be enough to assure that no further general revenues will be necessary in the next five years. This is because pension payments are still below those provided by existing legislation (Law 18.037), so there will be growing pressure to restore benefits to levels required by law. Technical staff of the Social Security Secretariat estimate that an increment of 17.2 percent in social security funds' revenues, beyond those already at hand, would be needed to pay for the legally-mandated benefits package.

Long-Term Financial Viability

3.10 The changes enacted to-date have not established an equilibrium between pension obligations and the income from wage taxes. The aging of the Argentine population, with its implications for an increasing dependency burden on the labor force indicates that wage taxes will be too narrow a base for expanding pension obligations. The dependency ratio, in this case the over-60 population divided by the population aged 20-59 multiplied by 100, is expected to rise from 24.1 percent in 1980 to 28.6 percent in 2020. Imbalance between wage-based revenues and pension payments are certain to plague the systems because the burden per worker must grow.

3.11 The problems of financing benefits for the elderly will be considerable in the next century, but it will already be a serious burden in the 1990s. Nearly 19 percent of all Argentines will be eligible for pension and related benefits in 1990. There will be about 2.4 workers obligated to pay wage taxes to finance the benefits of each retiree. But because of the potential for evasion, even fewer workers can be expected to contribute to pay for benefits. Further recourse to general revenues or special, additional taxes are inevitable if the Government desires to maintain benefit payments.

Inequality of Fund Resource Distribution

3.12 Political demands to extend coverage of social security benefits to poor and vulnerable groups, who now face health care costs and old age with no effective safety net, are likely to intensify. About 40 percent of the potentially eligible population, i.e., those old enough to retire and ill enough to require health care services, are outside the existing pension funds and social funds. They are eligible for public health care, but these services are widely regarded as inferior to those available through the social funds; most of these people come from the same poverty groups that do not qualify for retirement benefits.

3.13 Extension of the current Argentine system to provide 100 percent coverage could cost an additional three percentage points of GDP, increasing public-sector obligations for pension benefits and health care by 25 percent over current levels. The achievement of equity in caring for the elderly cannot be based on a simple extension of the current system.

3.14 Because there are three main funds, each with its own system of pensions and benefits, there is unequal distribution of payments and benefits. According to Annex Table 3.4 below, the Industry and Trade Fund subsidizes those covered under the self-employed fund. The industry and trade fund collected about 66 percent of all revenues collected by the three funds but paid out, in 1983 only 44 percent of the benefits paid by the three funds. The fund for the self-employed collected 17 percent of revenues but received 29 percent of benefit payments. These differences may reflect cross subsidization that achieves some equity objectives, but there are no data addressing such a possibility. Moreover, such large cross subsidies may in part explain efforts by many workers to avoid or evade the wage taxes that pay for heavy cross-subsidy burdens on those who do pay.

Annex Table 3.4: ARGENTINA - DISTRIBUTION OF INCOME AND EXPENDITURES BETWEEN FUNDS
(Percentage, for 1982)

	Payments to Revenues	Beneficiaries
Industry & Trade	64	48
Public Employees	17	26
Self-Employed	19	26
Total	100	100

Source: Golbert and Lo Vuolo, op cit, p.28.

Options for Pension Reform

3.15 Future governments can be expected to search for new sources of revenue, to consider means to reduce obligations to current and future beneficiaries, and to assess options for new approaches to providing basic social security protection on terms that are sustainable without excessive recourse to the limited power of the state to tax its citizenry. The Government might be attracted to adjustments aimed at mobilizing more resources for the current system: increasing formal-sector employment, increasing the wage tax on the self-employed, and reducing the current transfer of 10 percent of social security income to the health insurance fund for retired persons. None of these options, nor some combination of them, offer much hope for increasing system revenues. And even if they did, they would leave the system with its current features of inefficiency and inequity that at some point will have to be addressed.

3.16 A different pattern of benefits, based perhaps on a different theory of the function of social security, with sharply reduced pension obligations at higher income levels, may have to be introduced at some point, to substitute for entitlements that will be unsustainable in the future.

3.17 Benefits could be reduced in a way that could actually increase both the efficiency and equity of the benefits package, the following options could be considered:

(a) Increasing retirement age by 5 or 10 years, so that years of contribution would rise relative to years of receiving benefits;

(b) Reducing the rate of salary replacement from 82 percent to about 40 percent or, as an alternative, to a low, basic rate of salary replacement, with voluntary purchase of additional coverage; and

(c) Using an extended modest salary base for calculating pension rights (not the last salary but, for example, a ten-year real average).

A gradual phasing in of benefit changes for future beneficiaries, as was done in the United States' social security reform of the early 1980s, might offer an acceptable way of reducing otherwise unsustainable obligations.

An Alternative Pension System

3.18 Another option would be to de-link pensions from wage; to offer a basic stipend only, and to enact legislation to facilitate private contractual savings institutions under public regulation.4/ According to a recent report, this alternative would provide a monthly benefit of about US$100 to each of 3.6 million qualifying persons because of age, disability, or widowhood.5/

4/ A cut in wage-based taxes could reduce the bias against use of labor in manufacturing and could also encourage production of tradeable goods. System changes could also favor provision of basic benefits to poor and vulnerable groups now left outside prevailing arrangements. These medium-term operational changes would of course be dependent upon more broad-based reforms of the macroeconomic setting. These options are discussed among Argentine experts who favor a mixed system of a basic benefit and voluntary complementary insurance such as has been recommended by C. Mesa-Lago for Costa Rica (Schultess 1988; Golbert and La Vuolo, op cit).

5/ (Golbert and Lo Vuolo, op cit, pp.36-45).

3.19 The estimated cost in 1990 would be about 5.2 percent of a GDP of US$80 billion compared to the current system costing over 8 percent of GDP, yet leaving 40 percent of the aged poor without coverage. Under the proposed arrangement, a basic stipend could be provided to all the elderly, and that stipend could be financed with a wage tax (assuming no evasion) of 16 percent of the salary bill. The current system, as noted above, collects 23.5 percent of the salary bill, but it suffers from high rates of evasion, inequity, and low coverage of those most in need of a social safety net of income security.

3.20 In converting from the present arrangement, which provides pension entitlements to many persons at a level far above the assumed basic monthly stipend of US$100, the Government would presumably have to finance transition costs of an estimated US$720 million in 1990 (and even larger amounts in subsequent years) to provide for these higher acquired benefits that, for public employees at least, are virtual deferred compensation. But even with these additional costs, this more equitable package would demand less public resources than the present arrangement.

C. The Social Funds and Health Services

3.21 Health insurance operated by the social funds is separate from the pension funds, except for transfer from those funds that pay for health care provided to retirees. The disparate programs offer variable benefit packages depending on their financial capacity, but they offer at least minimal coverage to three out of four Argentines, i.e., including eligible retirees, survivors, and the dependents of affiliated workers. Some social funds are powerful entities with more than a million affiliates, such as those of commerce employees, rural workers, and metal workers, and others cover only a few thousand members. The 12 largest social funds account for half of all social funds' affiliates. A government institute, INOS, publishes some data on the social funds, but has limited supervisory authority over them.

3.22 Social security health care is, unlike pension benefits, almost entirely financed by wage taxes; the Treasury's only contribution is an earmarked subsidy for retiree health care, equivalent to 5 percent of total financing. A tax of 7.5 percent (3.5 points deducted from the employee's salary packet, 4 points added to the employer's wage bill) is levied at the source, and paid to the employee's social funds. An additional one percent or more of wages is deducted for employees who have dependents, (depending on the number of dependents). A further one percent is deducted to finance retirement health care. An amount equal to 2.9 percent of wages is transferred from social security revenues to the designated social funds. Each social fund is required in turn to transfer a share of its revenues to PAMI, the social fund created especially for retired persons and pensioners.

3.23 About two-thirds of receipts of the social funds are used to pay for the current health care of workers and their dependents; one-third of the receipts flows through PAMI to pay for the health care of retired persons and pensioners. There are about three million affiliates of PAMI.

This large group, a tenth of the population but possibly a third of all voters, obviously constitutes an important interest group that presses for maintenance of a high level of benefits.

3.24 The National Health Insurance proposal recently enacted by the Argentine Congress seeks to extend minimum coverage to all; currently, one in four Argentines is not covered by this prepaid health insurance system. This new law will centralize the monitoring of premiums to improve accountability through banks by assigning separate accounts for the inflow and outflow of resources; it will consolidate the resources available and transfer a greater share of those resources to the poor.

3.25 The burden of health care costs will grow faster than income because of the rise in the incidence of chronic diseases, such as cancer, stroke, and heart ailments. Argentina spends about 7 percent of GDP on all public and private health care, perhaps the highest percentage in Latin America and higher, as a percentage of GDP, than the UK. The burden of an aging population could be noted especially in growing costs for health care through the social funds.

Recommendations

3.26 The burden of rising health care costs in the 1980s fell progressively more heavily on the social funds as they gradually replaced private sector funding of health services. Without further change, the burden will grow further in the 1990s, eventually forcing the Government of Argentina, which has ultimate responsibility for funding public health care, to consider some system to ration health care services. The rising incidence of chronic disease and its high cost will in any case pose difficult choices in resource allocation, especially between the needs of essential basic health care for all citizens and the claims of social security beneficiaries on curative hospital services.

3.27 The Government's decision to incorporate INOS into the Health Secretariat late in 1988 was a useful step toward a rational process of obtaining value for money in health care in the future. Further steps could include the introduction of co-payments and deductibles to make users of health services aware of the costs of treatment and hence to curtail unnecessary demand.

3.28 Extension of health insurance to all people may be the only way to assure adequate basic care to poor and vulnerable groups. Argentina already spends more on public health care than many countries so that coverage extension should be traded for cuts in nonessential services, particularly costly curative care that middle-income groups could finance from their own resources. In extending coverage, policies could be developed that would encourage the social funds to compete among themselves for additional clientele by varying their benefit packages and payment schemes. Currently, each social fund has a designated, industry-based clientele, so that the change to a more open system of choice would represent a major shift in institutional arrangements. Despite the difficulties that such changes would present, the injection of consumer choice and competition into health insurance could have more than adequate compensating benefits.

ANNEX CHAPTER IV: FINANCIAL SECTOR

A. Overview

4.01 Argentina's industry and agriculture need an efficient financial sector to mobilize and allocate financial resources to modern, job-creating capital formation. The productive economy requires flexible access to modern, cost-competitive financial services, including commercial and working capital credit and investment finance. Because of Argentina's heavy external debt burden, international financial markets are unlikely to provide any significant part of these services in coming years. An efficient domestic financial system will therefore be more necessary than ever to meet the productive economy's needs.

4.02 The current macroeconomic crisis has subjected Argentina's fragile financial system to severe stress. Deposit withdrawals have generated liquidity problems throughout the financial system. Because many firms are finding it difficult to service their debt, doubts have arisen about financial institutions' solvency. The monetary authority has found itself compelled to maintain liquidity to the banking system at a time when hyperinflation would prescribe monetary contraction. Depending upon its evolution, the crisis could severely decapitalize the financial system. The system may then be forced to consolidate and reorganize in ways that cannot easily be predicted.

4.03 The crisis is partly the consequence of the financial system's structural problems. Persistent macroeconomic instability has distorted and impeded financial development. High, fluctuating and uncertain inflation, devaluation, and interest rates; the public sector's demand for credit; and lack of external finance have all combined to discourage financial applications, so that the financial system is now small relative to the economy's size. Frequent policy changes have compounded the problem. The narrow base of the financial system has meant that monetary policy has had little room for maneuver, and this has contributed in turn to macroeconomic instability. This vicious cycle has frustrated financial reform efforts.

4.04 The financial system's core structural problem is the peculiar Central Bank intermediation arrangement that has evolved to channel subsidies and to fund the public sector. In effect, the Central Bank borrows funds through the "forced-investment" mechanism from the private commercial banks, and passes these on through "rediscounts" to public sector financial institutions--mainly to the National Mortgage Bank (BHN), the National Development Bank (BANADE), and to the provincial banks. In recent years this problem has been aggravated because the Central Bank has been using the same mechanism to provide credit to the Treasury, to enable the Treasury to retire its maturing bond issues. This intermediation arrangement is the central obstacle to the financial system's medium-term development. Its alleviation is a prerequisite for other financial reforms.

4.05 This arrangement works badly for the following reasons:

(a) The resources forcibly lent by the commercial banks to the Central Bank are diverted from the productive private sector. The consequent shortage of commercial and working capital credit contributes to high real interest rates, and is detrimental to productive activity;

(b) The intermediation process--which channels funds from commercial bank depositors to BHN and BANADE borrowers--is inefficient, since it takes resources away from high-yielding production credit applications to subsidized, generally inefficient (and loss-making) housing and investment applications; and

(c) The Central Bank's intermediation stance compromises its capacity to carry out monetary policy, since it has virtually no assets that it can use for monetary absorption.

Not only is it inappropriate for a central bank to hold so large a quantity of unmarketable assets; the assets in question appear in fact to be value-impaired.

4.06 The Central Bank was called upon to provide credit to the public banks because the macroeconomic instability has made it impossible for them to raise sufficient funds to finance their activities. Longer-term housing and investment finance have become unprofitable: because of the unstable macroeconomic conditions, most borrowers cannot pay the higher real interest rates that the same conditions induce wealth-holders to demand. While the private sector therefore withdrew from these activities, the public banks continued in them, decapitalizing themselves as a result, and effectively subsidizing their borrowers. The problem was compounded by the inefficiency of the public banks' operations. Once the macroeconomy is firmly stabilized, private financial institutions would provide more, if not all, of the housing and industrial finance activity. Even before then, however, BHN, BANADE, and the provincial banks need to be restructured and reduced in scope, so that they carry on no activities beyond what they can finance without draining resources from the Central Bank and the commercial banks.

4.07 The Alfonsin Government has carried out a number of fundamental financial system reforms. The liberalization of interest rates in October 1987 implied that interest rates now reflect the true cost of credit. This will encourage financial applications and, once the Central Bank reduces its demand for credit, promote better allocation of credit resources. The Central Bank has curtailed the flow of new credit to public sector banks (although it has had to capitalize the interest owed it). Since early 1987, the Central Bank has acted more resolutely in dealing with problem banks, quickly intervening in bankrupt private banks and suspending overdraft facilities to provincial banks that abused them. Important steps have been taken to introduce a deposit insurance system. These steps should more rapidly cleanse the system and reduce the costs to the system of bank failures. The Government has also taken initial steps to improve portfolio management in BHN and BANADE. In addition, the managements of these institutions have begun to consider restructuring.

Macroeconomic instability, fiscal imbalance, and the Central Bank's intermediation role prevented these reforms from having immediate positive effects, but they will contribute to financial sector effectiveness over the medium-term.

B. The Present Structure of the Financial System

4.08 The banking system constitutes the bulk of Argentina's institutional financial system. The number of commercial banks seems relatively large for the size of the market, although in recent years there has been some consolidation of privately-owned banks (the fall in the number of privately-owned institutions between March 1987 and September 1988 shown in Annex Table 4.1 reflects this trend). Seeking to clear up problem bank portfolios and to promote efficiency, the Central Bank has supported consolidation. The decapitalization resulting from the present crisis is likely also to hasten this consolidation process. As of August 31, 1987 there were 20 financial institutions under intervention and another 180 in liquidation.

4.09 In March 1988, the 36 publicly owned banks accounted for 45 percent of the banking system's total deposits and provided 71 percent of all bank credit. They include two national commercial banks. Of these the Banco de la Nacion is the nation's largest, accounting in September 1988 for 15 percent of total deposits. There are 24 provincial banks, of which the Banco de la Provincia de Buenos Aires alone accounted for more than 9 percent of total deposits. The Banco de la Nacion and the Banco de la Provincia de Buenos Aires have the nation's largest branch networks.

4.10 The BHN (National Mortgage Bank) was founded in the late 1800s to provide term finance for industry, agriculture, and housing through mortgage bonds. Through the first four decades of this century BHN was a prestigious institution capable of raising finance in international financial markets. In the 1940s, however, BHN was drawn into financing lower cost housing, and it slipped into increasing reliance on Central Bank funding. This was because its loans came to incorporate substantial subsidies that made it unable to pay for funds from the financial markets. BHN held 5 percent of total deposits in September 1988, although public institutions account for a large proportion of these deposits. BHN is now virtually the only source of housing finance for the middle class. In addition, the Government's National Housing Fund (FONAVI) provides subsidized housing finance intended for the poorest 40 percent.

4.11 BANADE was formed in 1970 from the Banco Industrial (originally founded 1944), to serve as an autonomous government-owned industrial development bank. It is now the only significant domestic source of term lending for Argentine industry. BANADE has been chronically troubled since its establishment, partly because macroeconomic instability has made it difficult to secure finance and has made its assets excessively risky, but also because its management has never succeeded in streamlining and modernizing its operations. In addition, successive governments have used the institution to channel subsidies and "to rescue" failing industrial enterprises. In recent years it has come to rely on Central Bank rediscounts for its funding, since it has been unable to secure a sufficient quantity of deposits or foreign credit.

Annex Table 4.1: ARGENTINA - FINANCIAL INSTITUTIONS

	March 1987	September 1988
Banks:	189	172
Commercial Banks:	183	166
Publicly Owned:	31	30
Federal (BNA)	2	2
Provincial	25	24
Municipal	5	5
Privately Owned:	151	135
Domestic Capital	121	103
Foreign Capital	31	33
Investment Banks:	2	2
Publicly Owned:	1	1
Federal	1	1
Privately Owned	1	1
Development Banks:	2	2
Publicly Owned:	2	2
Federal	2	2
Mortgage Banks:	1	1
Publicly Owned:	1	1
Federal (BHN)	1	1
Savings and Insurance Bank:	1	1
Publicly Owned:	1	1
Federal	1	1
Finance Companies	64	40
Savings and Loan Associations	11	7
Credit Unions	30	23
Total	294	242
Total Branches	4,471	
of which Commercial Banks	4,377	
of which, BNA		543
of which, BPBA		330
Total Employment	144,000	142,000
Percentage Employed by:		
Official Banks	56	58
Private banks	42	40
Other Institutions	2	2

Source: Central Bank of the Republic of Argentina, Departmento de Expansion y Servicios de Entidades

4.12 Non-bank financial institutions, including finance companies, savings and loan associations, and credit cooperatives, accounted for 2.6 percent of the system's total deposits and 2.2 percent of the system's total credit. There were 70 such institutions in March 1988, compared with 100 in June 1987 and 243 in December 1981.

4.13 Apart from banks, there are several other kinds of financial intermediaries, including finance companies and insurance companies. They are quantitatively relatively unimportant. There are five stock exchanges. There is also an active inter-firm financial market, in which firms with spare resources lend to other firms. This disintermediated market arose largely as a consequence of the formal financial system's inability to meet corporate credit needs, partly because of interest-rate regulations, and partly because the public sector takes a large proportion of the system's available resources. Furthermore, there are informal markets that carry on over-the-counter trading in securities and provide various other kinds of financial services, such as linking companies with excess funds to others requiring funding and providing guarantees.

C. Financial Liberalization and Centralization

Background

4.14 Over the past decade and a half Argentina's financial policy has shifted between a centralized, controlled approach and a liberalizing approach. From 1973 to 1976, the Government effectively nationalized the banking system's deposits by means of a 100 percent reserve requirement. Under the system that resulted, commercial banks effectively passed their deposit proceeds to the Central Bank and provided credit on the basis of Central Bank rediscount allocations, all at regulated interest rates. The Government intended to use the Central Bank to direct financial mobilization and allocation. This approach worked poorly, because the credit allocation had no rational price basis and because the Authorities could not control inflation, which discouraged financial applications.

4.15 Since 1976, Argentina's governments have generally intended to liberalize, but have succeeded only partially. This was essentially because the persisting financial shallowness and the public sector's heavy borrowing requirement meant that anti-inflationary measures either tended to force real interest rates excessively high or required the Government to force the financial system to lend to it. In 1977 a new government enacted a liberalizing reform that returned banking to a fractional reserve system, freed interest rates in the formal financial system, removed barriers to entry into the banking system, and liberalized foreign financial inflows. The required reserve ratio was 45 percent at first, because the Authorities feared that a lower ratio would prove too inflationary. The Authorities, therefore, instituted payment of interest on bank reserves, to enable banks to maintain relatively low spreads between asset and liability rates.

4.16 Beginning in 1977, the banking system expanded its operations in response to the policy liberalization. The private sector took on growing volumes of external debt. Unfortunately, the macroeconomic policies at this time--centered on the maintenance of an unsustainably overvalued exchange rate (see Chapter 1)--required high domestic interest rate levels. In March 1980 a bank crisis took place as a result of the excessively rapid expansion of bank activities. In mid-1982 a deeper crisis resulted from the combination of the sharp devaluations carried out over the preceding fifteen months, persisting high interest rates, uncertainties resulting from the 1982 conflict with Great Britain, and finally the onset of the world debt crisis. The devaluations and high interest rates drove domestically and externally indebted private firms--and consequently their banks--into virtual bankruptcy. The financial system had to provide funding for the public sector deficit, which had widened under the same pressures.

4.17 To cope with the crisis, the Authorities reluctantly took policy measures that effectively reversed the liberalizing reforms. To relieve the pressure on the private sector, the Central Bank effectively assumed the private sector's external debt. The Government enacted a blanket rescheduling of private sector loans to domestic commercial banks for a sixty-month period at negative real interest rate. To ensure that this did not decapitalize the banks, controls were reintroduced on deposit interest rates. These measures permitted the private sector to recover financially, and enabled the banking system to maintain its capitalization. The cost, however, was that the financial system was reestablished with high reserve requirements and interest rates inadequate to attract new deposits. The private banking system had to narrow its lending activities. Disintermediation resulted, taking the form of a reinvigorated inter-firm financial market.

4.18 The Alfonsin Government tried to resume financial liberalization, and undertook a number of fundamental reforms, including the complete liberalization of interest rates in October 1987. Unfortunately, the public sector's credit needs remained substantial. In particular, the large public sector banks--the BHN, the BANADE and some of the provincial banks--came to require heavy financing after 1985. In the absence of other funding sources, the Central Bank was called upon to supply financing to the public banks, through such mechanisms as provision of "rediscount" credit to public sector banks, advances for social security payments, and, during 1988, amortization of the Government's outstanding bond issues.

4.19 After 1985, the Central Bank came to rely on "forced investments" of the commercial banks as its principal tool of monetary absorption. The Central Bank's heavy use of forced investments has contributed to the commercial banks illiquidity, and has severely impaired the commercial banks' capacity to lend. Over 1987 and 1988, the Authorities made determined efforts to reduce both the provision of rediscounts and hence the forced investments. However, despite the liberalization of interest rates, the Authorities failed to relieve the strain on the banking system caused by the public sector's heavy demand for credit.

4.20 Despite the banking system's central role in the system, inflation and macroeconomic uncertainty generally have discouraged the Argentine public from holding monetary assets. Annex Table 4.2, which provides basic monetary indicators, indicates the unusually low money holding with which the economy has operated in recent years. Holdings of narrow money (M1), have been on the order of only 5 percent of GDP since early 1986. They rose to 7 percent of GDP only when the 1985-1986 Austral Plan repressed inflation temporarily. During 1988 holdings fell below 4 percent of GDP. Holdings of M4, the broad money aggregate incorporating remunerated deposits, reached 19 percent in early 1987, but then slipped to 14 percent in the third quarter of 1988. Holdings of M4 rose slightly toward the end of 1988, when the stringent monetary policy associated with the Primavera Plan elevated real interest rates to unprecedented heights. For the most part, M4 holdings have fluctuated between 14 and 18 percent of GDP, following a pattern broadly similar to that of M1.

Annex Table 4.2: ARGENTINA - MONETARY AGGREGATES: 1986:IV - 1989:I
(Billions of Dec. 1988 Australes, Percent of GDP)

	1986 IV	1987 IV	1988 I	1988 II	1988 III	1988 IV	1989 I
	Billions of December 1988 Australes (average of daily values)						
M1:	88.7	62.8	61.2	50.3	43.8	51.7	57.9
M2:	157.9	89.7	85.8	69.8	60.6	70.8	78.7
M3:	260.9	211.7	222.0	204.8	192.2	226.4	257.6
M4:	262.8	212.9	222.7	204.9	192.3	226.5	257.7
M5:	264.4	213.2	223.0	205.1	192.5	226.7	258.0
M6:	264.4	213.2	223.0	205.1	192.5	226.7	258.0
	As a percentage of GDP:						
M1:	6.3%	4.5%	4.4%	3.6%	3.2%	3.6%	
M2:	11.2%	6.4%	6.2%	5.0%	4.4%	4.9%	
M3:	18.6%	15.2%	16.1%	14.7%	14.0%	15.6%	
M4:	18.7%	15.3%	16.1%	14.7%	14.0%	15.7%	
M5:	18.8%	15.3%	16.1%	14.7%	14.0%	15.7%	
M6:	20.1%	18.1%	19.0%	16.9%	15.9%	17.0%	

M1 = Currency plus demand deposits,
M2 = M1 plus regulated time deposits,
M3 = M2 plus unregulated time deposits and acceptances,
M4 = M3 plus adjustable deposits,
M5 = M4 plus private bills,
M6 = M5 plus Austral-denominated bonds.

Source: Carta Economica

4.21 Annex Table 4.3, which shows the monetary structure of the Central Bank's liabilities, indicates the degree to which the Central Bank absorbs commercial bank resources. While the reserve ratio of the basic liquidity aggregate, M4, to the conventionally-defined monetary base is relatively low, the implicit reserve ratio against the Central Bank's overall monetary liabilities--virtually all of which are the "forced investments" of the commercial banks--is debilitatingly high.

Annex Table 4.3: ARGENTINA - CENTRAL BANK MONETARY LIABILITIES
(Billions of December 1988 Australes)

	Dec 87	Mar 88	Jun 88	Sep 88	Dec 88	Mar 89
	Billions of December 1988 Australes					
Conventional Monetary Base a/	65.0	51.4	45.5	43.1	58.0	51.5
Effective Monetary Base b/	74.9	57.9	51.0	47.2	60.9	54.4
Monetary Liabilities c/	155.8	140.1	132.1	141.6	170.8	157.3
Conventional Bank Reserves	25.8	21.1	21.0	19.9	24.2	20.1
Forced Investments	80.9	82.2	81.1	94.4	109.9	102.9
Broad Monetary Liabilities d/	155.8	143.6	136.3	151.4	170.9	157.3
Conventional money multiplier e/	3.5	4.3	4.4	4.7	4.2	4.9
Conventional reserve ratio f/	13.9%	11.1%	12.1%	11.1%	11.5%	9.1%
Broad money multiplier g/	1.4	1.5	1.5	1.3	1.4	1.6
Broad "reserve" ratio h/	62.6%	59.8%	64.4%	71.1%	65.2%	57.2%

a/ Currency plus bank reserves.
b/ Conventional monetary base plus "imputations," mainly advances for social security payments.
c/ Conventional monetary base plus "forced investments."
d/ Including Central Bank bills.
e/ M4/(conventional monetary base).
f/ (M4-currency)/Conventional reserves.
g/ M4/(monetary liabilities).
h/ (M4-currency)/(res. + fcd. invmts.).

Source: Carta Economica

4.22 The figures in Annex Table 4.4 regarding deposits and loans are also relevant to the financial sector's central issue. Because of the heavy "forced investments" private banks are required to make in the Central Bank, their lending capacity is considerably below their deposit base. Their loan-deposit ratios have fallen from 64 percent in the third quarter of 1985 to 42 percent in the third quarter of 1988. On the other hand, the loan-deposit ratio of all publicly owned banks taken together rose over the same period from 77 to 121 percent. These figures are heavily influenced by the BHN: largely because of its Central Bank support, the BHN managed to have a loan-deposit ratio exceeding 800 percent in mid-1988.

Annex Table 4.4: ARGENTINA - BANKING SYSTEM LOANS AND DEPOSITS
(Billions of December Australes)

	1985 IV	1986 IV	1987 IV	1988 I	1988 II	1988 III
Loan-deposit ratio:	69.5%	79.1%	90.3%	92.0%	81.8%	78.4%
Privately-owned banks	62.4%	57.0%	52.2%	50.9%	44.1%	42.3%
Publicly-owned banks	74.5%	96.2%	130.8%	139.2%	124.8%	120.8%
	Billions of December 1988 Australes.					
Banking-system deposits	200	234	198	203	200	188
Privately-owned banks	83	102	102	109	107	101
Publicly-owned banks	117	132	96	95	93	86
Banking-system loans	139	185	178	188	165	147
Privately-owned banks	52	58	53	56	47	43
Publicly-owned banks	87	126	125	132	118	104

Source: G.F. Macroeconomia

D. Principal Financial Sector Policy Issues

Hyperinflation

4.23 The hyperinflation will inevitably damage the financial system's liquidity and its solvency. Uncertainty regarding inflation, devaluation and interest rates will encourage deposit withdrawals, inducing illiquidity. Demand for money and other financial assets is likely to erode further from already low levels. The deteriorating economy is not only making it increasingly difficult for borrowers to make cash payments on their loans, thereby contributing to financial system illiquidity; it is also impairing the value of the loans, thereby contributing to financial system insolvency. At the same time, the monetary authority is likely to have no choice but to create money in order to relieve illiquidity and to prevent financial panic.

4.24 The financial system's illiquidity has been aggravated by the fact that deposits have been growing at much a slower rate than the interest paid by banks on deposits. The increase in deposits was only 50 percent in April 1989 and 30 percent in May of the total interest capitalized into bank deposits. Time deposits grew 23 percent in April and 36 percent in May, against average interest rates of 35 and 103 percent respectively. As a result, the Authorities announced restrictions on cash withdrawals from banks. In addition, toward the end of May the monetary authority announced that it would inject funds into the system by releasing almost 10 percent of the stock of forced investments.

4.25 The financial system's assets have also become increasingly illiquid, since private debtors have found their repayment capacity deteriorating. Many have found it necessary to draw down their deposit balances. Rising real interest rate levels have added to the pressure on private firms, particularly in view of the sharply diminishing real economic activity. At present the economy's real indebtedness to commercial banks remains relatively low in real terms, perhaps 50 percent of its record level in early 1981 and 70 percent of its level in early 1982. Nevertheless, real interest rates are so high that the real indebtedness level could reach a record level through interest capitalization by the end of July. Moreover, dollar-denominated indebtedness is believed already to be near an historic high. This would matter to the debtors if the interest on the debt were fully capitalized, but that would leave the banks extremely illiquid.

4.26 In an effort to reduce the pressure on financial institutions' solvency, the Central Bank has increased the rate of return on remunerated bank reserves. Nevertheless, independent analysts in Buenos Aires estimate that the typical Argentine bank may have suffered the loss of 10 to 20 percent of its capital position in April and May. The monetary authority is clearly hoping that, by releasing funds into the system in early June, it will succeed in reducing interest rates, and thereby reduce the pressure on bank solvency. The risk, of course, is that inflationary pressures will intensify.

4.27 While the precise outcome of the crisis cannot be predicted, it is clear that the financial system will emerge in worse condition. Public as well as private financial institutions are likely to find themselves severely decapitalized. If the hyperinflation persists, the financial system could be so decapitalized that it will require complete recapitalization and reconstruction when the hyperinflation finally "burns out." It is therefore useful to look beyond the crisis, to consider the kinds of reform (or reconstruction) that will be required for the medium-term development of an effective financial system. The three core issue areas will be (i) the role of the Central Bank, (ii) the public sector banks, and (iii) prudential regulation. The Central Bank's intermediation role is discussed in Chapter IV; the present discussion is therefore confined to the two remaining issue areas.

Public Banks

4.28 The same circumstances that discourage private long-term finance have contributed to decapitalization of the public sector's longer-term financial institutions, the BHN and the BANADE. Savers have been unwilling to place funds with the BHN because it could not remunerate them competitively, essentially because its mortgage holders have not been paying sufficient debt service to cover the institution's funding costs. Loan arrears have become significant. The BANADE, which has relied on public sector and external finance, has also had difficulties with arrears, partly because the economy has performed poorly. In addition, both BHN and BANADE have had difficulties effectively managing loan recovery. Both the BHN and the BANADE have received substantial Central Bank financing.

4.29 BHN is now insolvent by reasonable standards. Its losses totalled US$367 million in 1987, US$329 million more than the previous year. Net equity was negative US$216 million. This recorded performance was better than the actual performance, since BHN has tended to accrue interest on many loans that are not genuinely performing. Even where loans are placed on non-accrual, the BHN has tended to avoid making provisions against them on the presumption that recovery will be possible through foreclosure. BHN's charter allows foreclosure without BHN having to go through the regular judicial process. Nevertheless, BHN has been reluctant to initiate foreclosure proceedings. BHN's basic problem is that the interest and adjustment on its loan portfolio, which is relatively old and incorporates considerable subsidies, is insufficient to pay current Argentine market interest rates. (This problem is worsened by the fact that adjustment indices applied to loans tend to lag behind actual inflation rates.) BHN's performance is worsened by its current overstaffing: almost three fourths of BHN's operating budget of US$52 million during 1987 went to staff. The staff totalled 4,200 (1,000 of whom were over the age of 50).

4.30 At the end of 1987 BHN's loan portfolio, constituting 78 percent of its total assets, amounted to about US$1.3 billion. There were about 250,000 borrowers, so the average loan was about US$5,200. The average contractual real interest rate was only about 2 percent per year and the average maturity about 30 years. (In fact, given the way in which BHN interest charges are calculated, the real interest rates are negative on average.) Loan delinquency was nonetheless severe: about 25 percent of the loan portfolio was in arrears exceeding three monthly payments. In view of the context of high interest rates, this amounts to a subsidy that no one has authorized BHN to provide. To the extent housing subsidies are regarded as justified for poorer people, it is appropriate that they be channelled through the National Housing Fund (FONAVI), which was created with the intention of providing such a subsidy.

4.31 The Central Bank was BHN's principal funding source. Credit outstanding by the Central Bank to BHN totalled about US$1.7 billion at the end of 1987. The pressure on BHN intensified over 1988, since it had made a large number of new loan commitments during 1987. BHN's competitive disadvantage in attracting deposits sharpened after October 1987, when commercial bank deposit rates were freed. Interest rates due the Central Bank were on the order of 8 percent, well above rates of return on the loan portfolio. Since BHN lacked the cash flow with which to pay the Central Bank, what it owed was capitalized, which was one of the main causes of the high growth rate of BHN's obligations to the Central Bank.

4.32 Like BHN, BANADE has effectively been decapitalized through a combination of the troubled macroeconomic context and inadequate management of its loan portfolio. BANADE's present management recognizes this problem, and has made a preliminary restructuring proposal. BANADE is now Argentina's only source of term credit for the private sector, because the macroeconomic instability has made such activities impossible for the private sector. Like BHN, BANADE effectively channels a disguised subsidy to industrial enterprises fortunate enough to secure access to its credit, since long-term industrial credit would cost far more if it were available from private sources. Nevertheless, BANADE's recent performance indicates that it is unable to serve this function adequately. Despite several organizational restructurings and recapitalizations undertaken by the Alfonsin Government, BANADE has been unable to reduce its reliance on inappropriate Central Bank financing. With the suspension of Central Bank rediscounts during 1988, BANADE was forced to halt its lending operations. Since at least 1985, BANADE's cash flow has been negative in real terms and the institution has been insufficiently profitable to reverse its decapitalization.

4.33 At present its staff numbers approximately 3,000, spread through a main office, 33 branch offices and 8 mining area offices. Its assets total approximately US$5 billion and its loans about US$3.5 billion to a total of 4,300 different clients, although provisions total about 10 percent of the portfolio. In mid-1988, external funding accounted for about 70 percent of BANADE's resources, Central Bank rediscounts about 20 percent, and private deposits only about 5 percent. Since private domestic deposits are only available at high interest rates for relatively short terms, such deposits are an inappropriate source of finance for BANADE.

4.34 BANADE's difficulties revolve around its loan portfolio problems, which derive in part from the troubled macroeconomic conditions. BANADE's loan portfolio quality is also affected by political interference in lending decisions, a pervasive perception by borrowers that they need not repay, lack of financial standards for BANADE managers to follow, and lack of an effective supervisory agency to monitor BANADE performance. The total value of BANADE's loan portfolio as of June 30, 1988 was US$3.4 billion. About 38 percent of the loan portfolio is to public enterprises and 58 percent to private enterprises, with the remaining 9 percent going to mixed enterprises. About 52 percent of this loan portfolio is classified as problem loans, 21 percent being in arrears of more than one year. There is an inadequate level of provisions for the bad debts, in the sense that problem loans not covered by guarantees net of provisions are about twice the institution's equity position.

4.35 In the years 1986-1988, BANADE failed to achieve a positive cash flow. Over the first six months of 1988 the cash flow deficit was about US$136 million, which had to be covered through borrowing at high short-term interest rates. New lending operations ceased when the Central Bank refused further rediscount credit. The share of cash income to capitalized (i.e., accrued only) financial income was only 5.37 percent in the first six months of 1988, down from about 10 percent in 1987. Collection efforts have been relatively unsuccessful, in part because few uncollectable loan cases go to judicial action, fewer are quickly resolved in BANADE's favor, and fewer still are executed. These problems are undoubtedly affected by the concentration of BANADE's portfolio: the fifty largest debtors accounted for about 75 percent of total loans and 75 percent of all the problem loans. BANADE's equity position is clearly deficient: long-term unsecured debt was 8.8 times equity; total unsecured debt was 11.6 times equity; and total long-term debt was 27.1 times equity.

4.36 The provincial banks have received substantial quantities of "rediscount" finance (in many cases, through overdraft facilities) from the Central Bank, particularly since 1982. In this way, the provincial banks have often served effectively to channel Central Bank resources to provincial governments. The Central Bank has tried to curb this credit flow, and largely succeeded in doing so during 1988, although the outstanding credit of provincial banks to the Central Bank continues to grow through the effects of inflation adjustment and interest capitalization. Toward the end of 1987, for example, in the face of political opposition, the Central Bank temporarily suspended some provincial banks' overdraft facilities on the grounds that they had abused them.

The Regulatory Environment

4.37 Prudential regulation of Argentina's financial institutions is less vigorous and efficient than it could be. A superior regulatory system would help instill public confidence in financial institutions. In recent years, a better bank examinations system would assuredly have resulted in significant savings for the Central Bank, since it would have permitted speedier identification of problem banks, hence timely corrective measures or intervention at a less costly stage.

4.38 The Central Bank carries out the superintendency function for all financial enterprises. Broadly speaking, it appears to have the laws and regulations it requires to set and enforce regulatory standards, with the important exception that the Central Bank's legal capacity to regulate provincial banks is unclear. Banks are subject to conventional rules governing such matters as the largest allowable loans (as ratios to callable capital), lending to enterprises judged to be "linked," minimum capital requirements, appropriate asset portfolio composition, and so on. In addition, banks must meet the Central Bank's minimum reserve and forced investment requirements. The Central Bank's examinations staff is charged with verifying compliance with these standards.

4.39 The Central Bank has a standard loan classification system for financial institutions. There appears to be a generalized view that the criteria for incorporating loans in the categories are insufficiently precise. A more important problem is that, in reality, it is the banks themselves who determine the classification of their assets. Moreover, the banks themselves determine whether loans should be placed on nonaccrual status. This is because the Central Bank's overstretched examinations staff has generally been able to do no more than verify that banks are complying with reserve requirements. It has been unable adequately to examine loan classification, nor whether lending limits to linked institutions have been exceeded.

4.40 The Central Bank has the power to (i) require that banks present plans to regularize reserve deficiencies, (ii) authorize fusions and consolidations; and (iii) carry out "interventions"--i.e., appoint delegates to substitute for a troubled banking institution's board of directors. It may initiate judicial bankruptcy and liquidation proceedings. In order to facilitate a particular bank restructuring, the Central Bank Directorate has the power to permit exceptions to its normal rules for temporary periods. An important problem with the intervention proceeding is that, once initiated, it requires the Central Bank to cover all deposits, whether insured or not. A further problem is that the judicial proceedings have tended to be lengthy, stretching in certain instances over several years.

4.41 The Central Bank has been more willing to initiate interventions in recent years. Nevertheless, since the public has concluded from experience that intervention is followed sooner or later by liquidation, announcement of intervention tends to induce deposit runs. Liquidation also imposes severe costs on the Central Bank, since the Central Bank is required to assume the payroll, legal proceedings, severance pay, and so on, as well as the funding for depositor compensation. Central Bank "rediscounts" to intervened institutions have accordingly tended to be quite substantial in recent years.

4.42 The Central Bank has considerable difficulty regulating publicly owned banks, especially the banks owned by provincial governments. The provincial banks are often established legally under provincial legislation, and in some instances this is understood to limit the Central Bank's regulatory power. (The looser regulatory environment for public banks amounts to discrimination against private banks.)

4.43 Relief of macroeconomic instability is a prerequisite to reestablishment of longer-term financial markets. Once the public believes that stability has been attained, private longer-term financial activity may develop spontaneously, beginning with housing finance. Until then, the management of the public sector banks will be an issue of critical importance, because the macroeconomic instability makes their activities almost inherently loss-making. In addition, relief of macroeconomic instability and restoration of control over the public sector deficit is the only ultimately sound basis for reducing the high real interest rates now burdening the financial system.

E. Recommendations

4.44 It is essential that, while the Government deals with the present crisis, it also look beyond it to work out a medium-term financial development strategy. If the hyperinflation persists, the financial system will have to be reconstructed; in any case, it will need to be restructured, to ensure that it meets the needs of development over the coming decade and that it becomes a force for stabilization rather than destabilization. The restructuring process will be easier, of course, to the extent macroeconomic performance improves and to the extent the external constraint is relieved--in particular, the more closely the public sector borrowing requirement can be limited and monitored, and the more external financing can be obtained for the public sector.

4.45 The essential core reforms can be grouped simply in three categories: those involving the Central Bank and the commercial banks, those involving the large public sector financial institutions, and those involving prudential regulation. First, a package of reforms is needed simultaneously (i) to remove the value-impaired assets from the Central Bank's balance sheet, in exchange for a genuinely marketable asset that the Central Bank can use in open-market operations, i.e., use to back its currency issues; and (ii) to relieve the commercial banks' forced investment burden. The Central Government could carry out this recapitalization by gradually issuing bonds to buy the value-impaired assets gradually from the Central Bank. The recommended reforms involving the Central Bank and the commercial banks are discussed in more detail in the monetary policy chapter.

Public Banks

4.46 Where the BHN and the BANADE are concerned, it is increasingly clear that the public sector can no longer maintain these costly institutions. Since any serious stabilization effect must involve elimination of Central Bank credit flows to these institutions, it is likely that they would be forced into some sort of receivership, during which they would be drastically restructured or even liquidated. Current macroeconomic conditions make it impossible for the private sector, let alone the public sector, to carry on housing and industrial finance activities profitably. Eventually, once stabilization takes hold, the private sector should gradually begin to carry on efficient longer-term financing.

4.47 This suggests a restructuring strategy that would sharply reduce their size and improve their operating profit flow to the point of making them self-financing, at least until the macroeconomy stabilized sufficiently to restore the profitability of longer-term financial operations. The two institutions could do this by acquiring relatively small "core" portfolios of high-yield assets whose yield covers their operating expenses. This would require that their operating expenses be reduced to the minimum necessary to manage the "core" portfolio and to reduce losses on the existing portfolio. Recapitalization ought to be carried out only to the extent necessary to enable them to go into this "survival mode." Such recapitalization ought to be conditioned on thorough independent audits of assets and liabilities, agreement with the Central Bank regarding their obligations to that institution, more vigorous loan collection efforts, and measures to improve operating efficiency. At the same time, the Government needs to take hard decisions about these institutions' future role. To the extent private initiative becomes willing to carry on their functions, it should be allowed and encouraged to do so.

4.48 The provincial banks' future role needs to be examined more closely, although given the nation's federal structure this will have to be accomplished largely on a province-by-province basis. Provincial governments should be encouraged to re-examine the role their banks play in their economies. The provinces' specific circumstances vary considerably, and there is scope for different approaches and objectives. In any case, the Central Bank's future relationships with provincial banks will have to be re-examined. While it is legitimate for provincial banks to have access on the same basis as any other bank to Central Bank liquidity rediscounts, the Central Bank cannot permit its funds to be used by provincial banks to finance their provincial government deficits. In any case, provincial governments should regard it as in their interest to have their banks take advantage of high-quality prudential regulation (see below). This could be made a condition of future access to Central Bank liquidity rediscounts.

The Regulatory Environment

4.49 Along with the need to revitalize the private financial system, there is a collateral need to strengthen both the superintendency and deposit-insurance functions. It is advisable that both be made independent of the Central Bank. Independence of the superintendency would ensure that its determinations are not influenced by considerations of the short-run intervention costs. An independent superintendency, with full powers to examine financial institutions as necessary and carry out interventions as required on technical grounds, would go a long way toward reestablishing public confidence in the financial system and ensuring that small problems are dealt with expeditiously before they become large problems. In any case, the superintendency function needs to be strengthened by providing it with the personnel and computer resources to enable it to monitor all aspects of bank activity. In particular, loan classification needs to be made fully technical. Establishment of a separate deposit-insurance fund should eventually make it possible to deal with bank failures without the Central Bank directly having to create money.

4.50 Even before 1989, many financial enterprises were severely weakened, and the current hyperinflation has worsened this tendency. The regulatory staff may therefore be compelled to increase its activity. Since the financial crises of the early 1980's, many banks and non-bank financial institutions suffered decapitalization and weakened loan portfolios. Lax regulation has implied, among other things, that many financial institutions have made inadequate provisions and have accrued too much income from doubtful assets. A large number of institutions continue under Central Bank intervention. The weakness of the system has made it vulnerable, and increased the dangers and costs of tightened monetary policy. The Central Bank's recent policy of encouraging consolidations will almost certainly have to continue.

Annex Table 4.5: ARGENTINA - FINANCIAL INSTITUTIONS' NET ASSET POSITION WITH THE CENTRAL BANK

	March 87	September 88
Public National Banks, B. Prov. de B.A.		
Net Liabs. to the Central Bank/Deposits	-20.3%	-11.2%
= Rediscount Credit/Deposits	53.2%	68.1%
- Reserves/Deposits	-20.6%	-25.0%
- Forced Investments/Deposits	-52.9%	-54.3%
Deposits/Total System Deposits	35.4%	25.0%
Provincial and Municipal Banks		
Net Liabs. to the Central Bank/Deposits	64.8%	9.0%
= Rediscount Credit/Deposits	65.7%	62.3%
- Reserves/Deposits	0.0%	-22.1%
- Forced Investments/Deposits	-0.8%	-31.2%
Deposits/Total System Deposits	10.3%	14.1%
Private Banks, Argentine Capital		
Net Liabs. to the Central Bank/Deposits	-50.0%	-49.5%
= Rediscount Credit/Deposits	23.1%	16.0%
- Reserves/Deposits	-15.9%	-11.5%
- Forced Investments/Deposits	-57.2%	-54.0%
Deposits/Total System Deposits	28.7%	32.4%
Private Banks, Foreign Capital		
Net Liabs. to the Central Bank/Deposits	-54.6%	-51.1%
= Rediscount Credit/Deposits	12.5%	18.7%
- Reserves/Deposits	-15.2%	-15.8%
- Forced Investments/Deposits	-51.9%	-54.0%
Deposits/Total System Deposits	14.3%	17.0%
Cooperative Banks		
Net Liabs. to the Central Bank/Deposits	-75.7%	-52.8%
= Rediscount Credit/Deposits	5.5%	16.6%
- Reserves/Deposits	-18.8%	-19.0%
- Forced Investments/Deposits	-62.5%	-50.5%
Deposits/Total System Deposits	8.7%	9.1%
Other Financial Institutions a/		
Net Liabs. to the Central Bank/Deposits	-43.0%	-34.8%
= Rediscount Credit/Deposits	34.5%	12.7%
- Reserves/Deposits	-5.6%	-7.0%
- Forced Investments/Deposits	-71.8%	-40.6%
Deposits/Total System Deposits	2.7%	2.4%

Source: BCRA

a/ Finance companies, savings and loan associations, credit companies and investment banks.

ANNEX CHAPTER V: EMPLOYMENT AND LABOR

A. Introduction

5.01 The most important medium-term objective of Argentine economic policy is to ensure adequate employment and to increase wage levels. Average wage rates have fallen since the mid-1970s: 1987 real wages were only 60 percent of their 1975 value and 75 percent of their 1980 value (Annex Table 5.1). Because of anemic growth in the 1980s, employment growth has stagnated and unemployment has begun to drift up in recent years.

5.02 Two related features dominate employment and labor patterns. First, labor productivity growth has been extremely disappointing. Although overall productivity rose more than 7 percent between 1985 and 1987, it has apparently stagnated since then, and is now only 18 percent above its 1962 level and is still 9 percent below its 1980 level (Annex Table 5.1). Second, during the 1980s the public sector has been the principal source of net employment creation. Public sector employment rose 16.5 percent between 1980 and 1988, while the overall labor force participation rate remained roughly unchanged at just under 40 percent. The present crisis clearly indicates, however, that the public sector will have to reduce its employment over coming years. That is, labor must be efficiently reallocated, out of the public sector and into what must become an increasingly productive private sector.

5.03 A genuine increase in labor productivity is the only basis on which private sector employment and real wages can increase over the medium term. In order for the private sector to increase its demand for labor, private capital formation must increase and labor markets must be made more efficient. Wages need to be flexibly associated with productivity at the level of each firm. In particular, wage setting needs to be decentralized, so that wage determination is associated more directly with productivity. The job tenure system needs to be relieved, to enable employers to hire workers without having to assume that they must keep them indefinitely, and to permit workers to move freely in response to changing conditions and incentives. Provision of labor security must become less of a burden to private firms. More precisely, taxation of labor use must be relieved.

B. Productivity, Wage Trends, and Wage Determination

5.04 The overall stagnation of labor productivity (Annex Table 5.1) masks some differences across industries (Annex Table 5.2). In agriculture, productivity has risen despite declining growth rates. In manufacturing labor productivity has remained broadly constant over time, with output variations apparently not reflecting technical change nor changes in capital intensity. In the non-tradeable construction and services sectors there has been a declining productivity trend. However, from 1983 through 1987 productivity rose in the services sectors, mostly because of slowing employment growth.

Annex Table 5.1: ARGENTINA - LABOR MARKET INDICATORS

	LFP	E	U	U*	W/P	RER	WT/WN	LG	PRO
1962	45	42	7.3	n.a.	83	125	0.83	96	100.0
1970	44	42	5.3	n.a.	100	100	0.89	100	106.2
1975	40	39	3.5	6.5	115	82	0.92	128	120.0
1980	39	38	2.8	5.1	93	46	0.89	121	126.4
1984	38	36	4.8	7.4	99	107	0.96	122	117.7
1985	38	36	6.4	9.8	81	133	0.97	124	110.9
1986	39	35	5.2	8.9	76	106	0.96	132	118.4
1987	39	35	5.6	9.9	67	107	1.04	137	119.0
1988	39	34	6.5	10.4	70*	111	1.03*	141	117.5

Source: Sanchez (1988); Lopez & Riveros (forthcoming)

LFP = Labor force participation rates;
E = Total employment divided by total population (1986-1988 are preliminary);
U = (Urban) open unemployment rates;
U* = Underemployment rates (see text);
W/P = Real consumption wages (deflator: CPI) Index 1970 = 100;
RER = Real exchange rate: nominal exchange rate adjusted for changes in domestic and foreign prices (increase denotes depreciation);
WT/WW=Ratio of wages in tradeables (Agriculture and Manufacturing) to wages in non-tradeables (Services and Construction, excluding government);
LG = Employment in the Public Sector (Central and Local Governments and Parastatals) Index 1970 = 100;
PRO = Overall labor productivity (Real GDP divided by LFP times estimated population).
* Provisional (based on first quarter data). For tradeables we used wages in manufacturing and for non-tradeables, wages in construction.

Annex Table 5.2: ARGENTINA - CHANGES IN SECTORAL OUTPUT AND EMPLOYMENT
(Percent p.a.)

	1970-1975	1975-1983	1983-1987
GDP (Factor Costs)	2.9	0.1	1.5
Employment	2.4	0.8	0.5
Agriculture			
Output	2.8	2.2	0.2
Employment	-0.6	-1.1	-0.7
Manufacturing			
Output	3.4	-1.6	0.9
Employment	3.9	-1.8	0.9
Construction			
Output	1.0	-2.8	-4.1
Employment	2.1	-2.7	2.8
Services			
Output	2.6	0.3	2.4
Employment	2.7	2.7	0.3

Source: IBRD (1988)

5.05 The poor productivity performance is the underlying reason for the poor growth rates of real wages. Between 1940 and 1985 real wages never rose nor fell for more than three consecutive years (Annex Table 5.3). In general, periods of decline tended to follow brief periods of increase. Real wages were only 61 percent higher in 1985 than in 1945, implying annual growth of only about 1 percent. This poor longer run real wage performance has been characterized by intense short-run fluctuations. These result in part from the interplay of Argentina's macroeconomic constraints and a centralized wage determination system (discussed below) that fails to reflect productivity conditions.

5.06 Nominal wages in most industrial sectors are determined either by institutionalized procedures for bargaining at the national level or set by the Government. Since the late 1960s the Government has generally announced the periodic wage adjustments applicable to most workers under formal contracts.1/ The Government's wage adjustments are compulsory for virtually all workers and firms in the formal sector.

5.07 The centralized power of the trade unions, job security regulations, and the protection granted certain industries from domestic and external competition have together meant that the wage adjustment has become an influential macroeconomic policy instrument, with a significant effect on prices, short-term economic activity, and the balance of payments. Unfortunately, it has limited influence on medium-term real wage levels. Whenever wage rates have exceeded their productivity basis, the consequence has tended to be additional inflationary pressure and balance of payments problems. For example, the broad real wage increase that took place in 1984 contributed to the burst of inflation that preceded the Austral Plan. The more modest real wage increase that took place in 1986 contributed to the resurgence of inflation (although there was an accompanying short-term productivity increase as a result of higher capacity utilization). The broad lesson appears to be that, while the centralized wage determination process is capable of increasing wages in the short-term, long-term productivity improvement is the only genuine basis for a sustained increase in real wages.

5.08 Moreover, centralized wage setting has troublesome microeconomic consequences. It is probable that the inability of firms to set their own wage increases has contributed to a longer-term tendency to more capital-intensive investment. Since the wage adjustments apply essentially to formal activities in urban centers, the system effectively sustains a

---

1/ From 1958 through 1966 nominal wages were increased every year through national bargaining between unions and employers. In mid-1967 this system was suspended in the context of a stabilization plan. An attempt to restore national bargaining was made in 1975, which resulted in wage increases of more than 100 percent and contributed to hyperinflation. The Government returned to dictating wage adjustments until 1979, when a short-lived system of wage bargaining at the level of the firm was created. This ended in the 1982 crisis, and the Government returned to dictating the adjustments.

Annex Table 5.3: ARGENTINA - REAL WAGES
(Index 1970=100)

	Real Unskilled Wages	Real Skilled Wages	W Skilled/ W Unskilled	Minimum W/ W Unskilled	Real Minimum W
1970	100	100	100	100	100
1971	103	97	99	102	108
1972	91	90	99	96	96
1973	99	97	98	103	112
1974	111	108	96	111	137
1975	99	101	102	90	103
1976	61	61	101	72	53
1977	64	65	102	75	51
1978	67	76	113	62	42
1979	75	90	121	67	47
1980	95	116	123	68	56
1081	81	100	123	68	64
1982	52	72	138	97	57
1983	72	75	104	97	85
1984	77	80	105	97	92
1985	60	63	104	80	62
1986	59	61	104	78	59
1987	61	66	105	85	65

Source: Sanchez (1987); Lopez & Riveros (forthcoming) and IBRD data (BESD)

Note: Wages of unskilled and skilled labor correspond to manufacturing and are deflated by the price of tradeable goods (weighted average of the GDP deflator for agriculture, mining and manufacturing). The nominal monthly minimum wage is deflated by consumer prices. Figures for 1986 and 1987 are provisional.

protected segment in the labor market. Because wage adjustment applies to the nation as a whole, relative wage movements fail to reflect differential regional labor market conditions. Finally, the fact that wage adjustments tend to apply more rigorously in urban centers has encouraged internal migration.

5.09 Wages set through a centralized process obviously cannot reflect labor productivity, since labor productivity is a matter for each individual firm. A firm that must adjust its wages according to a government guideline and follow restrictive job tenure policies (see below) cannot adjust its staff to changing conditions; this partly explains why the slow employment growth noted above has accompanied erratic output growth since 1970.

C. Employment and Unemployment Trends

5.10 Despite huge fluctuations in economic activity, Argentina's open unemployment has remained relatively low. The main reason for this has been that the public sector has continued to create employment. From 1974 to 1980, while GDP grew at a real average annual rate of only 1.9 percent, average unemployment (for Buenos Aires, 20 capital cities and 5 non-capital cities) remained at only 3.4 percent. GDP fluctuations in the 1980s produced an average unemployment rate of 5.1 percent, the same level observed in 1981 and 1982, two consecutive years in which real GDP declined.

5.11 Annual employment growth averaged only 0.7 percent from 1975 through 1987. In manufacturing and agriculture total employment declined over that period. There was no significant change in the construction sector. Employment expanded in services, particularly in the public sector; it undoubtedly expanded as well in the economy's extensive informal sector.

5.12 Labor force participation rates2/ have been falling and the working age population has declined as a share of total population, and these have contributed to keeping unemployment relatively low. Public sector employment is nevertheless the main reason why unemployment has remained relatively low. Public sector employment grew at an annual average rate of 3.5 percent from 1960 to 1980 and 3.3 percent from 1980 to 1987. By comparison, wage employment grew in the 1960-1980 period at a yearly average of only 1.4 percent and total employment (wage plus non-wage employment) grew at 1.3 percent (Llach and Sanchez 1984); from 1980 to 1987 total employment grew at an annual average rate of only 0.8 percent.

5.13 The most rapid increases in public employment have been in regional and local governments: employment was 28 percent higher in 1988 than in 1983 (see Annex Table 5.4). The central administration's employment was 20 percent higher. Employment in state enterprises was 6 percent higher in 1988 than in 1983.

2/ The labor force participation rate is defined as the number of employed plus unemployed divided by working age population (15 years and older).

Annex Table 5.4: ARGENTINA - EVOLUTION OF PUBLIC EMPLOYMENT, 1971-1988
(In Thousands)

	Administration a/	Public Enterprises b/	Regional, Local Government c/	Total
1971	572.7	399.4	483.7	1455.8
1972	581.7	407.5	500.9	1490.1
1973	601.5	414.3	518.0	1533.8
1974	625.2	429.1	564.5	1618.8
1975	637.4	441.0	630.8	1709.2
1976	638.0	476.7	646.7	1761.4
1977	643.1	438.2	661.1	1742.4
1978	582.7	418.9	688.6	1690.2
1979	564.5	387.4	726.9	1678.8
1980	557.1	374.4	721.7	1653.2
1981	573.5	350.1	724.2	1647.8
1982	574.5	334.7	725.0	1634.2
1983	548.3	325.8	747.2	1621.3
1984	605.5	348.8	720.1	1674.4
1985	622.4	360.3	752.3	1735.0
1986	637.5	351.0	795.2	1783.7
1987	650.2	341.5	910.5	1902.2
1988e	658.0	346.1	956.2	1960.2

Source: Secretaria de Hacienda

a/ Central Administration, Special Accounts, and Decentralized Agencies.
b/ Including the official Banking System.
c/ Permanent Staff in Provinces, MCBA and Tierra de Fuego.

D. Labor Market Regulation and Efficiency

5.14 Excessive labor market regulation and intervention has contributed to the disappointing wage trends as well as to inflexible intersectoral labor mobility. The centralized wage setting discussed above is, of course, one aspect of labor market regulation. Three additional aspects of labor regulation influence labor market efficiency: the job tenure system, heavy payroll taxation and the minimum wage legislation.

5.15 Employees are entitled to prior notice of dismissal and to a severance indemnity, both scaled to length of service. The severance indemnity now amounts to one month's current salary, up to a maximum equivalent to three times the officially established minimum wage, for each year of service. Seniority in formal sector firms averages about 12 years. For an average worker, accordingly, the total maximum dismissal indemnity is about 36 times the monthly minimum wage. This is not high (given the current real level of the minimum wage), although it adds to the problems of firms undertaking adjustment that need to reduce staff. Probably more important, businessmen find they must deal with powerful labor unions and undertake intricate legal procedures in order to carry out such dismissals. In all, this effectively increases labor costs, discourages employment and encourages more capital intensive investment than would otherwise be warranted over the long run.

5.16 Like other Latin American countries, Argentina has a system of payroll taxes that fund a variety of welfare programs. Employers' contributions go to pensions (12.5 percent of monthly earnings), housing (5 percent), family allowances (9 percent), unemployment (8 percent), and health (4.5 percent). In addition, there is a "thirteenth month's" wage payment each year and legal minimum vacations (which add another 3 percent to total wage costs. In total, non-wage labor costs to employers amount to 46 percent of wage costs3/ This ratio is high relative to other developing countries, particularly if compared in terms of per capita income. Employees pay part of their pension (another 11 percent), health insurance (another 3 percent), and certain other levies (about 2 percent), so that payroll levies on wages now total 58 percent.4/ There is a large difference between what firms pay for labor and what employees receive, which inevitably produces inefficiency. The benefits provided by the social security system are considered extremely deficient, particularly in view of their high cost. Workers who can afford to do so make supplementary contributions to private pension plans.

5.17 The real minimum wage has fluctuated considerably during the past 15 years (see Annex Table 5.3). Real levels are lower in the 1980s than they were previously. The wage structure in formal sectors is generally substantially above the minimum wage. Given the fact that the minimum wage

---

3/ These data have been obtained from the Price-Waterhouse report: "Doing Business in Argentina" (1984).

4/ In 1985, the non-wage costs as a percentage of per capita income was 64 in Argentina, 42 in Brazil, 50 in Mexico, 24 in Chile, 21 in Korea, 44 in the United States and 17 in Japan.

is a "floor" for formal sector wages, and that it still serves in some measure as an indicative index, its increase may shift the entire wage structure up, thus affecting inflationary trends.5/

5.18 The increasing importance of the informal sector, which is partly a response to the burden of labor regulation, has reduced the effective coverage of the regulation in the economy.

E. Protective Regulations and Equity

5.19 As noted earlier, Argentina's labor market comprises protected and an unprotected sectors. The protected sector is covered by mandated wage adjustment and labor regulations. It comprises government, formal services, import substituting industries, large agricultural firms and some export industries. The unprotected sector, which comprises small firms, informal activities and producers of agricultural goods for internal consumption, is not covered by mandated wage adjustments and labor regulations.

5.20 The simultaneous existence of protected and unprotected sectors has the obvious distributional implication that workers in the unprotected sectors are in a relatively poorer position than those in the protected sectors. This is supported by the observed ratio of wages of skilled labor to unskilled labor (see Annex Figure 5.4 and Annex Table 5.3), which displays a clear increase between the mid-1970s and 1982 amid huge fluctuations in aggregate real wages.

F. Recommendations

5.21 The Government should consider vigorously cutting its inefficient activities and dismiss workers as necessary, not only with the severance pay to which they are entitled, but with programs of income maintenance, retraining, and placement. Private sector industrial federations might be asked to participate in these programs. In addition, early retirement programs might be introduced. The short-term costs to the public sector can be expected to be compensated by lower public sector costs and higher operating efficiency; over the medium term, as workers are hired in more productive private sector jobs, their increased productivity should ultimately provide the public sector with more tax revenue.

5.22 The present centralization of wage-setting needs to be reversed, so that wage rates can be set by each firm in accord with its own productivity conditions. That is, some loosening must be permitted of the present system under which firms revise their wage scales up with the centralized wage indicator.

5.23 The present system of wage taxation needs to be relieved. The private sector must be allowed to provide a larger share of the benefit programs, since it could assuredly do so more effectively at lower cost. This would in turn reduce the Government's costs. The heavy labor taxation

---

5/ This argument has been empirically substantiated in Paldam & Riveros (1988). See also Sanchez & Giordano (1988).

burden must gradually be worked down over time, to reduce the cost to the private sector of hiring workers. Measures to relieve the stringency of job tenure would also be helpful to encourage the private sector to hire workers.

5.24 More innovative approaches that might usefully be considered over the medium term would include encouraging profit-sharing between owners and their labor forces. This would permit labor costs to vary with firms' performance. It should also have the effect of giving the labor unions and their members a more direct stake in the profitability of the enterprises in which they work.

ANNEX CHAPTER VI: TRADE POLICY

A. Trade Performance

6.01 Argentina's postwar experience of persistent macroeconomic instability and a declining rank among countries according to per capita GNP was preceded by the closure of its economy to foreign trade. Prior to the Great Depression the share of imports (GNFS) in GDP kept close to 50 percent. This indicator of openness fell to 5 percent at the end of World War II, and again in the mid-1950s. Since then, the import share has remained near 10 percent without ever exceeding 11.5 percent in a single year (Annex Figure 6.1). Exports as a share of GDP have followed the general trend of imports, but have fluctuated heavily as a result of volatile international commodity markets, stochastic shifts in domestic agricultural output supply and variations in domestic aggregate expenditures. Since the onset of the debt crisis, export earnings have exceeded import expenditures in all years, but have continued to oscillate. As a result, the resource balance has moved between a high of 7.5 percent of GDP in 1985 and a low of .2 percent in 1987. A new upswing occurred in 1988 (3.2 percent) and appears to continue into 1989.

6.02 The structure of Argentina's imports and exports reveals the dominating influence of the import substitution strategy (Annex Table 6.1). Tariff and nontariff protection has kept consumer good imports at less than 6 percent of total import expenditures, except for an average of 14 percent in 1979-81, when a trade liberalization episode coincided with a period of easy access to foreign credit. The remainder is taken up by primary inputs, intermediates and capital goods. Reflecting lower investment, the share of capital goods imports has declined from 20-24 percent before the debt crisis to 12-16 percent since. On the export side, primary products and agro-based manufactures contribute about 75 percent to the country's export earnings. The policy of promoting the exports of non-agro-based manufacturers ("industrial manufactures") has not until recently been successful in diversifying exports. Their export share has hovered between 13 percent and 21 percent before, in 1987, it jumped to 27 percent as a result of tumbling agricultural earnings. In 1988, however, the share increased further despite a strong recovery of agricultural export prices and supplies. While some of this improvement may stem from the recession beginning in mid-1988, this growth in nontraditional exports is also a response to a substantive reform of the export regime that began in early 1987.

B. Import Substitution Strategy

6.03 The overall contraction of international trade resulting from the Great Depression and World War II was largely responsible for the initial loss of Argentina's exports and imports. Of more lasting importance, however, were the inward oriented trade and industrial policies initiated during the 1930s, but maintained and intensified in the 1950s, when other countries were increasingly removing their external trade barriers and a rapid expansion of international trade flows took place. Successive governments opted for the import-substitution strategy hoping to: (i) foster investment and productivity growth through accelerated domestic industrialization; (ii) achieve a higher degree of macroeconomic stability by reducing the country's exposure to volatile world commodity markets; and (iii) support both targets with a policy of selective industrial export promotion. These hopes have been frustrated.

Annex Figure 6.1 - ARGENTINA-EXPORTS AND IMPORTS (GNFS), 1925-1987

(Percentage of GDP)

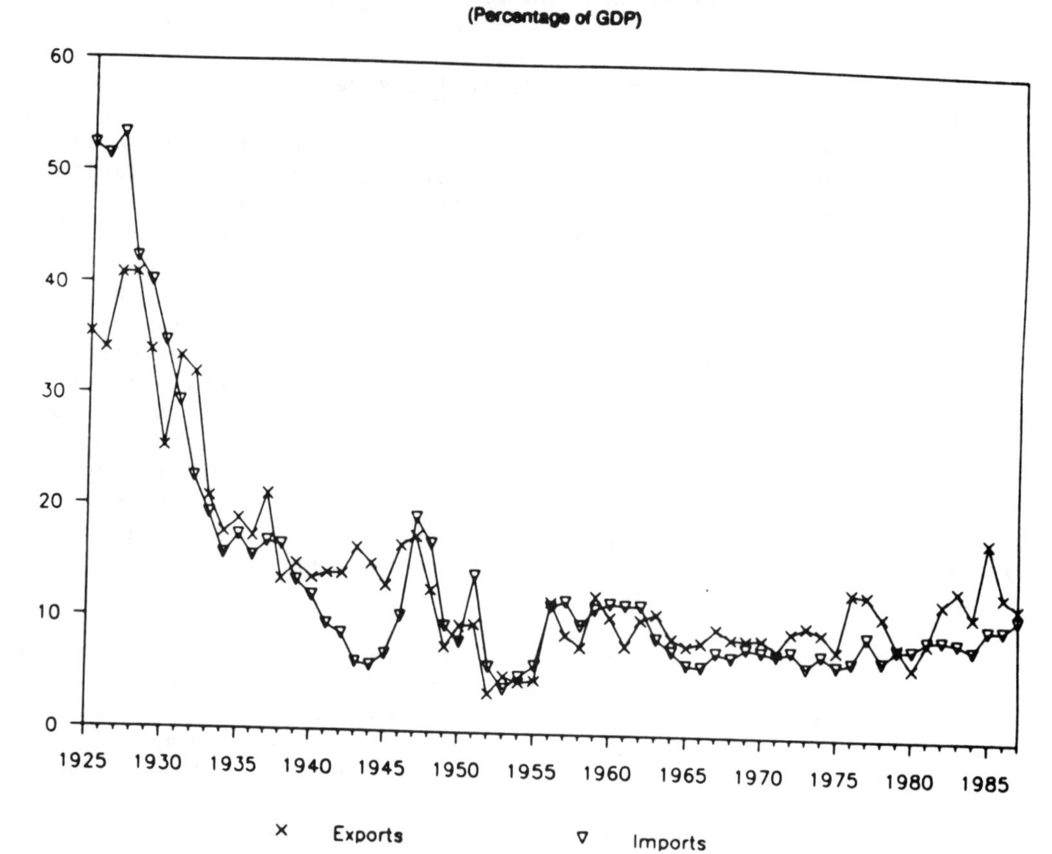

Sources: 1. (1925 - 1984) Y. Mundlak, D. Cavallo and R. Domenech, Agriculture and Economic Growth, Argentina 1913 - 1984. Forthcoming.
2. (1985 - 1987) World Bank, World Tables, 1988 - 1989 Edition.

Table 11.1 ARGENTINA - COMPOSITION OF EXPORTS AND IMPORTS, 1977-88
(in Percent)

	1977	1978	1979	1980	1981	1982	1983	1984	1985	1986	1987	1988*
EXPORTS	100.0	100.0	100.0	100.0	100.0	100.0	100.0	100.0	100.0	100.0	100.0	100.0
Primary Products	40.8	41.1	40.4	39.8	45.5	39 8	48.6	46.5	43.8	36.9	27.5	41.0
Agro-based Manufactures	39.0	36.8	41.0	36.8	31.5	32.6	33.7	35.4	30.9	39.2	44.4	27.3
Industrial Manufactures	19.6	21.3	17.9	18.8	16.1	20.4	13.3	13.8	18.4	21.7	26.5	30.0
Fuels	0.5	0.8	0.6	3.5	6.8	7.2	4.4	4.3	6.7	2.1	1.4	1.7
Unclassified	0.1	0.0	0.1	1.1	0.1	0.0	0.0	0.0	0.2	0.1	0.2	0.0
IMPORTS	100.0	100.0	100.0	100.0	100.0	100.0	100.0	100.0	100.0	100.0	100.0	100.0
Capital Goods	20.5	23.7	19.9	21.5	22.5	14.7	14.2	13.2	15.7	11.7	14.7	15.7
Intermediates	60.4	59.2	53.5	50.3	53.6	65.9	70.5	72.4	68.6	73.3	68.3	69.0
Fuels	16.3	12.3	16.4	10.1	10.4	12.5	10.2	10.5	11.9	8.9	11.3	10.9
Consumer Goods	2.8	4.7	10.2	18.0	12.6	6.5	3.3	3.8	3.5	6.0	5.5	4.4
Unclassified	0.0	0.1	0.0	0.1	0.9	0.4	1.8	0.1	0.3	0.1	0.2	0.0

Source: INDEC
* Bank Staff Estimates

(a) The first, easy stage of inward-oriented industrialization (foodstuffs, textiles, clothing and leather products) was already completed in the late 1940s. Import substitution in the second, more capital intensive stage, would have required substantial dynamism to avoid foreign exchange shortages. Instead, recurrent external crises, followed by recessions have stifled industrial expansion. As a share of GDP, manufacturing in the 1980s has returned to where it had been four decades earlier. Investment has reached high levels (28-30 percent of GDP) only in the late 1970s, but is hovering between 10 percent and 13 percent since 1983. Most disappointing has been the contribution to productivity growth. A forthcoming study indicates that, over the long run, total factor productivity in the nonagricultural sectors (excluding government) has grown at less than two thirds of the rate achieved in the agricultural sector.1/ Indicators for the industrial sector suggest that total factor productivity has actually declined during the 1970s and early 1980s (para. 1.11). The momentum of industrial productivity has stalled as: (i) scale economies which tend to become increasingly important beyond the first phase of industrialization could not be realized on the narrow domestic markets; and (ii) import protection and domestic industrial policies have relaxed the pressures of competition that force managers to allocate resources efficiently and that tend to drive the process of innovation.

(b) The closure of the Argentine economy, instead of reducing macroeconomic instability, has arguably contributed to the secular increase in the inflation rate and to the instability of the real exchange rate. The share of exports and imports in GDP turned out to be less important for the country's exposure to trade shocks than the mechanisms that enable the open economy to cope with them. A price elastic aggregate import demand and capital flows tend to absorb a good part of the shock impact on the real exchange rate and thus on relative prices, while the discipline of intense competition keeps relative price changes from turning into a source of inflation. These mechanisms have been weakened. Increasing the import share of non-substitutable primary inputs and intermediates rendered aggregate import demand price inelastic; intermittent balance of payments crises impaired the country's access to external credit; and the creation of sheltered domestic markets fostered a pricing behavior that has added an inertial component to inflation. To overcome the inertial forces, governments have repeatedly and unsuccessfully, in the context of stabilization programs, relied on fixing the nominal exchange rate, and have in the end increased the instability of the real exchange rate.

1/ Y. Mundlak, D. Cavallo and R. Domenech, Agriculture and Economic Growth: Argentina 1913-1984, forthcoming.

(c) To relax the import capacity constraint and to reduce the instability of export earnings, Argentine governments have at various times since 1963 applied industrial export promotion incentives. Until 1987, however, neither the share of exports in total industrial production (fluctuating between 6 percent and 10 percent with the exception of 1985 when it reached 13 percent) nor the share of industrial manufacturers in total export earnings (8-9 percent) have exhibited an upward trend. Two major reasons can be distinguished: (i) A selective promotion policy faces inherent difficulties that consecutive administrations have found impossible to solve: (a) the Government lacks the information to clearly identify the potential winners on export markets and, in the Argentine political environment, is unable to forge a lasting public consensus on the selection made. The incentives, therefore, remain subject to change, i.e. cannot influence the long-term investment decisions required for the expansion of export capacity and the access to export markets; and (b) selective promotion measures invite countervailing duties (CVD) on the side of importing countries, particularly the United States and the European Community. (ii) Equally important for the reluctance of Argentine industrialists to undertake irreversible export investments has been the volatility of the real exchange rate for exporters which, in turn, has resulted from the interaction of inappropriate trade and macroeconomic policies.

In addition, trade and industrial policies have contributed to macroeconomic instability through their impact on the public sector deficit. Trade tax revenues have been an unstable part of the budget while the revenues foregone through industrial promotion incentives and overcharging on public sector purchases resulting from the "buy national" obligation of public procurement have added up to 5 percent of GDP to the deficit. To regain macroeconomic stability, therefore, the Government has correctly emphasized that trade and industry reform not be postponed. The reform can support the stabilization as it impinges both on the underlying fiscal cause and on the inertial forces that have worked to give inflation a life of its own.

C. Policy Reform

6.04 A political consensus has finally been established in Argentina that the import-substitution strategy has run its course. Beginning in early 1987, the Government has transformed the consensus into an active reform policy. To have started and continued the reform is a major achievement considering not only the resistance of powerful vested interests, but also the legacy of an aborted previous liberalization attempt caught in the collapse of the 1979-81 stabilization program. The experience of the previous export promotion and import liberalization policies is reflected in characteristic elements of the current reform: (i) a strong emphasis on the free trade status for exporters early in the reform process, coupled with a phase-out of various specific export promotion measures; (ii) a negotiated sector-by-sector approach to import liberalization; and (iii) an attempt, not always successful, to maintain a competitive exchange rate for trade transactions. In addition, recognizing the complementary relation between foreign trade and domestic industrial policy instruments, the Government has taken care to coordinate the trade reform with changes in the industrial policy. These will be discussed in Annex Chapter VII.

The Export Regime

6.05 The supply of Argentine products to world markets has in the past been restrained by the export regime, in particular: (i) export taxes and equivalent differentials in the exchange rates for export and import transactions; (ii) export license requirements aimed at giving domestic market supplies priority over export supplies; (iii) tariffs and quantitative restrictions on imported inputs and capital goods used for export production; and (iv) the imposition of domestic indirect taxes on exports which amounts to double taxation as generally importing countries raise domestic indirect taxes in addition to import duties. Since late 1987, the Government acted on these barriers as follows:

(a) The average export tax rate (on a production weighed basis) was reduced for agricultural goods from 18.2 percent to 4.1 percent, and for manufacturing goods from 9.5 percent to 0.6 percent. The production share of positions of the export nomenclature (NADE) which are subject to export taxation was reduced from 85.9 percent to 34.5 percent in agriculture and from 43.4 percent to 6.3 percent in manufacturing. However, after August 1988 the export tax equivalent of the multiple exchange rate system for trade transactions more then compensated for the direct export tax reduction. In the fourth quarter of 1988 the export tax equivalent came to 17.4 percent for agricultural commodities and to 12.7 percent for manufactures (including agro-based manufactures).

(b) The Government has simplified the export licensing process through a one-stop-window arrangement for exports (except grains) shipped through Buenos Aires. However, it has not yet reduced the number of NADE positions that are subject to license requirements. The production coverage for these positions stands at 54.5 percent for agricultural commodities and 17.7 percent for manufactures. These controls continue to cause both uncertainty and resource costs for exporters.

(c) The duty free admission of imported goods for use in export production (temporary admission regime-TAR) has been extended to all primary inputs and intermediates with small exceptions. At the same time, the general import liberalization measures have removed the control of domestic producer associations over the import licensing process of these goods, and have made import authorization fully automatic for about 85 percent of the imports affected by the TAR. The new TAR complements the older drawback scheme under which exporters can apply for the reimbursement of duty payments after exports have been shipped. In the inflationary environment of Argentina, a comprehensive and automatic ex-ante TAR is the key instrument for removing the direct effects of protection for imported inputs. The duty exemption part of this measure has reduced the average production cost of manufactured exports by 3 percent, while the cost equivalent of the automatic access is possibly higher. In addition, the Government has introduced a TAR for capital goods called ARGEX. Under this regime, applicants have to enter multi-year contractual export commitments with the Government. Almost all firms have chosen to avoid such commitments, and have instead taken advantage of Decree 515/87 under which the Secretary of Industry and Foreign Trade can grant duty exemptions on capital goods on a case-by-case basis.

(d) The reimbursement of indirect domestic taxes has been extended to the export of most industrial manufactures, though not to agricultural and mineral commodities and to agro-based manufactures. The average rate of reimbursement (export-weighted) for industrial manufactures increased from 6 percent to 13 percent, but stayed below 1 percent for all other goods. While agricultural production is generally exempt from the value-added tax, a substantial share of mineral and agro-based production is not, i.e. the extension of indirect tax reimbursement has stopped short of being complete.

6.06 Before the turmoil in the foreign exchange market rate put a halt to further reform measures in February 1989, the Government had been working on a sweeping reform of export licensing for industrial goods (including agro-based manufactures). The reform would have removed all license requirements that currently enable the authorities to give domestic supplies priority over export supplies, except for purposes of national defense which are specified by law. Export licenses would only have been required for goods that are subject to: (i) quality and sanitary controls; (ii) environmental restrictions; and (iii) international agreements on export restraints like the Multi-Fiber Agreement and bilateral voluntary export restraints.

6.07 The Government has reduced the selective export promotion granted through: (i) subsidies on incremental exports in the context of the PEEX program that calls for multi-annual contracts between the Government and the exporting firms; and (ii) Central Bank rediscounts of first-tier banks pre-shipment loans and of foreign trade bills (post-shipment financing) at subsidized interest rates. Since end-1987, the Government is grandfathering the PEEX program, i.e. neither entering new nor extending old contracts. The Central Bank has raised the interest rates on rediscounts in several steps from 1 percent (pre-shipment) and 4.5 percent (post-shipment) to 8 percent for both in August 1988. Commercial banks offer pre-shipment financing at LIBOR plus about 2.5 percent. The measures of the Central Bank have thus substantially reduced, but not removed the subsidy element.

6.08 Abstracting from the tax equivalent of the exchange rate, by end-1988 the Government had made impressive progress towards extending free trade status to Argentine exporters; it has already started to move away from inefficient selective export promotion policies. The reform of the export regime, however, is less than complete. The major issue is the extension to agricultural and agro-based industrial exports. The export taxation for these goods to some extent compensates for a weakness of domestic indirect taxation (Chapter III), and this needs to be addressed.

Beyond that, previous arguments for a general discrimination against agricultural exports have been discredited by Argentina's experience: (i) the country's vulnerability against terms of trade fluctuations has increased rather than abated in the wake of reduced agricultural exports; (ii) the rate of productivity growth in the agricultural sector has been higher than in the rest of the economy; and (iii) the static optimal tax argument for commodities has lost relevance, as the country's share in the world exports of its major commodities has fallen below 10 percent. In a number of specific cases, furthermore, prohibitive export tax rates (up to 38 percent) have served to create rents in user industries. Such rents are costly for the economy as a whole.

6.09 Since February 1989, the Government has been faced with a run out of the national currency, which has rapidly been transmitted into hyperinflation. The Government has taken a succession of short-term exchange rate and export tax measures and has temporarily suspended the reimbursement scheme and the export financing facility of the Central Bank. Despite the various tax and tax-equivalent measures, however, the short-term incentives for exports have dramatically improved due to the nominal devaluation of more than 1,000 percent within three months. These are, of course, side effects of a hyperinflation that will eventually be stopped. It is important that the disinflation package not undo the reform of the export regime already achieved, but set the stage for the completion of the reform.

Recommendations

6.10 Recommendations for the export component of the stabilization package include:

(a) the imposition of an across-the-board export tax be used to abolish all product-specific export taxes;

(b) export tax payments and indirect tax reimbursement claims be credited against value-added tax payments if and when made;

(c) imports and exports be transacted at the same exchange rate;

Recommendations for the further reform steps, once the inflationary pressure has abated, include that:

(d) the reform measures prepared for early 1989 be enacted;

(e) the subsidy element in the export financing facility of the Central Bank be removed; allowing new exporters and small and medium-scale firms access to the facility while charging an adequate spread to cover the higher risk;

(f) the access of indirect exporters to the TAR, the indirect tax reimbursement and to export financing be improved; and

(g) the GATT code on subsidies and countervailing duties be signed as would become possible with the phase-out of the PEEX and other minor export subsidy programs and the removal of the subsidy element in the Central Bank export credit facility; as a signatory to the code, the Government would be able to contest the countervailing duties currently imposed by the United States and the European Community against a number of imports from Argentina.

The Import Regime

6.11 The Argentine import regime has in the past ascribed to the executive branch ample discretion over the issuance of import licenses, changes in the tariff schedule and duty exemptions on a project-by-project basis. The discretionary element has been complementary to the corporatist character of government-industry relations, most visibly in the work of honorary commissions through which domestic producer associations advise the Government on the issuance of import licenses to the private sector. The "prior intervention" of these commissions has developed into the equivalent of quantitative import restrictions (QRs) for goods that compete with domestic supplies. Importing by the public sector (including public enterprises) has only been possible with the approval of the Compre Argentino Commission that has been dominated by domestic suppliers. The Compre Argentino law stipulates that import requests are to be denied when domestic substitutes are available at "reasonable prices." In reforming the import regime the Government has rightly emphasized the reduction of the discretionary element. However, the Government has not considered it feasible to impose its concept of an optimal reform. Instead, it has negotiated the reform sector-by-sector, gaining leverage by bringing to the table the organized interests of both sides, i.e. the competing domestic suppliers and the industrial users, though not the unorganized consumers. The Government has in this way made impressive progress towards a more rational import regime, though the outcome, as had to be expected, leaves much room for improvement.

6.12 Between the second quarter of 1987 and the fourth quarter of 1988, the Government has:

(a) abolished the prior intervention of honorary commissions required for the issuance of licenses for the import of goods under all but 2,033 of the 11,780 positions of the import-tariff nomenclature (NADI); thereby reducing the production coverage of QRs from 62 percent to 18 percent for manufactures and from 29 percent to 2 percent for agricultural commodities;

(b) made the import authorization process fully automatic for 8,029 NADI positions (out of 9,747 positions that are no longer subject to the prior intervention of Honorary Commissions) by rescinding the discretion of the Secretary of Industry and Energy Trade, and transferring the authorization process to the commercial banks;

(c) implemented the GATT code on import valuation, abolishing official reference prices, rescinding the discretion of the National Customs Administrator to set normal prices and, instead, basing the calculation of duty payments on the invoice;

(d) dissolved the supplier dominated Compre Argentino Commission without, however, appointing a new Government dominated commission; meanwhile "urgent" import requests are decided by the Secretary of Industry and Foreign whereas "non-urgent" imports are delayed;

(e) reduced the level of tariff protection by abolishing the 15 percent across-the-board surcharge of the Plan Austral and installing a new schedule with the following characteristics (old regime in parentheses): (i) average production weighted tariff rate for manufactures of 28 percent (43 percent) and for agricultural goods of 20 percent (29 percent); (ii) tariff band of 0-40 percent for NADI positions with 98 percent production coverage (0-53 percent for positions with 95 percent production coverage); (iii) maximum tariff rate of 50 percent (115 percent); and (iv) the dispersion of effective protection (standard deviation of effective rates of tariff protection at the five digit ISIC level) has come down from 39 percent to 28 percent;

(f) abolished the discretion of the Secretary of Industry and Foreign Trade to grant duty exemptions for capital goods imports on a project-by-project basis, and, instead, appointed a commission to advise on permanent tariff rate changes on a position-by-position basis.

6.13 In addition, prior to the turmoil on the foreign exchange market, the Government had been planning for the second quarter of 1989 to:

(i) further reduce the production coverage of QRs for manufactures from 18 percent to 15 percent; and

(ii) expand the fully automatic import authorization process to all NADI positions not subject to QRs; this would also increase the share of fully automatic TAR positions from 85 percent to 100 percent.

These measures were postponed as the Government became absorbed by the unfolding crisis.

6.14 After the core measures had been enacted in October 1988, the Government weakened the reform with three steps:

(i) The Government imposed specific tariffs on 300 NADI positions with a production coverage of 2 percent. Since these tariffs have been set at import prohibiting levels, they have the same effect as QRs, thus directly undoing a part of the October 1988 reform. Perhaps more importantly, introducing specific tariffs has compromised the credibility of the program in the eyes of the Argentine public.

(ii) In the context of the tax reform law of December 1988, the Congress reestablished the discretionary power of the Secretary of Industry and Foreign Trade to grant duty exemptions on capital goods imports on a project-by-project basis. This distortion has in fact become more serious than it had been before the reform because the tariff reform had placed the rates for a large number of capital goods at the upper end of the tariff band, expecting that the Tariff Commission (para 10.10 (f)) would soon revise the whole schedule for capital goods. So far the Commission is inoperative.

(iii) The Government has opened six anti-dumping cases, thus departing from its long established and well-founded policy of utmost reservation against this instrument. Since the injury test under Article 717 of the Codigo Aduanero is particularly undemanding (and inconsistent with the GATT anti-dumping code), the experience of successful anti-dumping cases could easily lead to a flood of new applications.

6.15 In less than two years, the import reform has come a long way considering the political constraints under which the Government had to operate. Though superior to the previous regime, the new regime is not fully consistent with the Government's target of integrating Argentina into the world economy, allowing the citizens to detect and capture their dynamic comparative advantages. That target would be best served by a fully automatic import regime, for both the private and the public sector, without quantitative restrictions and with a flat moderate import tariff except for goods used in export production. The reform process, however, has had a distinct influence on the public debate, raising the awareness in all major political parties that the remaining QRs, the new specific tariffs and the procurement practice under the Compre Argentino law do not serve the national interest well. The Government has thus opened the way for further progress that might eventually lead to the complete removal of QRs and to the repeal of the Compre Argentino law.

Recommendations

6.16 The priority for disinflation requires that elements of the trade reform process be strengthened that can support the stabilization process. In addition, further reform steps should be planned so as to aid the private investment response to the trade reform. Recommendations include that:

(a) the structural reform package include the removal of all QRs (including fully automatic import authorization) that were exempted from the 1987-88 reform in conjunction with a temporary tariff surcharge; and a subsequent tariff adjustment bring ad valorem tariff rates into the band established in 1988, IV;

(b) the structural reform package include the removal of specific tariffs; only for seasonal products should specific tariffs at non-prohibitive levels be considered;

(c) the tariff band be narrowed immediately from 0-40 percent to 10-40 percent; and work commence on a further narrowing of the band to 10-20 percent together with a much lower average tariff of about 15 percent; and

(d) the Government use its decree power to restrict the application of the Compre Argentino law to the minimum required by the law, i.e. preference for local suppliers at prices equal to import prices including duties and domestic taxes; and

To support the private investment response to trade reform, attention should be given to the import and domestic production of capital goods. It is recommended that:

(e) all capital goods not produced in the country be placed at the lower end of the tariff band;

(f) specific tariffs on components and inputs of capital goods produced in the country be replaced by ad valorem tariffs at the lower end of the tariff band;

(g) the "Informatica Program" (Resolutions ME 978/85 and ME 418/86) be repealed; the program has severely hampered the technological development of the country, particularly of the domestic capital goods industry; and

(h) a program to unify the tariff rates for capital goods produced and not produced in the country be developed.

ANNEX CHAPTER VII: INDUSTRIAL POLICY

A. Sector Performance

7.01 Rapid industrialization has been the central target of the twin policies of import protection and industrial promotion. For nearly two decades, however, the policies have failed to even prevent the relative decline of the industrial sector (Annex Table 7.1). Domestically, industrial value added has grown dismally slow and at a lower rate than GDP at factor cost. Internationally, the sector's contribution to GDP growth has been significantly lower than the average of relevant country groupings (Latin America, middle-income and highly indebted countries). Though the stock of fixed capital has apparently grown rapidly before the negative trend shift of domestic investment that occurred in 1982, industrial employment has fallen dramatically, implying a rapid increase in capital intensity. The value added, labor and capital stock figures in Annex Table 7.1 strongly suggest that total factor productivity has declined through the 1970s and early 1980s and has since recovered slowly from a low level.

Annex Table 7.1: ARGENTINA - SELECTED INDICATORS OF INDUSTRIAL PERFORMANCE
(Average Annual Growth Rates in Percent)

	1970-82	1982-87	1970-87
GDP (at factor cost)	1.06	1.60	1.22
Value Added a/	-0.52	2.85	0.46
Employment a/	-2.59	-0.47	-1.97
Fixed Capital b/	5.97	0.35	4.29
Contribution to GDP growth c/	0.13	0.74	0.31
Memo items			
Contr. to GDP growth c/			
- LAC d/	1.90	0.64	1.53
- MIC e/	2.25	1.24	1.95
- HIC f/	2.33	0.42	1.76

a/ Manufacturing sector.
b/ Stock of nonagricultural machinery and equipment.
c/ Industrial sector.
d/ Latin American and Caribbean.
e/ Middle-income countries.
f/ Highly-indebted countries.

Sources: World Bank, World Tables 1988-89 edition.
INDEC.
S. Goldberg and B. Ianchilovici, "El Stock de Capital en Argentina." Desarrollo Economico, No. 110 (July-September 1988).

7.02 The failure of the industrialization drive is recognized by all political forces in Argentina. The conclusions for industrial policy, however, are a matter of debate because industrial policy impacts on industrialization only in an indirect way and in interaction with other policies and external events. Particularly, it is asked whether an active promotion policy would not be successful in Argentina as it is perceived to have been in several East Asian countries, if only the macroeconomic environment were more stable and the measures of promotion would be planned more competently. A growing evidence, however, supports another view, namely that the policies of protection and promotion have contributed to macroeconomic instability by: (i) fostering an export-shy but import-craving industrial sector (Annex Chapter V); (ii) allowing the fiscal burden of promotion to run out of control; and (iii) bringing about a non-competitive market structure inducive to inflation inertia and to cost-push dynamics. The State, furthermore, appears to have been weakened to an extent that investment incentives are bound to be appropriated by the incumbents of protected industries, allowing them to raise rents, expand market shares and avoid the need for adjustment and productivity enhancement.

B. Policy Impact

7.03 **Fiscal Burden.** Estimates of the fiscal costs of the numerous industrial promotion regimes are difficult to come by. In 1987, the so-called theoretical cost as a share of GDP amounted to 0.8 percent for the general promotion regime and 1.8 percent for the special regimes of four provinces (La Rioja, Catamarca, San Luis and San Juan). The true revenue foregone has probably been higher because the theoretical cost calculation is based on the downward-biased estimates at the time of approval. The fiscal cost for the Tierra del Fuego regime is estimated at close to 1 percent of GDP. No estimates are available for the small sector regimes (mining, forestry, fishery, shipbuilding, production and processing of wine, sugar, tobacco, yerba mate and cotton, and production of internal combustion engines). The core promotion instruments are tax exemptions and deferments that can last for up to 15 years. The practice of approving new projects on a massive scale has, for the foreseeable future, eroded the base for the value added tax and the profit tax. The temporary suspension of new project approvals under the general promotion regimes and the special regimes for four provinces, which is part of the tax law approved by the Congress in December 1988, has failed to restore the tax basis. The same is true for limits on the open-ended tax benefits which the tax administration found difficult to enforce. In addition, the December 1988 law has not been effective in preventing new project approvals as late as May 1989.

7.04 **Market Structure and Macroeconomic Instability.** A recent Bank Study on the Industrial Sector has shown that import protection and industrial promotion have been instrumental in bringing about a dualistic market structure, most commonly in subsectors producing intermediate goods, but also in various capital and consumer goods industries. The pattern is one of a small group of large firms that collude more or less openly in matters of price setting and sharing of investment incentives and supply contracts with the public sector. The oligopolists coexist with a fragmented fringe of small firms. The large firms enjoy substantial cost advantages, partly due to scale economies and partly to their ability to block the access of fringe firms to the investment incentives granted under the various industrial promotion regimes.

7.05 Without import competition, output prices under a colluding oligopoly tend to be more rigid downward than upward: a slack in demand might even push prices upwards as marginal costs increase because of scale economies or as firms follow a full-cost pricing rule. Frequent intersectoral demand shifts, which are typical for a country like Argentina exposed to volatile terms of trade, then ratchet inflation upwards. In addition, a colluding oligopoly tends to interact with strong unions in triggering wage-push inflation. In both cases, the Government is faced with the choice of either choking off the inflationary pressure through a severe recession or validating more inflation.

7.06 For the same reason, a policy of disinflation threatens to be extremely costly. The Government recognized this fact in the design of the Plan Austral when it complemented orthodox stabilization measures with direct price controls. In the Plan Primavera, the Government went a step further by directly soliciting the cooperation of industry leaders in the setting and overlooking of industrial output prices. The agreements, however, turned out to be inherently fragile as the industry leaders asked for concessions (*inter alia*, on public sector prices and on tax rates) that tended to undermine the orthodox component of the package. It is different to envisage how the Government can avoid such inconsistencies without exposing the key subsector to the potential entry of competing imports.

C. Recent Developments

7.07 In September 1988, the Congress passed a new industrial promotion law to replace the old general promotion regime and the regimes for La Rioja, Catamarca, San Luis and San Juan, though not the regime for Tierra del Fuego and the numerous small-sector regimes. The new law, which has not yet been enacted, would replace open-ended tax exemptions with a fixed amount of tax credits. The low-value added bias would be removed and the capital-intensity bias reduced. The fiscal effect would be uncertain. Most importantly, the acquired rights of beneficiaries of the old regimes, that last for up to 15 years, would not be curtailed. In the context of the December 1988 tax reform, the Government suspended the approval of new projects under the old regimes and imposed limits on the tax exemptions. These measures appear to have been circumvented.

7.08 At this point, policies of industrial promotion have been thoroughly discredited in Argentina as instruments of fostering the industrial contribution to growth and of raising productivity. In addition they have prominently contributed to the fiscal collapse of the State. A radical departure from the promotion policies is, therefore, in order. Indeed, it is difficult to imagine that a package for stopping hyperinflation could be credible without the cancellation of acquired rights.

Recommendations

7.09 It is recommended that:

(a) a stabilization package include (i) the cancellation of acquired rights under the old industrial promotion regimes (Laws 21608, 22021, 22702 and 22973), and (ii) the restriction of the Tierra del Fuego regime (Law 19640) to trade tax exemptions; to lend credibility to these measures, the Government might seek a constitutional amendment;

(b) the industrial promotion law of September 1988 not be enacted, but a more efficient approach be taken to the regional and national targets of that law as well as of the Tierra del Fuego law: (i) the approach would assign the task of regional promotion to the system of financial co-participation, leaving it to the Provinces to decide strictly within their budget limits on the allocation of resources between industrial promotion and public services; (ii) the approach would recognize that trade reform is essentially removing the rationale for the promotion of "priority projects" on a national level, except for the rare industrial cases of strong positive external effects; resources for the promotion of such exceptional projects could be sought in the Federal budget on a case-by-case basis, rather than by establishing ex-ante the amount of tax credits to be allocated annually to new projects;

(c) to study the implications of the small-sector promotion regimes (mining, forestry, fishery, shipbuilding, production and processing of wine, sugar, tobacco, yerba mate and cotton, and manufacture of internal combustion engines) with the aim of integrating these sectors into the general system.

Regulations

7.10 **Barriers to Entry and Exit.** Opening the industrial sector to international competition, it becomes increasingly important to enable entrepreneurs to rapidly respond to new opportunities and allow firms to exit operations that have become unprofitable. The suspended promotion laws had virtually preempted new entries by concentrating benefits on incumbents. In addition, they included capacity restrictions in the special sector regimes. Continuing these policies would compromise the flexibility of the industrial sector. The investment response of entrepreneurs to new opportunities would be delayed as they would wait for the approval of benefits, lest they be surprised by the cost advantage of a competitor. On the exit side, firms are hampered by labor and by bankruptcy regulations. Compensation requirements add to the cost of giving up a production line or a location. More restrictive appears to be an extended bankruptcy procedure that binds resources in non-profitable activities. The difficulty to exit discourages entrepreneurs from investing into export production which tends to be more risky than the production for a sheltered domestic market. Managers then tend to expand operations by incremental steps only, unwilling to shift resources to areas characterized by rapid demand changes and short product cycles. Furthermore, trapped in a certain activity, entrepreneurs are highly motivated to seek protection against changes in international market conditions.

7.11 **Price Regulations.** The Government has far-reaching powers of regulating prices and of intervening in the production and distribution of goods and services under the so-called supply law (Ley de Abastecimiento). Except for drugs and a small number of food items, however, these powers have not been used to pursue industrial policy targets. Price controls have intermittently been applied in the context of stabilization programs, though the controls appear to have been handled flexibly allowing for relative price adjustments.

7.12 **Public Procurement and Public Supplies.** Weaknesses in public procurement planning and internal procedures have led to a situation in which a small number of large domestic firms share the market for goods purchased by the public sector. The rents obtained by colluding suppliers have not only added to the public sector deficit, they have also increased the leverage of these suppliers in the respective domestic markets. YPF and Gas del Estado have been obliged to give priority to domestic customers without being allowed to freely negotiate prices. In general, the movement of their domestic supply prices has not reflected changes in international prices, resulting at times in substantial implicit input price subsidies to domestic user industries. The continuation of such subsidies would not only bear on the public sector deficit, it would also create new allocative distortions as the protection and the investment promotion of other sectors is reduced.

Recommendations

7.13 It is recommended that the Government:

(a) study, as a matter of urgency, possibilities for expediting bankruptcy procedures;

(b) strengthen public sector procurement planning and internal procedures to emphasize cost effectiveness and accountability concepts; and

(c) phase out the subsidy element in the supply prices of YPF and GDE.

As stated before, it is further recommended that the Government use its decree power to restrict the application of the Compre Argentino law to the minimum required by the law, i.e. preference for local suppliers at prices equal to import prices including duties and domestic taxes.

ANNEX CHAPTER VIII: THE ENERGY SECTOR

A. Overview

8.01 Argentina's energy resources are abundant and diverse, and include oil, gas, hydropower and uranium. Since 1984, about half of the energy produced was from oil, one-third from gas, six percent from hydropower and nuclear, and the rest from biomass, coal and other sources. These resources have been developed primarily by the public sector. Because of inadequate pricing and taxation, distorted investment and regulatory policies which have compounded inefficiencies within the State owned energy institutions, energy sector development has placed a heavy financial burden on the public sector.

8.02 **Energy prices** are generally set by the Government and are without the discipline of linkage to international prices (for hydrocarbon fuels) and marginal costs (for electricity). The Government is continually being pressured by many different special interest groups (private companies, labor unions, the provinces, public companies, etc.) to receive favorable prices to resolve their financial problems. In attempts to accommodate all these interests on an individual basis, complex distorted pricing, regulatory and taxation policies have evolved. Since prices are set by the Government, these interest groups have little incentive to reduce costs or save energy, but rather have a strong incentive to negotiate favorable prices and special treatment.

8.03 Energy prices received by producers have not reflected economic cost nor have they been high enough to cover financial costs. Public energy producing companies have not been able to recover their capital investments, to pay their operating costs or to receive a reasonable profit. Although producer prices are low, the high level of taxation on energy (particularly on oil products) tends to force final consumer prices of oil products above economic cost, while most consumer prices for gas and electricity remain below economic cost. Energy taxation is characterized by a high level of taxation, a complex system of specific taxes, inflexible earmarked funds, and multiple taxation at many stages. The overall impact of the pricing and taxation system is to create severe distortions for both producers and consumers, to create financial problems for public energy companies and to lead to wasteful energy use.

8.04 **Energy demand** is unusually high, reflecting wasteful consumption encouraged by low consumer prices for electricity, natural gas, and until the last few years, very low prices for oil products. Following the global oil price shocks of the early and late 1970s, most countries increased energy prices which promoted energy conservation. Argentina did not take these measures and continues to have an elasticity of energy consumption to real GDP greater than unity. (Countries which did adjust prices have elasticities less than unity).

8.05 The most notable structural shifts in demand have been the steady increase in final energy consumption of gas from 3 percent in 1960 to 27 percent in 1985 and the doubling of electricity use. Eleven percent of final demand for energy in 1985 was electricity, 27 percent gas, 50 percent petroleum products and 12 percent other fuels.

8.06 **Energy investment** has been undertaken primarily by the public sector and has been biased in favor of large scale power investment and generally against oil and gas investments. Although self-sufficiency in energy has been a major government objective, this goal has not been achieved. While production of electricity and gas has increased, oil production has fallen steadily since 1981, leading Argentina to become a net importer in 1987. Production rates in recent years have exceeded additions to the resource base; the country has depleted this most essential resource to a critical level from which it will be difficult to recover. A recent reassessment of gas reserves has shown that reserves are only three quarters of former official estimates. Future supply will be seriously limited if reserves are not discovered in the near future. Significant potential exists to increase oil production and find additional low cost natural gas, but fulfillment of this potential is frustrated by low natural gas prices, lack of financial resources, distortions in the regulatory framework that discourage private sector investment, and inefficiencies in the State enterprises.

8.07 Investment in electric power supply to meet rapidly growing electricity demand has been based heavily on hydropower. This strategy has been financially costly because of high capital cost and delays in construction of large hydropower (and nuclear) projects. It has left the country vulnerable to power shortages during periodic droughts, as occurred this past year. Efforts to control the public sector deficit have compounded the distortions in investment by cutting the smaller, more flexible oil, gas, and rehabilitation investments while continuing to fund large ongoing power projects. Unfortunately, the end result of the energy investment strategy over the last decade has been to increase the share of investment in large inflexible power generation schemes, which have had low productivity for the macroeconomy, and to decrease the share of investment in oil and gas exploration and production which could have provided valuable exports during the period of high world oil prices.

8.08 **Regulatory Policies and Institutional Structure** of the sector has resulted in overlapping responsibilities of many government agencies governing operations of the State energy enterprises. This has prevented the development of clear and consistent guidelines for operations and has discouraged public and private entities from pursuing the most profitable activities.

8.09 These pricing, investment and regulatory policies have imposed a heavy cost on the economy and will continue to do so, unless urgent reforms are undertaken. It is estimated that $10 billion in potential revenue for the Government and economy will not be realized over the next seven years, if no action is taken to: (i) restructure energy pricing and policies; (ii) redirect investment priorities; and, (iii) change the regulatory environment. Of primary importance is the need to reduce subsidies, increase producer prices and explicitly link hydrocarbon prices to international prices. To do this properly requires reform of the energy tax system in order to simplify structure and adjust tax rates. These reforms are crucial to reducing wasteful energy consumption. They will also improve the financial position of State energy companies, so that important ongoing investments can be completed (such as the Yacyreta hydropower dam) and productive oil and gas investments can be made. Concurrently with the price and tax changes, certain regulatory and institutional changes are needed to eliminate overlapping regulatory functions and, within each subsector, to: (i) improve operational efficiency; (ii) increase private sector investment; and, (iii) encourage competition. Described below are issues and recommendations to reform the system of subsidies, pricing and taxation, to increase oil and gas supply, to improve gas utilization, to improve refinery operations and improve operations and investment strategy in the power sector.

B. Energy Subsidies, Pricing, and Taxation

8.10 Many of the inefficiencies and problems in the energy sector, for both producers and consumers, derive from the current complex system of subsidies, pricing and taxation. Most producer prices received by State companies for crude oil, gas and electricity are below financial and economic cost; this exacerbates the financial problems of both these enterprises and the public sector. There are large subsidies to private sector entities and the provinces. The heavy taxation of energy fuels is overly complex and is misapplied; this creates financial as well as economic distortions. Accelerating inflation in 1989 has unfortunately reduced real energy price levels. However, the current crisis situation does provide an opportunity to make urgently needed reforms in the energy pricing, taxation, and subsidy system.

Subsidies

8.11 Many private sector entities that sell equipment and inputs to state enterprises, receive favorable prices above economic and financial costs. Those which obtain feedstock from state enterprises (such as private petrochemical companies and private refiners) pay prices below economic and financial cost. Financial distortions are particularly large in the hydrocarbons subsector, where it is estimated that subsidies cost at least $2,200 million per year. This figure includes approximately $1,300 million in subsidies to private sector entities, $327 million to the provinces as excess royalty payments, and $500 to 600 million for costs of the "Compre Argentino" policy. This financial drain has contributed to the severe financial problems of YPF and GdE and, as a result, the country.

Pricing

8.12 Retail prices paid by the final consumer for petroleum products during the fourth quarter of 1988 were generally much above economic cost (varying from 5 percent to 222 percent above, with gasoline being the highest) due to large taxes on petroleum products. Natural gas and electricity are also taxed, but final consumers pay prices which are significantly below economic cost (particularly residential gas consumers, residential electricity consumers, and other gas consumers). As shown in Annex Table 8.1, in the fourth quarter of 1988 all gas prices (without taxes) were very low, 41 percent to 67 percent of economic cost, while residential electricity prices were 51 percent to 63 percent of economic cost. These consumer price distortions encourage wasteful consumption of electricity and gas and distort oil product demand patterns lending to costly refinery imbalances. Unfortunately, the accelerating inflation and sharp devaluations in 1989 have eroded the real energy price levels and reduced ratios of domestic to international prices. A permanent erosion of real prices will have serious economic and financial consequences for the sector.

8.13 The indirect and direct costs of pricing distortions to the economy have been large, in excess of $3 billion over the last 10 years. This is additional net revenue which could have been earned through: (i) increased oil and gas production; (ii) reduced waste of energy, improved production incentives; and, (iii) more efficient energy use.

Taxation

8.14 Argentina imposes very high taxes on the energy sector. Taxes collected and paid by the energy sector have risen in the 1980s to account for a full 20 percent of national government tax revenue. This is unusual for a country that is not a significant exporter of energy. The rise in energy taxes has resulted in part from the deterioration in the collection of broad based taxes, such as VAT and income taxes at a time when the need to reduce to the consolidated public sector deficit is urgent. In addition to this high overall level, the structure of taxation has many serious problems. Numerous taxes at various levels distort incentives throughout the production process. There is a very complex inflexible system of ear-marked taxes. Little or no reliance is placed on corporate income tax.

8.15 This high level of taxation (which in most cases has not been completely passed on to the consumer) has had a negative financial impact on the public sector energy enterprises. The public energy sector (YPF, GdE, and national power companies) had an aggregate operating income of US$3.6 billion in 1987 and US$4.8 billion in 1988. However, after sub-tracting a myriad of sales taxes, fuel taxes, provincial taxes, federal taxes and royalties, the sector had a consolidated net loss of around US$2 billion each year. The cash flow for 1987 and 1988 is a negative US$0.9 billion. This has affected the ability of these companies to make invest-ments and has forced the Central Government to increase transfers to the sector. For example, in 1987 the Government received US$2.6 billion in taxes and royalties, but it also provided compensation to the sector of US$2.76 billion, leading to a net transfer to the sector of $0.16 billion. In 1988, after imposing large taxes on gasoline, the net transfer was a positive US$1.4 to 1.7 billion to the Government.

Recommendation on Subsidies, Pricing and Taxation

(a) Remove the subsidies to private sector entities as quickly as possible through a phased program of feedstock price increases and reduction in "Compre Argentino" costs.

(b) Increase prices of crude oil, natural gas, and refined pro-ducts as quickly as possible to cover economic costs, and explicitly link these prices to the international value of these products. (This implies elimination of preferential industrial diesel and fuel oil prices, increases in natural gas prices (before tax) of about 50 percent, and increases in LPG prices (before tax) by about 20 percent to reach economic levels). Increase electricity prices (before tax) to cover marginal cost (this implies an increase of about 20 percent). Preserve lifeline rates for poor consumers of gas and electricity. (The price increases referred to here are approximate increases, above inflation, from prices prevail-ing fourth quarter of 1988).

(c) Restructure the taxation of energy by: (i) merging all existing taxes (except VAT) into one *ad valorem* tax applied to the commercial price; (ii) reducing tax rates so as to create a more efficient and equitable structure; and (iii) applying VAT uniformly to all fuels. New *ad valorem* tax rates (as percent of commercial price, and also shown as "new taxes" in Annex Table 8) are as follows:

	Present	Proposed
Gasoline		
-Extra	193	165 to 180
-Regular	156	135
Gas-oil	105	105
Diesel	45	30
Kerosene	46	30
Fuel Oil	48	30
LPG	20	30
Natural Gas	44	30
Electricity	38	30

(d) All fuels would be subject to an ad valorem tax of 15 percent, for a single Energy Fund, a VAT of 15 percent, a road user charge for road funds of about 35 percent on transport fuels, and extra taxes on gasolines and gas-oil.

(e) Commercial prices, on which taxes are based would be: ex-refinery international prices for oil products, 90 percent of international fuel oil equivalent for natural gas and long run marginal cost (LRMC) for electricity. Oil products and gas prices should be explicitly tied to international prices. The ad valorem tax should apply to electricity to cover the higher financial costs of electricity (about 15 percent higher than LRMC), to contribute to investment and limit interfuel distortions. Also, income taxes and possibly an income surcharge tax should be applied to YPF, which will earn more income.

8.16 The potential net fiscal impact of the combined tax, price and subsidy recommendations would be $1,300 million per year, which should go to urgently needed investment and help reduce the fiscal deficit. (It would be proportionally less if subsidy reduction and price increases were phased in more slowly). While tax rates would decline, producer prices would increase and more tax revenue would be collected from income and windfall profits taxes. Revenues going to Energy and Road funds would be preserved at their present level. The impact on consumers would be minimized by the new lower tax rates and lifeline rates for poor consumers. Energy price increases should have a small impact on poor households due to the small monthly energy expenditures by the lower income groups.

8.17 In addition to numerous improvements in efficiency and incentives, that could lead to large savings for the country, the price increases would dampen demand for energy. The immediate short-run effect of raising prices (above inflation from fourth quarter 1988 levels) is estimated to be an annual savings in oil equal to about $70 million/year increasing to savings of $200 million/year by the mid-1990's. Energy prices have fallen in real terms during the rapid inflation in mid-1989. If, for example, all energy prices were instead reduced permanently to 30 percent below the levels of fourth quarter 1988 in real terms, the subsequent increase in demand is estimated to entail a loss of $180 million per year in the short run, growing to $560/million per year by the mid-1990's. Argentina simply cannot afford to pursue such a costly strategy of low real energy prices.

Annex Table 8.1: ARGENTINA - COMPARISON OF ENERGY PRICES AND COSTS

FUEL	PERCENTAGE RATIOS				PERCENT CHANGE
	Commercial Price w/o taxes to Economic Price w/o taxes	Commercial Price w/taxes to Economic Price w/o taxes	Commercial Price w/taxes to Economic Price w/new taxes	Commercial Price w/new taxes to Economic Price w/o taxes	(Increase+ Decrease-) to go from Commercial Price (w/taxes) to Economic Price (w/new taxes)
PETROLEUM PRODUCTS					
(A/cubic meter, 10/88)					
Gasoline					
-Regular	115	296	118	250	-15%
-Premium	109	322	115	280	-13%
Kerosene	96	140	107	130	-6%
Gasoil	94	194	92	210	8%
Diesel					
-Industry/Other	79	115	88	130	13%
-Power Sector	52	78	67	115	48
Fuel					
-Industry/Other	123-149	86-105	66-80	130	50 to 24
-Power Sector	78- 94	59-71	51-62	115	95 to 61
LPG (A/45 kg)	84	100	77	130	29
NATURAL GAS (US$/MCF, 08/88)					
-Residential/Commercial	46-41	71-62	54-49	130	83 to 103
-Industrial	67	100	79	130	27
-Power	62	94	82	115	21
-Refining	56	86	66	130	50
ELECTRIC POWER (USc/Kwh)					
Segba					
-Residential	51	72	56	130	79
-Commercial	158	247	190	130	-52
-Industrial	76	119	92	130	9
Average of 10 Major Utilities					
-Residential	63	87	67	130	50
-Commercial	182	269	207	130	-48
-Industrial	87	126	97	130	3

Commercial prices and tax rates as of October 1988; economic prices of gas linked to its opportunity value of the marginal fuel (fuel oil) plus differential distribution costs for residential; economic prices of electricity are based on marginal cost. New taxes refer to proposed new tax rates. Economic prices w/o taxes for petroleum products based on approximate international petroleum product prices plus distribution cost in fourth quarter 1988.

C. Petroleum and Gas Supply

8.18 Production of crude oil fell 14 percent from 1981 to 1987, and continues to decline (excluding a short-term surge in gas liquids production in 1988). YPF's production, which accounts for two-thirds of all crude oil production, fell 17 percent from 1983 to 1987. Fewer development wells were drilled during the period 1981-1987, largely because of lack of funds for new investments, although the total number of actively producing wells was higher by 1987 (up 18 percent), which meant that per well production had dropped about 27 percent between 1981-1987. Consequently, increasing investments would be required to maintain the same crude oil production, as per well production rates probably will continue to decline. If YPF's actual exploration and development investment continues to be only about US$500 million per year, oil production is likely to decline at 3 percent per year, leading to increasing crude oil imports (Minimum Supply Scenario as shown in Annex Figure 8.1). While new oil exploration and production contracts under Plan Houston have recently been signed, any additional oil production from Plan Houston is likely to occur only in the mid-1990's and is likely to be moderate. Petroplan, if implemented, would make only marginal additions to production.

8.19 Additional investments in oil and gas exploration are urgently needed to maintain and, if possible, expand production to avoid costly oil imports and gas shortages in the early to mid-1990's. Additional investments of about $300 to $500 million per year could lead to increasing oil production and even oil exports (Maximum Supply Scenario, Graph 13.1). Such an expanded program would have large benefits for the economy; even under very conservative assumptions (i.e., assuming a low world oil price projection of $11/bbl, and relatively high production costs), the net present value of such a program is near $1 billion. The net present value is substantially higher, over $6 billion, if projected future world oil prices are at higher levels ($16/bbl) and local production costs can be reduced by 40 percent. Production costs could feasibly be reduced by 40 percent through a combination of: (i) efficiency improvements; (ii) lower local costs, due to a lower real exchange rate; (iii) higher efficiency of private sector investment; and (iv) reduction in "Compre Argentino" (which alone adds extra costs of 40 percent).

8.20 Mobilizing such large additional investments is very difficult, if not impossible, from scarce public investment funds. More private sector investment is needed. Since the oil sector has an advantage over other sectors in mobilizing domestic and foreign private investment, petroleum policies should be oriented toward attracting as much private sector investment as possible. Investment undertaken by the private sector is also like to be lower cost and more efficient than public investment.

Recommendations on Increasing Petroleum and Gas Supply

(a) Increase investment in exploration and development for oil and gas, above levels of the last several years, through a combination of increased public and private sector investment.

Annex Figure 8.1

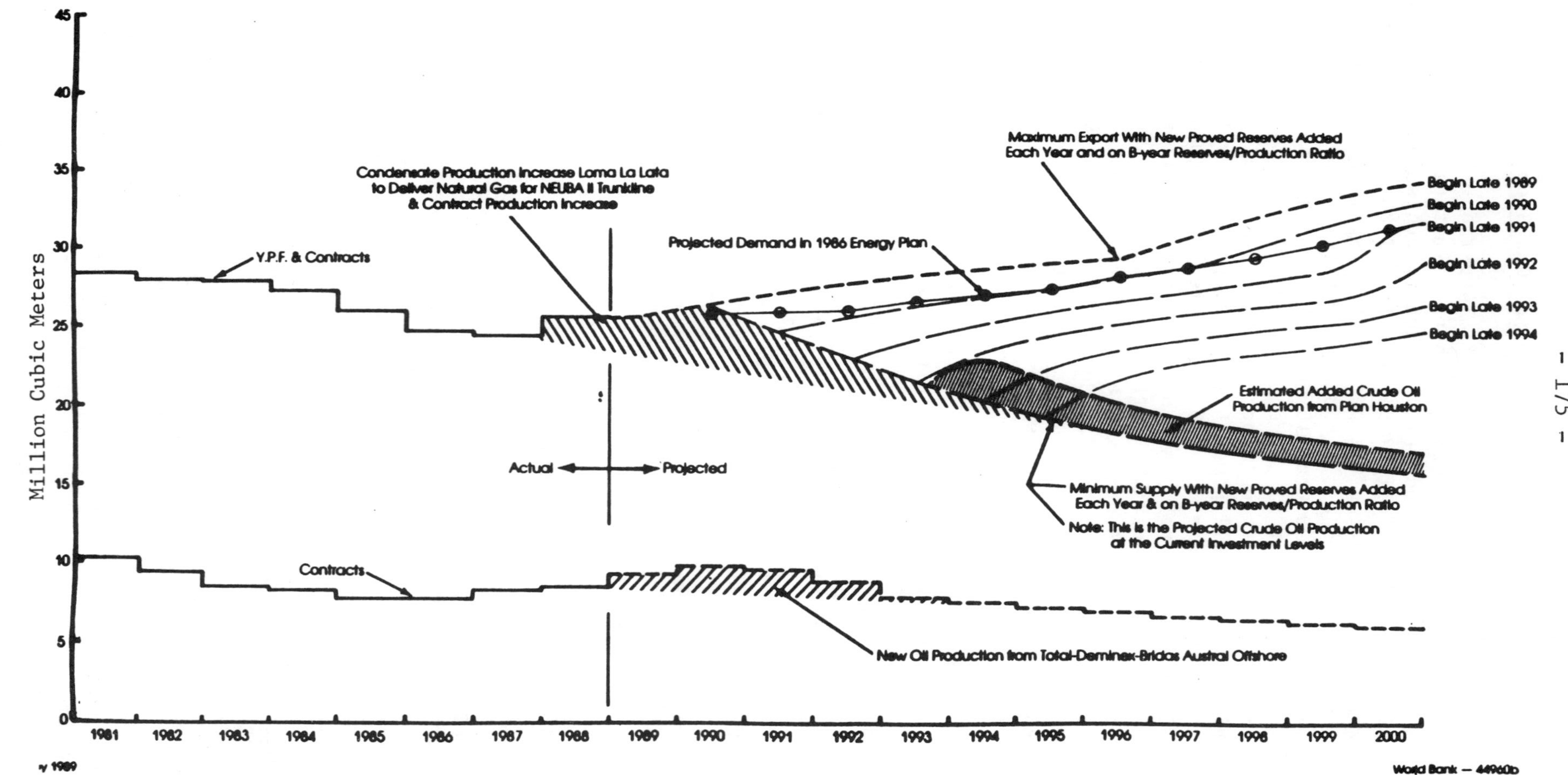

(b) Adopt incentives for increased private sector participation in petroleum exploration and production operations, as follows: (i) increase the prices to be paid at the wellhead for newly found crude oil and natural gas and for any incremental production of these hydrocarbons; (ii) eliminate obstacles for more rapid approval of contracts for exploration and production under the Houston Plan; (iii) increase exploration in new areas by offering additional technical information: (iv) implement the so-called Petroplan, thus allowing the private sector to operate some of the YPF marginal areas; (v) move towards wider private sector participation in YPF's operations, through joint venture or traditional service contracts.

(c) Improve efficiency of operations within YPF and GdE. YPF has already begun to undertake organizational changes and efficiency improvements. Key recommendations include:

(i) Change organizational structure to allow operating departments to function as true cost/profit centers.

(ii) Establish clear transfer prices between various operating departments.

(iii) Improve strategic planning capabilities and prepare a long-term investment program.

(iv) Improve efficiency in upstream and downstream operations through selected investments and corresponding technical assistance.

(v) Install management information system and increase training.

(d) Establish an effective regulatory mechanism for setting and implementing petroleum policy which: (i) eliminates the present unproductive overlap of responsibilities, (ii) sets clear guidelines for operations of public and private enterprises; and, (iii) provides equal treatment for both YPF and private companies.

D. Natural Gas Utilization

8.21 Natural gas is playing an increasingly important role in Argentina as a source of energy for power generation and industrial production and as a residential and commercial fuel for cooking and space heating. Over the past ten years, production has increased 5 percent per year, and it is projected to grow 4 percent per year through 2000. However, the policy of rapid development of natural gas was based on an overly optimistic estimate of proven natural gas reserves. A recent review of the proven reserve base reduced natural gas reserves by about 25 percent and brought into question the advisability of promoting natural gas use. Argentina's proven natural gas reserves are adequate to meet current needs, but as consumption increases new reserves must be discovered to maintain an adequate inventory. The existing pricing structure provides little incentive to explore for and develop natural gas. If the nation's goal of increasing the use of gas to replace exportable liquid fuels is to be fulfilled, incentives to explore for natural gas must be provided and an accelerated exploration and development program must be initiated.

8.22 An accelerated program to develop additional gas resources will have significant benefits for the national economy. Net benefits of such a program over the next 10 years are estimated to be US$4 to 6 billion through the substitution of gas for more expensive oil products and electricity. Analysis indicates that the economic cost of producing and delivering gas to consumers is less than the economic net-back value for all projected uses, except for methanol and fertilizer production. Natural gas is more efficient and lower cost than many other fuels. More use of gas in the residential sector would be desirable, as the cost per unit of useful energy is lower than substitute fuels (e.g. one third that of electricity). The financial cost of producing gas, including royalties, VAT, and income taxes, as well as a reasonable return on equity invested in gas production, is less than the internationally-based price of liquid products. An appropriate pricing structure based on the international value of the fuels that natural gas could replace would generate sufficient revenue to attract investment in gas exploration and development.

8.23 If no new reserves were developed, the economic cost of supplying natural gas for some petrochemical projects would exceed the net-back value. Therefore, gas should not be used for these purpose unless additional reserves are discovered (as seen in Annex Table 8.2). The existing pricing structure for petrochemical feedstocks is complex and the retention price received by the feedstock supplier may not be sufficient to encourage development of new natural gas reserves. Even if gas reserves are increased by 40 percent over the next ten years, the economic benefits of some petrochemical projects, notably the methanol project in the Austral region, are marginal. If proved gas reserves are increased by about 8 percent (40,000 MCM) there will be enough gas supply for all the power sector thermal stations which are scheduled for installation through 2000.

Annex Table 8.2: ARGENTINA - COMPARISON OF ECONOMIC COST AND NET-BACK VALUE ON

Sector	AIC Supply Cost $/MCM	Depletion Allowance (5) $/MCM	Economic Cost $/MCM	Net-back Value $/MCM
Residential	85	25	110	200
Comm/Inst.	83	25	108	203
Genl.Ind.	41	25	66	121
Cement	30-50	25	55-75	92
Power	38	25	63	79
Fertilizer	17	25	42	30
Methanol	17	25	42	15-25

8.24 Excess gas transport capacity is likely to become available in the NEUBA pipeline in the early to mid-1990's, so it is urgent to first accelerate exploration for gas in the Neuquen area (which is close to NEUBA). Policies must be implemented to accelerate exploration and production of gas in other regions as well. In addition to developing new reserves, the delivery infrastructure must be expanded to meet growing demand. The timing of the expansion of the north (and possibly the south) gas lines, is dependent on the size of potential resources in Neuquen. If the pipeline system is not expanded, the capacity of the system, including the expanded Neuba II pipeline, will reach a limit in 1996. Even with assumptions on relatively low growth in gas demand, there would be no opportunity for further growth in gas supply, and the available gas would have to be conserved for the highest value users. An interruption or slowing of the pipeline expansion program would have significant economic costs since the value of the fuels that additional gas could replace would be much higher than production and new pipeline costs. If the expansion program were delayed five years, the cost of the extra liquids fuels burned during the 1996-2000 period would be about $200 million in present value terms. This clearly illustrates the benefits of expanding the pipeline system to keep pace with the growing demand.

8.25 An additional problem of gas utilization is the utilization and pricing of natural gas and natural gas liquids (NGL's) in petrochemical projects. The current pricing system is inflexible and producer prices are too low to encourage development of new gas reserves. A new pricing formula is needed. A new formula could index the producer price to the international price of alternative feedstocks such as naphtha. This is somewhat similar to the method now used, but instead of the retention price and rebate structure, the supplier would negotiate directly with the buyer. A floor price for natural gas would have to be established, as well as the appropriate base price and indexation factors. For projects such as methanol or MTBE production, where dry natural gas is the feedstock, a product price based system would be preferred. The formula could provide for a fixed price during the start-up and market development period. Subsequently, if earnings exceeded a fixed level, the feedstock supplier would receive a share of the "excess" earnings.

Recommendations on Improving Natural Gas Utilization

(a) Implement a controlled transition to a new pricing system. During this transition period a new system of controlled prices would be put in place while needed studies and institutional arrangements are completed. This would entail: (i) gradually increasing the controlled price to just below the international level of fuel oil, for the industrial sector, plus additional distribution costs for the other sectors; (ii) eliminating arbitrary transfer prices for natural gas and natural gas liquids; (iii) establishing a system for regulating downstream gas operations as public utilities; and (iv) ultimately eliminating retail price controls except the margin for delivering gas from the wellhead to the consumer. The tax regime will have to be modified to assure that all types of energy which compete with natural gas are subject to the same unified energy tax (on an energy equivalent basis) as natural gas. The Government should also capture a larger share of the economic rent from the producer in the form of an income tax and windfall profits tax.

(b) Set the retail price of natural gas in the industrial and power market equivalent to, or nearly equivalent to, the price of fuel oil, the next best substitute and the marginal fuel displaced by incremental gas production. The price in the residential and commercial sectors should be set to cover the additional incremental transmission and distribution costs plus an excise tax, if needed, to reach kerosene equivalent.

(c) Unless additional gas is discovered, do not use gas for methanol or fertilizer production.

(d) Set up a regulatory system for natural gas transmission and distribution. Increase competition in gas supply by allowing producers to sell gas directly to consumers.

(e) Expand YPF exploration in the Neuquen area. Prepare a strategic plan for optimal least cost gas exploration/production and pipeline expansion options.

(f) Establish and implement a new pricing formula for natural gas and natural gas liquids used in the petrochemical industry based on alternative feedstock and profit sharing principles.

(g) Undertake studies on:

(i) Gas demand, which account for different demand growth rates, different price assumptions and cost/benefit analysis of gas use in each sector.

(ii) Cost of service and tariffication to establish long-run marginal cost of producing, transporting and distributing gas.

(iii) Institutional structure to regulate gas distribution and transmission.

(iv) Pricing options for NGL's used in petrochemical sector.

E. Refining Operations

8.26 In recent years large investments in refinery conversion have been made to convert lower value fuel oil into more valuable diesel and gasoline. It had been hoped that increasing gas production would displace fuel oil which could be refined in the new conversion units. However, declining investment in oil production has reduced oil supply for refining and demand for gasolines has gone down sharply due to higher retail prices (from new taxes). Therefore, because of both reduced crude supply and a distorted demand pattern for oil products, the refinery sector cannot operate in an optimal manner. At present there is an excess of refinery capacity: there are surpluses of gasoline (which cannot be exported due to low quality and logistical problems) and shortages of fuel oil. Steps need to be taken to profitably utilize excess refinery capacity, improve operational efficiency in refineries, and restructure the demand pattern for oil products.

8.27 The large installed capacity of Argentine refineries, almost twice domestic requirements, can be considered a resource waiting to be exploited. A US$1 per barrel marginal profit might be obtained from processing additional crude oil and exporting the products, which could earn US$35 million per year in revenue. In addition, the lack of optimization causes opportunity losses, worth possibly US$50 million per year, which could be earned if the products were to be brought up to international market quality and then exported. Marginal refinery capacity could be economically employed for generating exports and for arranging third party (foreign) processing agreements. However, the existing 10 percent tax on refined crude oil effectively kills the incentive to optimize this resource, in much the same way that unrealistic and unnecessary price controls distort demand and stifle efficiency of operations for the domestic market. Thus, a pre-requisite for optimizing the Argentine refining industry is to link crude oil and ex-refinery product prices to international values and eliminate of distortionary taxes, which reduce the effective use of this industry's potential.

8.28 Although the Argentine refining industry has moved to higher value production, there is further scope for improving yields, efficiency and profitability. Refining capabilities to produce these higher valued yields have not been fully utilized. The basic reasons for a consistent lack of optimization include the following: (i) import and export controls, which have reduced the refineries ability to balance crude slates and product yields with international market requirements and standards; (ii) industry reliance on government protected refinery margins; (iii) domestic price controls, which maintain overall prices at low levels relative to international markets, and which distort price spreads between different product grades; (iv) strategies by the private refiner to overcome, both the local regulations and distorted price structures, to increase private company profitability at the (partial) expense of YPF refineries; and (v) inability of YPF to act promptly as a profit seeking entity.

8.29 One solution to overcome these problems is to create an adequate operating environment for all the refineries, which would encourage better economic use of existing facilities, both to increase production to satisfy the domestic market and, at the same time, be able to export substantial volumes to maximize foreign revenue. These objectives may be reached by a combination of adequate regulatory changes and price and tax changes.

Recommendations on Refining Operations

(a) Improve refinery utilization. Options could include the award to refineries of throughput licenses to allow the utilization of various different kinds of domestic crude oils, to obtain the crude slates that can be utilized to maximize profits. The combined value of potential optimization of domestic product qualities is estimated at US$60 million per year.

(b) Expand physical port facilities and control infrastructure to expand import/export of crude oil and products. Upgrade marine terminals.

(c) Take measures to improve refinery efficiency, reduce losses, improve product blending, handle high sulfur crudes, handle heavier crudes and export excess naptha.

(d) Take steps to allow further increase of private sector participation in downstream operations, where the main options that can be considered include the following: (i) selling off specialty producing refineries and plants; (ii) disposal of the transport fleet; (iii) selling shares in the large YPF refineries and petrochemical plants; (iv) contracting out more of the drilling, well service and workover services; and (v) privatizing the pipeline transportation system for both crude oil and products.

F. Electric Power Sector

8.30 Argentina has traditionally enjoyed reliable and extended electricity service, provided originally by private companies and, since 1960, with increasing participation of the national utilities. In the last 16 years the country has increased its total installed capacity at a rate of 6.4 percent p.a. and developed a large National Interconnected System (NIS) which covers a substantial part of the country supplying about 90 percent of public electricity services. Currently, about 95 percent of the total population has access to electricity. Past investment has been biased toward large hydropower investments. This bias has put a heavy financial burden on the sector and has resulted in power supply being excessively vulnerable to power shortages during droughts (as in 1988/89). Some large plants are not yet completed but are near completion. One major issue is how to make the power sector more financially self sufficient (through tariff increases and cost reduction) so as to be able to finish ongoing projects. However, the sector faces important problems arising from: (i) an inadequate legal/institutional framework, (ii) lack of consistent planning, (iii) deterioration of generation and distribution facilities as a result of the reduction in investment and poor operating and maintenance practices, and (iv) a weak financial situation characterized by a heavy foreign debt burden and low level of internal cash generation, due mainly to rate levels which do not reflect costs. The economic difficulties now faced by the country call for an increased effort to conduct power activities efficiently and economically, thus reducing the sector's reliance on government support. To achieve this objective corrective measures should be promptly taken to: (i) improve the sector organization, (ii) ensure that the sector expansion follows principles of economy and efficiency, (iii) improve the efficiency in the operation of the existing facilities, (iv) promote energy conservation, (v) reduce technical losses and electricity theft, and (vi) reduce the level of investments and improve sector finances.

8.31 In the context of the Bank's lending for power, the Government is carrying out a program of reforms which will address the main sector issues: (i) a study on the sector organization and efficiency of public utilities, expected to provide recommendations to improve the institutional structure, is underway, (ii) the SE has committed itself to adhering to least cost principles in the updating of the expansion plan, (iii) the SE, SEGBA and AyE are preparing a program for the rehabilitation of thermal facilities, (iv) SEGBA, the largest national utility, is carrying out a program for reducing losses and electricity theft, and (v) the Government began to implement a Financial Rehabilitation Program (FRP) to recover the sector's financial health. The following recommendations are possible ways of addressing issues which are not fully covered in the existing Bank/Government agreements.

Recommendations on Improving Power Sector Supply

(a) Expansion Planning:

(i) On the basis of Bank analysis of forecast methods, the SE should improve forecasting models.

(ii) Expansion plan should be based on a low demand scenario. (Potential economic savings in investment plus operation costs due to a lower level of demand may amount to about US$2 billion in the period 1990-2000.)

(iii) Expansion should be based on more flexible, robust solutions which could save investments and adapt better to demand fluctuations.

(iv) Reliance on large capital intensive hydroprojects should be reduced.

(b) Improve development program:

(i) Government policies should reflect least-cost principles. Consideration should be given to the use of Combined Cycle Plants and Rehabilitation of existing thermal facilities. Complete ongoing study on thermal generation options on time and implement recommendations.

(ii) Eliminate new nuclear plants (after Atucha II) from development plan. Reinforce planning team at the SE.

(iii) Priority should be given to completion of ongoing works.

(c) Improve Operations:

(i) Complete ongoing studies on status of thermal installations and prepare rehabilitation plans. Speed up implementation of SEGBA's loss reduction program. This could reduce losses in SEGBA's system from 22 percent currently to 13 percent by 1992.

(ii) Expedite execution of the SEGBA V Project.

(iii) Evaluate the actual market for coal in power generation, to define whether expansion of YCF's mine in Rio Turbio is justified.

(d) Improve organizational and institutional structure:

(i) Complete study under preparation by the special commission on sector organization and efficiency of power utilities and implement recommendations.

(ii) Increase private sector participation by modifying regulatory framework to provide a sound environment to private investors. (There is a potential for substituting US$200 million to US$400 million from public to private investment.)

(e) Complete LRMC study and apply recommendations and increase electricity prices up to economic cost. (Impact on utilities finances of a cost recovery pricing policy could reduce financial deficit of national utilities from US$600 million in 1987 to zero by 1993).

G. Energy Planning

8.32 The 1986 Energy Plan was a very useful integration of supply, demand and investment. However, the Plan is now out of date and many objectives have not been fulfilled. At present there is no official integrated investment program. Approximate investment levels in oil, gas, refining, and power for 1989-2000 based on analysis in this report range from a low of US$15 billion, to a high of US$27 billion for the 1989-2000 period. Power investment should be based on the low demand scenario. Oil and gas exploration and development investment figures for 1989 through 2000 range from an investment level to reach the Minimum Supply Projection of US$6 billion (assuming a continuation of recent investment trends) to an investment level of US$14 billion (from increased public and private investment), to reach Maximum Supply Projection. The other investments in gas and petroleum infrastructure are partially dependent on the level of oil and gas investment and optimal planning. The potential role of the private sector is large, between US$4 to US$12 billion of total energy investment from 1989 through 2000.

Recommendations on Energy Planning

(a) Update annually the Energy Plan, including alternative investment plans to meet different energy demand growth projections, impact of price changes and interfuel substitution.

(b) Evaluate the costs and benefits of each investment program and rank investment priorities according to economic criteria.

(c) Evaluate energy conservation options, to complement price changes in each subsector.

ANNEX CHAPTER IX: AGRICULTURE

A. Agriculture in the Economy 1/

9.01 Agriculture is crucial to the recovery of Argentina's economy. Historically, export earnings from agricultural and agroindustrial products have ranged from 70 to 80 percent of total export earnings. Though this percentage has fallen somewhat in recent years (Annex Table 9.1), the capacity of the agricultural sector to generate increased export earnings, far exceeds that of any other sector.

Annex Table 9.1: ARGENTINA - KEY AGRICULTURAL STATISTICS

Period	Agriculture as % of GDP a/	Ag.Exports as % of Total Exports b/	Level of Direct and Indirect Taxation of Agric.Sector as % of Ag. GDP c/	Argentine Grain Exports as % of World Grain Trade
1971-75	13.2	76.2	18.0	11.7
1976-80	13.5	76.5	25.4	10.6
1981-85	15.1	74.2	45.6	10.9
1986	14.8	67.2	-	9.0 d/

a/ Source: Table 2.1-A, CEM '85
b/ Source: Table 3.6, CEM '87, Categories I -- VI, VIII, & X
c/ Source: Stutznegger, Background Paper, Table 17b
d/ Average for 1985-87.

9.02 Agriculture contributes to GDP approximately 15 percent, and agro-industry approximately 20 per cent. In contrast to the pattern observed in many countries, agriculture's share in GDP has actually increased in the last 15 years. While industry stagnated during this time, agriculture and agro-industry expanded despite declining real prices for foodstuffs in international markets. In 1970, agriculture contributed about 13 percent of GDP; by 1985, this share had increased to about 15 percent despite an adverse movement in the agricultural terms of trade. By contrast, industry's share declined from about 36 percent to under 30 percent during the same period.

B. Agricultural Potential

9.03 The potential for rapid expansion both in the area planted and the productivity of Argentine agriculture, given favorable incentives, is enormous. Most agricultural products are produced in the central-eastern region of the country, one of the most fertile regions in the world for

---

1/ A comprehensive review of the Argentine agricultural sector is contained in the recent World Bank report Argentina Agricultural Sector Review, Report No. 7733-AR, June 26, 1988.

temperate agriculture. Over 56 percent of the soils of this region are classified as Type I or Type II, that is, requiring little fertilizer under normal crop rotation and demanding only normal conservation measures. This natural productivity contributes to some of the lowest production costs in the world. Variable production costs in 1987 were US$35 per ton for wheat, US$27 for corn, and US$80 for soybeans, or roughly two-thirds to one-half the production costs in the United States.

9.04 The strong comparative advantage of Argentine agriculture is demonstrated in Annex Table 9.2. Using economic prices, estimates have been made of the domestic resource costs of production, or the amount of resources necessary to generate a dollar of foreign exchange. These estimates indicate the high economic returns associated with agricultural production: the foreign exchange generated is 2 to 3 times the cost of domestic resources used in production.

Annex Table 9.2: ARGENTINA - COEFFICIENTS OF COMPARATIVE ADVANTAGE a/ IN GRAIN PRODUCTION

Product	1981/82	1982/83	1983/84	1984/85	October 1986
Wheat	0.38	0.34	0.43	0.48	0.76
Corn	0.34	0.31	0.35	0.36	0.65
Grain Sorghum	0.43	0.41	0.50	0.39	0.64
Soybeans	0.31	0.14	0.32	0.18	0.32
Sunflower	0.33	0.41	0.31	0.31	0.60

Source: F. Cirio and M. Regunaga, 1986

a/ Domestic resource cost-coefficients divided by the official exchange rate.

9.05 Although the fertile plains of the pampa will most likely remain the center of Argentina's productive base in agriculture, a great diversity of micro-climates and soils permits the production of a range of crops from wheat, rice, and oilseeds to grapes and other fruits and even tropical products. Many of these crops can be produced at costs competitive with production costs elsewhere.

9.06 Though agriculture has grown faster than the economy as a whole, its performance has been substantially below potential. Investment in the sector has been minimal over recent years. Furthermore, its share of total world trade in grains has gradually declined (Annex Table 9.1). The unavoidable uncertainty arising from dependence on international markets has for many years been compounded by excessive explicit and implicit taxation. Domestic policy has tended to use the agricultural surplus to finance industrial development. Estimates of this transfer indicate that, in the early 1980s, the agricultural sector conferred on average 45 percent of its GDP to the public sector and to other parts of the economy.2/

2/ See World Bank, *loc. cit.*

C. Disincentives to Agricultural Investment

9.07 This has been brought about by a combination of export taxes, price controls at the wholesale and retail levels, official prices for exports and export quotas. Indirect negative effects on the sector have resulted from an overvalued currency and protection afforded the industrial sector. Some compensatory mechanisms have been in effect at various times through subsidized credit and tax exemptions on the purchase of machinery. However, these compensatory subsidies have been insufficient by far to redress the high level of explicit and implicit taxation on output.

9.08 Other factors have contributed to reduce incentives to the sector. The infrastructure that supports agriculture has deteriorated and the price differential between FOB and farmgate price is excessive. This differential reflects an inefficient transport system, high storage costs and a lack of transparency in the cash and forward grain markets. Furthermore, margins between Argentine FOB and international prices have been wide during the 1980s (Annex Table 9.2). Part of this is due to the ending of the US grain embargo on the USSR which had given Argentina's grain a premium. Another part of it is explained by the high cost of shipping grain from Argentine ports and the necessity to use smaller ships.

9.09 High margins can also be explained by the fact that the incentive system and the financial power in the market weigh heavily in favor of the major buyers. The bulk of agrarian exports are shipped by companies whose profits are maximized on a global basis, and the existence of export taxes (recently reduced), combined with foreign exchange controls and relatively easy access to pre-export finance by registered exporters, creates strong incentives for these companies, both national and multinational, to minimize declared prices in the FOB market. It is common practice for grain shipments to be traded several times off-shore before reaching their final destination. Of the 100 or so operators in the FOB market, no more than 15 are major on-shore purchasers of grain.

9.10 The financial power of the major grain buyers is strengthened by their access to pre-export finance. This is available to exporters up to 210 days before the harvest and is seen by the Authorities as a significant positive inflow of foreign resources into the financial system. Approximately US$2-3 billion are brought into Argentina annually for this purpose, but only a small proportion actually enters the forward purchase market. Financial resources for post-harvest storage by producers are equally scarce. Consequently, the bulk of the grain crop is sold for cash at harvest time, a situation which confers enormous market power on the major buyers.

9.11 The consequence of the disincentives described above has been the development of a low-cost extensive form of agriculture, with little capital stock (the average age of the tractor fleet is 18 years) and little debt. Of all the major agricultural producers in the world Argentina's agriculture is by far the least protected (in fact, it suffers strong negative net protection). Nevertheless, it has continued to grow. The implication is that with stronger incentives, the potential for new investment and growth is exceptional. Given the importance of the sector in GDP and exports and the positive coefficients of comparative advantage shown in Annex Table 9.2, a strategy of maximum growth for agriculture is clearly called for.

Annex Table 9.3: ARGENTINA - DIFFERENCES IN PRICE OF EXPORT BETWEEN FOB ARGENTINA AND FOB GULF a/

Year	Wheat	Maize	Sorghum	Soya
1980	27	38	22	-24
1981	24	-11	-21	-31
1982	4	-7	-21	-14
1983	-17	-4	-17	-14
1984	-18	-6	-18	-18
1985	-32	-10	-11	-17
1986	-27	-16	-14	-3
1987	-24	-2	-13	2

Source: National Grain Board

a/ In US$ 1980 per ton, at peak commercial periods.

D. Principal Policy Requirements

9.12 **Macroeconomic Framework**. The single most important element in the incentive framework for agriculture is the exchange rate. This applies to all subsectors (traditional and nontraditional) and to agro-industry. Growth depends exclusively on the export markets. Maintenance of a stable competitive real exchange rate is crucial. There is also the potential for substantial macro shocks from price changes.

9.13 **Cost Reduction**. Investment incentives are adversely affected by the high costs associated with marketing agricultural commodities. These can be broken down into four categories, in all of which the Government has a key role to play:

(a) Transport costs. Action by the Government in this area would include direct investment. At the same time, it is necessary to encourage a process which is already taking place whereby private, integrated systems of storage and transport are developing, covering the shipment of grain from farmgate to export terminal. Key elements in encouraging this process would be demonopolization of port facilities, privatization of rail car operations and privatization of storage facilities.

(b) Storage costs. The Junta Nacional de Granos charges US$5 per ton for terminal storage. The private sector can perform the same service for US$2 per ton. The modernization of the Junta's terminal port facilities, their privatization and the demonopolization of the facility at Bahia Blanca are urgent priorities. Recommendations to that end have been made by a joint private/public sector commission and these have been endorsed by the Government. Intensive discussions are currently under way among the Government, the JNG, the exporters and the country elevator operators to create an independent entity capable of taking over and operating the port elevators currently owned by the JNG.

(c) Marketing margins. Reduced marketing margins require a lowering of perceived risks throughout the trading system and an increase in the market power of producers *vis-à-vis* the main buyers. Key to meeting these ends are development of more transparent forward markets, development of tradeable forward purchase securities, expansion of the operations of the Bolsas, enhanced access to futures markets (both foreign and incipient domestic markets) for hedging price risks, and increased flows of pre-export finance to producers. The Government has a powerful regulatory influence over all these possible developments. A reorganized Junta Nacional de Granos (see para. 14.15) below would also have a key role to play.

(d) Financial costs to farmers could be reduced by ensuring a much larger flow of dollar-denominated pre-export financing to producers through forward purchase markets. This implies direct intervention to regulate use of pre-export funds by exporters and also requires development of more transparent, forward purchase markets themselves, as proposed above. Specific proposals to this end are currently under discussion in Argentina and should be given high priority.

9.14 **The Junta Nacional de Granos**. Restructuring the Junta is a critical element in improving the grain marketing system. There is widespread acceptance of the need to divest the Junta of its role as an operator of terminal elevators. At the same time, proposals are well advanced for restructuring the Junta based on an *a priori* acceptance of the need for a public sector institution involved in the grain trade. It is essential that the new Junta created be professionally strong and financially independent. This would be a dramatic change from the present situation, but there is an emerging view in Argentina that such a change is necessary.

9.15 Such an institution should be autarkic and self-governing. Its management should be professional, appointed and promoted on merit, and remunerated competitively with the private sector. It would act as an independent export trader, with access to financing on its own strength. As a pre-condition for this, it would operate under conditions that make it competitive, in principle, with private traders. This applies in particular to its relation with off-shore companies and external sources of financing. It would play a major role in the development of a market for traded forward purchase instruments referred to above and in the use of futures markets.

9.16 A continued role for the Junta as an executing agency for government programs (support for marginal farmers; domestic wheat purchases etc.) should be considered only under conditions of contract. Program offices should be established with contract personnel, hired temporarily for the management of the program in question. The Junta should account to the Government for its management, and should be able to refuse programs it considers flawed, for instance because of insufficient funding.

9.17 **Agricultural Taxation**. The substantial lowering of export taxes, which occurred in 1986, was a very positive move. As suggested in Chapter II, export taxes, while justifiable in the context of a short-term stabilization effort, need to be phased out over time. In the interim, any use of export taxation should be uniform across products and sectors, and accordingly not discriminate against agriculture as in the past. At the same time a reexamination of the overall agricultural tax issue is needed. The introduction, as an emergency measure in 1988, of a dual exchange rate penalizing agricultural exports once again gave strong disincentive signals to the agricultural sector. A more stable, production-neutral system of agricultural taxation remains an urgent priority. The difficulties encountered in introducing a Federal Land Tax in 1987-88 should not deter efforts to reform the fiscal role of agriculture. These efforts should concentrate on the strengthening of the Provincial Land Tax (with possible Federal revenue sharing) and on improving income tax collection in both rural and urban sectors.

ANNEX CHAPTER X: ANALYTICAL APPENDICES

A. PROJECTIONS AND MACROECONOMIC CONSISTENCY

10.01 The projections that form the basis of Chapter V are derived from a two gap model that integrates a public sector and financial sector with the Bank's revised minimum standard model (RMSM). The model consists of: (i) a set of consistent macroeconomic statistics for the base year 1987 that includes national income accounts, balance of payments, the financial sector, and external debt commitments; (ii) a series of exogenous variables and parameters, which consist of assumptions based on past performance or targets; and (iii) equations describing the relations among the variables that permit the computation of a consistent set of accounts.

General Description

10.02 The general purpose is to project requirements for external financing and financing of the public sector, given targets for GDP, growth in monetary aggregates and inflation, and increases in international reserves. The model contains several characteristics that differ from the conventional RMSM, and these are the focus of the discussion that follows.

Prices

10.03 Selected prices are given exogenously as part of the Bank's global projections exercise. These include movements in LIBOR, the manufacturing unit values, and the prices of commodities. Prices of both exports and imports are given by the World Bank's Commodity Division on the basis of global supply and demand models.

10.04 The model establishes targets for increases in the domestic price level in the form of the combined index of monthly inflation (page 120). After the hyperinflation of 1989, it was assumed that the price level under strong economic management could drop rapidly as given in the text.

10.05 With the benefit of the changes in external inflation given by the MUV and domestic price movements, it is possible to project a nominal exchange rate. A change in the real exchange rate can be entered exogenously as was the case for 1988 to reflect the slight appreciation that year against the US dollar; for 1989, we have assumed that the massive real devaluation already evident involves overshooting, and will end the year with a 40 percent movement. This real rate would then prevail for the projection period.

Balance of Payments

10.06 As with the RMSM, the model exogenously assumes that for a given policy regime and growth in the world economy export volumes can be expected to expand by the amounts shown on page 121. These were discussed in detail with sectoral experts in the Bank and in Argentina. Imports are projected on the basis of exogenous sectoral growth rates of the economy; import elasticities, shown on page 122, are consistent with the experience of the 1970-87 period, if somewhat lower to reflect the higher exchange rate. The ensuing trade balance, together with the assumed LIBOR times the debt stock to produce interest expenses, produces a current account balance. The current account plus assumed changes in reserves and net capital flows is then balanced through the unidentified finance line.

10.07 Arrears through 1988 are assumed to be capitalized as part of the loans from unidentified finance (the gapfill loan). In line with the rest of the capital account presentation, the unidentified finance line is shown on a net basis, so that future amortizations (assumed to begin in the fifth year after the loan is made) are assumed to be rescheduled.

10.08 Note the Government is assumed to have a sufficient number of future standbys with the IMF that its net exposure position remains constant.

Public Sector Accounts

10.09 The base year numbers for the public sector, were adjusted forward insofar as Bank staff had information, and are reasonably accurate through 1988. One problem is the lack of consistency with past time series because of the absence of recent information on the provinces, which the Government has assumed to be in balance since end-1987. A second problem is the absence of budget basis accounts for 1988; the cash accounts were used to estimate the budget accounts for 1988, by adjusting expenditures up by the same proportion as in 1987 (about 1 percent of GDP). A third problem is that the inflation of 1989 and changes in tax regime make it difficult to obtain accurate estimates of the worsening of the public sector deficit in that year.

10.10 The projected financing needs of the public sector are then derived by assumed strong policy actions in the nonfinancial public sector that would reduce financing requirements to those consistent with available foreign finance. By specifying targeted levels of domestic credit to the public sector consistent with expansion of monetary aggregates and hence inflation, and knowing the foreign credit to the public sector available from the balance of payments, it is possible to formulate a view on the necessary degree of deficit reduction for the combined public sector to achieve the inflation reduction target. The exact apportioning of the improvements as between revenues, the noninterest operating surpluses of the public enterprises, and expenditure reductions is a matter of judgment; the objectives were specified based on policy changes suggested in Chapters III, V, VI, and VII, and allowing for a modest increase in public investment.

10.11 The external financing component is derived from the flows expected to the public sector as coming from the balance of payments plus 100 percent of the unidentified finance. This includes 100 percent of disbursements of bonds, IDB, IBRD, and financial markets minus 100 percent of amortization to these creditors as well as commercial banks and bilaterals. (Short-term and nonguaranteed private flows were assumed to finance the private sector). These are then converted at the nominal exchange rate.

10.12 A new feature of this model is the illustration of the impact of increased foreign debt, changes in international interest rates and real devaluation on public finances. The foreign interest bill of the public sector is given by consolidating the interest of the central administration and public enterprises into the current expenditure account of the nonfinancial public sector. The projected external indebtedness of the public sector as taken from below the line in the public sector accounts and from the balance of payments is then incorporated into the future public sector accounts.

10.13 The model also permits some treatment of the quasi-fiscal deficit of the Central Bank. This includes the external component of the quasi-fiscal deficit, given by the net foreign interest earnings on the Central Bank's foreign assets; the internal component of the quasi-fiscal deficit is assumed to be in balance, a fairly conservative assumptions under the presumption of medium-term price stability in the model. The net foreign interest earnings are equal to earnings on reserves as derived from reserve stocks in the balance of payments less 22 percent of the total non-IMF interest bill (the 1988 share of the Central Bank in the total public sector debt) plus IMF charges. Consolidating the quasi-fiscal deficit with the nonfinancial public sector allows for a consistent view of total financing requirements of the public sector. It also provides for simplicity in modeling since it circumvents the problem of apportioning the increases in debt stocks to the various sectors of government, including the Central Bank.

National Accounts

10.14 National accounts are determined in the fashion of the RMSM. Given the growth in GDP, an investment parameter determines investment levels. Private investment is the residual of the total minus public investment. Subtracting total investment and public consumption from GDP produces private consumption. Gross national savings of the private sector are derived from investment less foreign savings (the current account) and public savings.

10.15 In a purely technical sense, the long-term growth prospects of a country are determined by its ability to improve its capital formation and productivity. Present levels of investment are among the lowest in the last 20 years, and must be reversed for output growth to resume and accelerate; this requires adequate foreign savings and increasing domestic savings. The model has taken incremental capital outputs ratios consistent with those of the 1970s (although the ICOR concept has limited application in an economy dominated by agriculture). Growth rates in GDP are consistent with historical experience, allowing for some improvement in the efficiency of investment associated with recommended policy changes toward the real sector and with price stability.

Financial Sector and Monetary Aggregates

10.16 The financial system on the asset side is given by changes in net international reserves as taken from the balance of payments, and the expansion in domestic credit is determined as the residual of the combined public sector deficit less available foreign financing. On the liability side, the change in M1 plus other interest bearing liabilities are given by an assumption about projected remonetization consistent with past experience under conditions of medium-term price stability; to this are added net increases in private foreign liabilities, which are the residual of increased indebted as shown in the balance of payments less external financing to the public sector. The balancing item is domestic credit to the private sector.

External Debt

10.17 Stocks of external debt and amortization schedules were adjusted with information from the Government of Argentina through end-1988, and from the pipeline of amortization and interest flows. To these were added new commitments as described in the text. These amounts are those reasonably expected under conservative assumptions about expected flows under conditions of medium-term price stability.

Exchange Rates, Prices and Interest	1987	1988	1989	1990	1991	1992	1993	1994	1995	1996	1997	1998
ER Real Index (% +=devaluation)		-9.7%	48.0%	8.0%	0.0%	0.0%	0.0%	0.0%	-5.0%	-5.0%	-5.0%	0.0%
ER Real Index (1987=100)	100.0	90.3	133.6	144.3	144.3	144.3	144.3	144.3	137.1	130.3	123.7	123.7
ER Nominal Index (1987=100)	100.0	438.2	25231.5	31004.2	31018.3	31024.6	31011.2	31005.5	29458.2	27992.8	26610.0	26595.3
ER (australes per 1US$)	2.1	9.4	540.0	663.5	663.8	663.9	663.6	663.5	630.4	599.0	569.5	569.1
ER (US$ per 1 austral)	0.5	0.1	0.0	0.0	0.0	0.0	0.0	0.0	0.0	0.0	0.0	0.0
ER Base Year	0.5	0.5	0.5	0.5	0.5	0.5	0.5	0.5	0.5	0.5	0.5	0.5
Monthly Domestic Int. (real)	0.44%	1.30%	1.20%	0.80%	0.52%	0.47%	0.49%	0.46%	0.44%	0.43%	0.40%	0.38%
Annual Domestic Int. (nominal)	216.7%	511.8%	4519.9%	39.5%	11.1%	10.3%	10.6%	10.2%	9.9%	9.9%	9.6%	9.2%
LIBOR	7.3%	8.5%	9.0%	9.8%	9.1%	8.8%	9.1%	8.7%	8.4%	8.4%	8.1%	7.7%
LIBOR (IBRD estimate)	7.3%	8.1%	9.8%	9.8%	8.4%	9.2%	8.9%	8.4%	8.4%	8.3%	7.9%	7.4%
Population Growth	1.5%	1.5%	1.5%	1.5%	1.5%	1.5%	1.5%	1.4%	1.4%	1.4%	1.4%	1.4%
Population (Mln.)	31.497	31.969	32.449	32.936	33.430	33.931	34.440	34.922	35.411	35.907	36.410	36.919
GNP per capita	2,486	2,362	2,240	2,266	2,310	2,351	2,407	2,477	2,547	2,616	2,699	2,787
MUV (Annual <>)		8.0%	2.9%	11.5%	4.4%	4.3%	4.3%	4.3%	4.3%	4.4%	4.5%	4.3%
MUV Index (Int. Deflator)	100.0	108.0	111.1	123.9	129.3	134.8	140.6	146.7	152.9	159.6	166.7	174.0
Average Monthly Inflation	9.6%	14.8%	36.0%	2.0%	0.4%	0.4%	0.4%	0.4%	0.4%	0.4%	0.4%	0.4%
Domestic Index (Annual <>)		424.0%	3903.7%	26.8%	4.4%	4.3%	4.3%	4.3%	4.3%	4.4%	4.5%	4.3%
Domestic Index	100.0	524.0	20979.6	26607.3	27779.7	28969.2	30209.6	31503.1	32852.0	34299.7	35854.0	37389.2
Investment Index (Annual <>)		424.0%	3903.7%	26.8%	4.4%	4.3%	4.3%	4.3%	4.3%	4.4%	4.5%	4.3%
Investment Index	100.0	524.0	20979.6	26607.3	27779.7	28969.2	30209.6	31503.1	32852.0	34299.7	35854.0	37389.2

1987 Real Exports (Annual <>)		1988	1989	1990	1991	1992	1993	1994	1995	1996	1997	1998
Exports of Goods		18.0%	-13.7%	8.3%	5.9%	5.8%	5.1%	5.5%	5.3%	5.3%	5.4%	5.4%
Exports of G&NFS		12.2%	-10.3%	8.2%	5.4%	5.3%	4.9%	5.1%	5.0%	5.0%	5.1%	5.1%
1. Livestock		9.0%	-27.3%	2.0%	2.0%	2.0%	2.0%	2.0%	2.0%	2.0%	2.0%	2.0%
2. Cereals		-18.8%	-33.8%	6.0%	1.0%	1.0%	1.0%	1.0%	1.0%	1.0%	1.0%	1.0%
3. Other Agriculture Goods		14.3%	-56.5%	15.0%	4.0%	3.5%	3.5%	3.5%	3.5%	3.5%	3.5%	3.5%
4. Fats & Oils		17.5%	6.6%	8.0%	5.0%	3.5%	0.5%	3.5%	3.5%	3.5%	3.5%	3.5%
5. Manufac., Food & Bever.		4.6%	1.5%	8.0%	6.0%	6.0%	6.0%	6.0%	5.0%	5.0%	5.0%	5.0%
6. Petroleum		79.9%	-4.4%	5.0%	5.0%	5.0%	2.0%	2.0%	2.0%	2.0%	2.0%	2.0%
7. Chemicals & Plastics		48.4%	3.9%	10.0%	8.0%	8.0%	8.0%	8.0%	8.0%	8.0%	8.0%	8.0%
8. Leather & Wool		-2.4%	-9.8%	5.0%	5.0%	5.0%	5.0%	5.0%	5.0%	5.0%	5.0%	5.0%
9. Other Manufactures		46.0%	-12.4%	10.0%	8.0%	8.0%	7.0%	7.0%	7.0%	7.0%	7.0%	7.0%
Non Factor Services		-5.0%	2.2%	8.0%	4.0%	4.0%	4.0%	4.0%	4.0%	4.0%	4.0%	4.0%
			16.11%									

Exports in 1987 Mln. US$		1988	1989	1990	1991	1992	1993	1994	1995	1996	1997	1998
1. Livestock	655	714	519	529	540	551	562	573	585	596	608	620
2. Cereals	747	606	401	425	430	434	438	443	447	452	456	461
3. Other Agriculture Goods	626	716	312	358	373	386	399	413	428	443	458	474
4. Fats & Oils	546	642	684	738	775	802	806	835	864	894	925	958
5. Manufac., Food & Bever.	1,337	1,398	1,419	1,532	1,624	1,722	1,825	1,935	2,031	2,133	2,239	2,351
6. Petroleum	125	225	215	226	237	249	254	259	264	269	275	280
7. Chemicals & Plastics	466	692	718	790	853	922	995	1,075	1,161	1,254	1,354	1,463
8. Leather & Wool	419	409	369	388	407	427	449	471	495	519	545	573
9. Other Manufactures	1,439	2,102	1,840	2,024	2,186	2,361	2,527	2,703	2,893	3,095	3,312	3,544
Non Factor Services	2,112	2,007	2,051	2,215	2,304	2,396	2,492	2,591	2,695	2,803	2,915	3,031

Export Price Indices (Annual <>)		1988	1989	1990	1991	1992	1993	1994	1995	1996	1997	1998
1. Livestock		4.2%	3.2%	-1.2%	6.1%	6.1%	6.1%	6.1%	6.1%	6.1%	6.1%	6.1%
2. Cereals		35.4%	0.9%	-13.6%	4.5%	4.5%	4.5%	4.5%	4.5%	4.5%	5.3%	5.3%
3. Other Agriculture Goods		46.3%	-1.0%	-9.9%	4.9%	4.9%	4.9%	4.9%	4.9%	4.9%	-1.3%	-1.3%
4. Fats & Oils		43.1%	1.7%	-1.2%	12.1%	12.1%	12.1%	12.1%	12.1%	12.1%	0.2%	0.2%
5. Manufac., Food & Bever.		39.7%	3.0%	-4.1%	3.3%	3.3%	3.3%	3.3%	3.3%	3.3%	2.0%	2.0%
6. Petroleum		-18.6%	10.7%	7.1%	5.6%	5.6%	5.6%	5.6%	5.6%	5.6%	9.7%	9.7%
7. Chemicals & Plastics		8.3%	6.3%	1.5%	3.6%	3.6%	3.6%	3.6%	3.6%	3.6%	4.6%	4.6%
8. Leather & Wool		8.3%	6.3%	1.5%	3.6%	3.6%	3.6%	3.6%	3.6%	3.6%	4.6%	4.6%
9. Other Manufactures		8.3%	6.3%	1.5%	3.6%	3.6%	3.6%	3.6%	3.6%	3.6%	4.6%	4.6%

Export Price Indices (1987=100)		1988	1989	1990	1991	1992	1993	1994	1995	1996	1997	1998
1. Livestock	100.0	104.2	107.5	106.2	112.7	119.6	126.9	134.6	142.9	151.6	160.8	170.6
2. Cereals	100.0	135.4	136.6	118.0	123.4	128.9	134.7	140.8	147.1	153.7	161.9	170.4
3. Other Agriculture Goods	100.0	146.3	144.8	130.5	136.9	143.6	150.6	158.0	165.8	173.9	171.6	169.4
4. Fats & Oils	100.0	143.1	145.5	143.8	161.2	180.7	202.6	227.1	254.5	285.3	285.9	286.5
5. Manufac., Food & Bever.	100.0	139.7	143.9	138.0	142.5	147.2	152.1	157.1	162.3	167.7	171.0	174.4
6. Petroleum	100.0	81.4	90.1	96.5	101.9	107.6	113.6	120.0	126.7	133.8	146.8	161.0
7. Chemicals & Plastics	100.0	108.3	115.1	116.8	121.1	125.4	129.9	134.6	139.5	144.5	151.1	158.1
8. Leather & Wool	100.0	108.3	115.1	116.8	121.1	125.4	129.9	134.6	139.5	144.5	151.1	158.1
9. Other Manufactures	100.0	108.3	115.1	116.8	121.1	125.4	129.9	134.6	139.5	144.5	151.1	158.1

Exports in Current Mln. US$		1988	1989	1990	1991	1992	1993	1994	1995	1996	1997	1998
1. Livestock	655	744	558	563	609	659	713	772	835	904	978	1,058
2. Cereals	747	821	548	502	530	559	591	623	658	694	738	785
3. Other Agriculture Goods	626	1,047	451	468	510	554	601	653	709	769	786	803
4. Fats & Oils	546	918	995	1,062	1,250	1,450	1,634	1,895	2,199	2,551	2,646	2,744
5. Manufac., Food & Bever.	1,337	1,953	2,042	2,114	2,315	2,535	2,776	3,040	3,297	3,576	3,830	4,102
6. Petroleum	125	183	194	218	241	268	288	311	335	360	403	451
7. Chemicals & Plastics	466	749	827	923	1,033	1,156	1,293	1,447	1,619	1,812	2,047	2,312
8. Leather & Wool	419	443	425	453	493	536	583	634	690	750	824	905
9. Other Manufactures	1,439	2,276	2,119	2,365	2,647	2,961	3,283	3,639	4,034	4,472	5,005	5,601

Export Totals in Mln. US$		1988	1989	1990	1991	1992	1993	1994	1995	1996	1997	1998
Exports of Goods (Constant)	6,360	7,503	6,477	7,012	7,426	7,854	8,255	8,707	9,167	9,655	10,173	10,723
Exports of G&NFS (Constant)	8,472	9,509	8,528	9,227	9,729	10,249	10,747	11,298	11,862	12,458	13,088	13,755
Exports of Goods (Current)	6,360	9,134	8,159	8,668	9,628	10,678	11,762	13,013	14,375	15,889	17,257	18,762
Exports of G&NFS (Current)	8,472	11,301	10,438	11,412	12,606	13,907	15,265	16,813	18,496	20,362	22,117	24,036

Import Elasticities		1987	1988	1989	1990	1991	1992	1993	1994	1995	1996	1997	1998
Imports of Goods (Annual <>)			-13.6%	-17.1%	6.0%	6.9%	6.1%	8.6%	8.6%	8.1%	8.2%	8.4%	8.5%
Imports of G&NFS (Annual <>)			-11.0%	-17.1%	5.2%	6.1%	5.6%	7.6%	7.6%	7.4%	7.5%	7.8%	7.9%
Import Elasticity (Imp. wrt GDP)			3.6	7.6	2.3	2.2	2.0	2.0	2.0	2.0	2.0	1.9	1.9
1. Food & Consumer Goods	GDP				1.0	1.0	1.0	1.0	1.0	1.1	1.1	1.1	1.1
2. Petroleum	GDP				0.1	0.1	0.1	0.1	0.1	0.1	0.1	0.1	0.1
3. Intermediate Goods	Industry				1.6	1.5	1.5	1.5	1.5	1.5	1.5	1.5	1.5
4. Capital Goods	Inv.				1.6	1.5	1.5	1.5	1.5	1.5	1.5	1.5	1.5
Non Factor Services	GDP				1.5	1.5	1.5	1.4	1.4	1.5	1.5	1.5	1.5

Imports in 1987 Mln. US$	1987	1988	1989	1990	1991	1992	1993	1994	1995	1996	1997	1998
1. Food & Consumer Goods	320	227	188	192	198	203	211	219	228	237	248	260
2. Petroleum	657	471	390	391	392	394	395	396	398	400	401	403
3. Intermediate Goods	3,987	3,425	2,838	2,916	3,073	3,239	3,516	3,817	4,143	4,497	4,888	5,314
4. Capital Goods	856	905	750	919	1,059	1,176	1,320	1,479	1,623	1,781	1,961	2,158
Non Factor Services	2,397	2,283	1,892	1,957	2,039	2,125	2,237	2,355	2,489	2,631	2,795	2,971

Import Price Indices (Annual <>)	1987	1988	1989	1990	1991	1992	1993	1994	1995	1996	1997	1998
1. Food & Consumer Goods		10.8%	4.0%	1.3%	4.0%	4.0%	4.0%	4.0%	4.0%	4.0%	4.0%	4.0%
2. Petroleum		-18.6%	10.7%	7.1%	5.6%	5.6%	5.6%	5.6%	5.6%	5.6%	9.7%	9.7%
3. Intermediate Goods		8.3%	6.3%	1.5%	3.6%	3.6%	3.6%	3.6%	3.6%	3.6%	4.6%	4.6%
4. Capital Goods		8.3%	6.3%	1.5%	3.6%	3.6%	3.6%	3.6%	3.6%	3.6%	4.6%	4.6%

Import Price Indices (1987=100)	1987	1988	1989	1990	1991	1992	1993	1994	1995	1996	1997	1998
1. Food & Consumer Goods	100.0	110.8	115.2	116.7	121.4	126.3	131.3	136.6	142.0	147.7	153.6	159.8
2. Petroleum	100.0	81.4	90.1	96.5	101.9	107.6	113.6	120.0	126.7	133.8	146.8	161.0
3. Intermediate Goods	100.0	108.3	115.1	116.8	121.1	125.4	129.9	134.6	139.5	144.5	151.1	158.1
4. Capital Goods	100.0	108.3	115.1	116.8	121.1	125.4	129.9	134.6	139.5	144.5	151.1	158.1

Imports in Current Mln. US$	1987	1988	1989	1990	1991	1992	1993	1994	1995	1996	1997	1998
1. Food & Consumer Goods	320	251	216	224	240	256	277	299	324	351	381	415
2. Petroleum	657	383	352	378	400	424	449	476	504	535	589	649
3. Intermediate Goods	3,987	3,709	3,268	3,407	3,720	4,062	4,568	5,137	5,777	6,497	7,387	8,399
4. Capital Goods	856	980	864	1,074	1,282	1,475	1,715	1,991	2,264	2,572	2,963	3,411

Import Totals in Mln. US$	1987	1988	1989	1990	1991	1992	1993	1994	1995	1996	1997	1998
Imports of Goods (Constant)	5,820	5,028	4,167	4,418	4,722	5,012	5,442	5,911	6,392	6,915	7,499	8,134
Imports of G&NFS (Constant)	8,217	7,310	6,059	6,375	6,761	7,137	7,679	8,266	8,881	9,546	10,294	11,104
Imports of Goods (Current)	5,820	5,324	4,700	5,083	5,642	6,217	7,009	7,903	8,869	9,955	11,321	12,874
Imports of G&NFS (Current)	8,217	7,789	7,274	7,507	8,278	9,081	10,154	11,356	12,675	14,155	15,982	18,042

Terms of Trade Index (1987=100)	1987	1988	1989	1990	1991	1992	1993	1994	1995	1996	1997	1998
Export Price Index	100.0	121.7	126.0	123.6	129.7	136.0	142.5	149.5	156.8	164.6	169.6	175.0
Import Price Index	100.0	105.9	112.8	115.0	119.5	124.1	128.8	133.7	138.8	144.0	151.0	158.3
Terms of Trade Index	100.0	115.0	111.7	107.5	108.5	109.6	110.6	111.8	113.0	114.3	112.4	110.5

Terms of Trade Index (Annual <>)	1987	1988	1989	1990	1991	1992	1993	1994	1995	1996	1997	1998
Export Price Index		21.7%	3.5%	-1.9%	4.9%	4.9%	4.8%	4.9%	4.9%	4.9%	3.1%	3.1%
Import Price Index		5.9%	6.5%	2.0%	3.9%	3.8%	3.8%	3.8%	3.8%	3.8%	4.9%	4.8%
Terms of Trade Index		15.0%	-2.9%	-3.8%	1.0%	1.0%	0.9%	1.1%	1.1%	1.1%	-1.7%	-1.6%

B. MACROECONOMIC CONSISTENCY FRAMEWORK

10.18 This appendix analyzes the consistency between the demands on the domestic financial system generated by the consolidated public sector deficits and the desired reduction in the rate of inflation.

The Budget Constraint of the Public Sector

10.19 Public sector deficits give rise to an increase in government financial liabilities; if the growth of the liabilities is inconsistent with the increase in the demand that can be expected of the public once external financing is discounted, then the underlying assumptions on growth, inflation or real interest rates are unlikely to be realized. The demand for money is a function of, among other things, expected inflation; given a desired (or feasible) accumulation of external and domestic debt, a target for the rate of inflation will, in the long run, be compatible with only one value of the consolidated public sector deficit. In the short and medium run, of course, cost-push shocks are probably a dominant factor in determining the dynamics of the rate of inflation.

10.20 The budget constraint facing the consolidated public sector provides the starting point to assess the internal consistency of a macroeconomic program. A deficit can in fact be financed in essentially three different ways: (i) issue of monetary base; (ii) increase in net external liabilities; and (iii) increase in voluntary and involuntary domestic borrowing. The latter two include any increase in domestic and foreign arrears. For ease of discussion, one can start from the financing of the non-financial public sector (NFPS):

$$1) \qquad DEF_t + r(1 + \pi)B_{t-1} + ei^*B^* \equiv \Delta b + e\Delta B^* + \Delta CR^g$$

where DEF is the primary deficit, r is the interest rate on real domestic debt (thus equal to $(i-\pi)/(1+\pi)$), B_{t-1} is the outstanding (beginning-of-period) stock of domestic debt, Δb is the real change in government debt (equal to $B_t - B_{t-1}(1+\pi)$),[1] e is the exchange rate, i^* is the foreign interest rate, B^* is the outstanding stock of net foreign liabilities of the NFPS, and CR^g is the credit of the Central Bank to the NFPS (a cross product term πrB is ignored for simplicity).1/ Eq. (1) follows the current Argentine methodology, i.e. including in the definition of the deficit only the real part of the domestic interest bill, on the grounds that the inflation adjustment only represents accelerated repayment of principal, and should not thus be included in the financing needs of the Government.2/

1/ Throughout the appendix, lower letters indicate real changes of the capital letter variables, i.e. $\Delta x = X_t - X_{t-1}(1+\pi)$.

2/ As argued by Olivera-Tanzi et al. (1988), this is only legitimate if the Government can indeed roll-over its stock of real debt at unchanged real interest rates. This has not been the case in Argentina during 1988 and 1989, but the issues will not be pursued any further here.

10.21 To arrive at the consolidated public sector financing needs, one must also include any profits/losses borne by the Central Bank. As discussed in detail in chapter 4, the deficit according to eq (1) has a logical counterpart in a definition of the quasi-fiscal deficit of the Central Bank encompassing the operating result (for simplicity net of operating costs that can be disregarded) minus the loss in value of net domestic assets:

2) $$QFD = iFD - ei^{*}NFA - iCR^{g} - iCR^{fs} - \pi/(1+\pi)\ (NW - eNFA)$$

where NFA are the net foreign assets of the Central Bank, CR^{g} is credit to the Government, CR^{fs} is credit to the financial sector, MB is the monetary base, equal to currency in circulation CU plus unremunerated reserves of the banking system, FD are involuntary investments of the banking system and NW is the net worth. This is equal to interest received minus interest paid minus the adjustment for inflation of the net worth (equal to net assets). The latter term is of particular importance, as it represents the inflation profit net of redistribution within the financial sector. Thus, defining the deficit in real terms implies considering the profits from the inflation tax as a revenue for the consolidated public sector, rather than a source of financing.3/

The deficit defined in (2) can be financed as follows:

3) $$QFD = \Delta fd - e\Delta NFA - \Delta cr^{g} - \Delta cr^{fs} + \Delta mb$$

i.e. through an increase in net external liabilities or in real net domestic liabilities (gross liabilities minus increases in assets). Note that this does not include non-operating changes in the balance sheet, such as valuation changes on the existing debt.

10.22 The consolidated public sector budget constraint (real definition) is obtained by aggregating (1) and (2):

4) $$DEF + (i-\pi)/(1+\pi)B + ei^{*}B^{*} + QFD \equiv \Delta b + e\Delta B^{*} + \Delta cr^{g} + \Delta fd - e\Delta NFA - \Delta cr^{g} - \Delta cr^{fs} - \Delta mb$$

or:

Primary deficit NFPS + real domestic interest payments
+ external interest payments
+ nominal quasi-fiscal deficit

3/ See L. Barbone and P. Beckerman, "Inflation, Monetary Policy and Quasi-fiscal Deficits: A Simple Model with Application to Argentina" dated March 20, 1989 for a discussion of the economic properties of alternative definitions of the quasi-fiscal deficit.

- net inflation tax/profit

equals

change in real domestic bonds in the hands of the public
- change in consolidated NFA of public sector
+ change in real monetary base
+ change in real net interest bearing liabilities of the Central Bank *vis-à-vis* the financial system.

Note that the term Δcr^g disappears, since it represents only an intra-public sector transfer, i.e. it does not involve a net creation of financial liabilities.

Argentina, 1988: How was the Deficit Financed?

10.23 The first application of the above framework consists in reconciling the ex-post accounts of the public sector. This is done in Table 1, where both the nominal and real definitions of the deficit are presented. The primary deficit (total revenues minus total non-interest expenditures) of the non-financial public sector amounted, for the whole year, to 0.9 percent of GDP. Real domestic interest of the non-financial public sector amounted to 0.6 percent of GDP, and (accrued) external interest payments of the total public sector (including interest payments of the Central Bank) would have amounted to 4.3 percent of GDP. Finally, the Central Bank is estimated to have registered a (nominal) surplus of 1.7 percent of GDP on account of interest operations. The sum of the above yielded a financing requirement of about 4.1 percent of GDP.

10.24 On the financing side, the inflation tax provided by far the largest source of support: it is estimated that, after correcting for the Central Bank's exposure to the financial system, the loss of real value of net monetary and quasi-monetary liabilities yielded the Government 4.9 percent of GDP, a tax equivalent to 23 percent of total revenues of the general government, or more than four times the total income tax collection for the year. The downside of this was of course the fact that, in order to collect the tax, the economy had to endure an average rate of inflation of over 12 percent monthly throughout the year.

10.25 Other sources of financing contributed positively on a net basis. Public external debt rose by about 1.7 percent of GDP (before valuation changes), representing about 50 percent of interest due on foreign debt.4/ The monetary base contracted sharply in real terms (minus 1.3 percent of GDP), as a result of the increased opportunity cost for holdings of domestic currency. Similarly, real holdings of government bonds with the public fell by 0.7 percent of GDP, despite increasingly higher real interest raters offered, and deep discounts granted. Central Bank liabilities, on the other hand, rose (albeit marginally). It should be remembered that most of the Central Bank's interest bearing assets are on an involuntary basis.

4/ The financing was, for the most part, involuntary, as Argentina suspended servicing of external debt to non-preferred creditors in April 1988.

10.26 As Table 1 shows, there remains a sizable unexplained statistical discrepancy (0.9 percent of GDP). Several factors may contribute an explanation, ranging from differences in recording of foreign credits, and the attendant changes in the real exchange rate, to the possibility that the deficit of the consolidated public sector may have been larger than reported. There are indications, in fact, that additional financing was accruing during the year to the Treasury via the non-payment to the Central Bank of australes corresponding to certain external interest payments made on behalf of public enterprises, for which the public enterprises had paid the Central Government.

10.27 The quarterly data shown in the table also offer interesting insights on the reasons why the Plan Primavera may have failed. As can be seen, the inflation tax fell sharply at the outset of the plan. From a high of 7.7 percent of GDP in the second quarter of the year, it was in fact reduced to less than 1.4 percent in the fourth quarter. However (in sharp contrast to what had occurred in the initial phase of the Plan Austral in 1985) the fall in the inflation tax was not compensated for by an increase in "legal" revenues. Indeed, if anything, tax and non-tax revenues registered a sharp fall, on the order of almost two percentage points over the second quarter. As a result, given the seasonal increase in expenditures associated to the end of the year, the financing requirements of the consolidated public sector (after collection of the inflation tax) rose by almost nine percentage points of GDP between the third and the fourth quarter, setting the stage for the subsequent hyperinflationary explosion.

Using the Model for Design of Stabilization

10.28 The above consistency equations can be used for assessing the internal consistency of stabilization programs. In order to do so, however, the model needs to be supplemented with a model of the dynamics of the Central Bank's balance-sheet, as well as with an estimate of the determinants of the non-interest deficit of the non-financial public sector. This section presents a re-evaluation of the design of the Plan Primavera, based on a minimal set of endogenous relationships. The inflation targets of the plan are taken as exogenous, as well as the expected NFPS deficit, the planned increase in net external debt and in domestic NFPS debt with the public, and, what is most important, the expected real rate of interest. On the other hand, the monetary sector's behavior is entirely endogenized, following the model discussed in Barbone and Beckerman (1989). In synthesis, this part of the exercise consists in checking whether sufficient allowance had been made for (voluntary) financing of the expected public sector deficit that would have prevailed during 1989.

10.29 The next section briefly discusses the main features of the model.

Demand for Assets and Evolution of the Quasi-fiscal Deficit

10.30 The starting point for the analysis is the assumption that the demand for money is a stable function of the rate of inflation; that there exists a feasible or desirable path for the growth of external indebtedness, and that there are limits to the (voluntary or involuntary) recourse to domestic bond markets for given real rates of interest. This sets a maximum to the financing sources of the deficit. The deficit, in turn, is a function of inflation and real interest rates. Lack of correspondence between the two sides of the equality indicates an inconsistency in some of the assumptions of the macroeconomic program.

Money Demand Function

10.31 The demand for monetary base is derived from the demand for various monetary aggregates, and the applicable reserve coefficients. For simplicity, the demand for base derived from aggregates other than M1 is disregarded.5/ The velocity of M1, in turn, has exhibited an upward trend throughout the 1970s and 1980s, as a result of financial innovation and of the secular dollarization of the economy. For purpose of the estimation of the model, the following regression for M1 was run with quarterly data over the period 1973-I 1988-I, constraining the income elasticity of the demand for money to one and proxying the secular increase in velocity through a trend (t-statistics in parentheses):

$$\ln(M1/GDP)_t = \underset{(2.27)}{-0.32} - \underset{(3.43)}{0.218}*INFL_t - \underset{(2.47)}{0.0043}*TREND + \underset{(9.84)}{0.75}*(M1/GDP)_{t-1} + \text{Seasonal Dummies}$$

$AdjR^2=.926$
Durbin's H =-.857
SEE = .112

The value of the long-run semi-elasticity with respect to inflation implied by the above estimates is 0.83.6/

In order to derive the demand for monetary base from the demand for M1, the simulations utilized a historic value of the M1 multiplier (approximately 0.8).

5/ The marginal reserve coefficients for aggregates other than M1 are negligible; the assumption made in the text is not utterly unrealistic.

6/ The value of semi-elasticity is more or less in line with other estimates, and it implies a unitary elasticity with respect to the rate of inflation at approximately 114 percent per quarter, or 29 percent per month. The unitary elasticity also maximizes the inflation tax; some authors have suggested even higher values for the revenue-maximizing inflation. Cf, for instance, Melvick (1988), Rodriguez (1989).

Annex Table 10.1: ARGENTINA - THE FINANCING OF THE PUBLIC SECTOR DEFICIT, 1988
(Percent of GDP - Real Definition)

	Total Revenues	Non-Int. Expend.	Primary Deficit	Domestic Interest Payments	NFPS Ext. Interest Payments	Other Interest Payments	Domestic QFD Nominal	Net Inflation Tax	Additional Financing Requirements
QI	26.95	28.98	2.03	1.21	3.16	1.30	1.49	-6.11	3.08
QII	28.79	27.46	-1.32	0.58	4.26	-0.41	-1.74	-7.65	-5.48
QII	26.43	27.08	0.65	0.01	2.08	1.75	-4.80	-4.59	-4.91
QIV	26.92	29.17	2.25	0.49	3.17	1.17	-1.61	-1.39	4.09
1988	27.27	28.17	0.90	0.57	3.17	1.16	-1.66	-4.94	0.80

Equals real change in:

	Public External Debt	Monetary Base	Central Bank Liab.	Govt. Bonds	Other Credit to Govt.b/
QI	5.48	-5.57	3.03	0.93	-0.79
QII	1.25	-3.00	-0.68	-0.63	-2.43
QIII	-2.66	-3.01	2.54	-3.48	1.70
QIV	2.58	6.30	-3.39	0.55	-1.94
1988 a/	1.67	-1.33	0.38	-0.65	-0.87

Source: Secretaria de Hacienda; BCRA; Bank Staff estimates

a/ Adjusted for intra-annual inflation.
b/ Includes statistical discrepancy.

ARGENTINA

STATISTICAL APPENDIX

TABLE OF CONTENTS

1. POPULATION

2. NATIONAL ACCOUNTS

3. BALANCE OF PAYMENTS

4. EXTERNAL DEBT

5. PUBLIC SECTOR

6. FINANCIAL SECTOR

7. AGRICULTURE

8. INDUSTRY

9. PRICES

Table 1.1: ARGENTINA - POPULATION, 1960-1988

	Census Population a/	Mid-year Population (Thousands)	Annual Growth Rate (%)	Five-year Growth Rate in the Projection
1960	20014 b/	20616	1.65	1.71
1965		22283	1.52	1.55
1970	23390 c/	23962	1.54	1.45
1971		24352	1.63	
1972		24764	1.69	
1973		25189	1.72	
1974		25621	1.71	
1975		26052	1.68	1.67
1976		26480	1.65	
1977		26912	1.63	
1978		27348	1.62	
1979		27789	1.61	
1980	27947	28237	1.61	1.61
1981		28694	1.62	
1982		29158	1.62	
1983		29627	1.61	
1984		30097	1.59	
1985		30564	1.55	1.58
1986		31030	1.52	
1987		31497	1.50	
1988		31969	1.50	

Source: INDEC.

a/ 1980 census data. Next census will be held in 1990.

b/ Census omissions are not included.

c/ Results obtained from sample.

June 1989

Table 1.2: ARGENTINA - DISTRIBUTION OF POPULATION BY JURISDICTION, 1980 a/

	Population (Thousands)	Area (Thousands of sq. kms.)	Density
TOTAL b/	27947	2780.2	10.1
Capital Federal	2923	0.2	14615.0
Buenos Aires	10865	307	35.4
Catamarca	208	101	2.1
Cordoba	2408	169	14.2
Corrientes	662	88	7.5
Chaco	701	99	7.1
Chubut	263	225	1.2
Entre Rios	908	79	11.5
Formosa	296	72	4.1
Jujuy	410	53	7.7
La Pampa	208	143	1.5
La Rioja	164	90	1.8
Mendoza	1196	149	8.0
Misiones	589	30	19.6
Neuquen	244	94	2.6
Rio Negro	383	203	1.9
Salta	663	155	4.3
San Juan	466	90	5.2
San Luis	214	77	2.8
Santa Cruz	115	244	0.5
Santa Fe	2466	133	18.5
Santiago del Estero	595	135	4.4
Tucuman	973	23	42.3
Tierra del Fuego	27	21	1.3

Source: INDEC.

a/ 1980 census data. Next census will be held in 1990.

b/ Not including Falkland Islands (Malvinas) and dependencies, South Sandwich Islands South Georgia, and Antarctic Argentina.

July 1987

Table 1.3: ARGENTINA - UNEMPLOYMENT AND UNDEREMPLOYMENT RATES, 1975-1989 a/
(Percent of Economically Active Population)

	1975		1976		1977		1978		1979		1980		1981		1982		1983		1984		1985		1986		1987		1988		1989
	Apr.	Oct.	Apr.	Oct.	Apr.	Oct.	Apr.	Oct.	Apr.	Oct.	Apr.	Oct.	Apr.	Oct.	Apr.	Oct.	Apr.	Oct.	Apr.	Oct.	Apr.	Oct.	Apr.	Oct.	Apr.	Oct.	Apr.	Oct.	May
TOTAL UNEMPLOYMENT b/	..	3.8	5.2	4.4	3.9	2.7	4.2	2.3	2.6	2.4	2.6	2.5	4.2	5.3	6.0	4.6	5.5	3.9	4.7	4.4	6.3	5.9	..	5.2	6.0	5.7	6.5	6.1	..
Gr. Buenos Aires	2.4	2.8	4.8	4.1	3.4	2.2	3.9	1.7	2.0	2.0	2.3	2.2	4.0	5.0	5.7	3.8	5.2	3.1	4.0	3.6	5.7	4.9	4.8	4.4	5.4	5.2	6.3	5.7	7.7
Gran Rosario	5.3	5.7	5.3	4.1	3.5	2.6	5.5	2.3	3.1	2.7	4.3	2.4	4.9	6.5	8.5	8.0	6.3	..	7.0	6.2	10.9	10.2	6.8	7.2	7.3	8.3	7.8	7.0	..
Cordoba	6.1	7.2	6.5	5.4	5.9	4.0	5.1	2.7	2.6	1.8	2.1	2.7	2.9	4.7	4.8	3.9	4.4	5.6	4.4	5.1	5.3	4.7	6.4	5.1	4.9	5.5	5.0	5.9	..
Gr. San Miguel de Tucuman	8.4	6.9	7.4	5.6	7.3	4.3	6.8	4.9	5.9	4.9	6.3	8.3	8.8	10.6	11.0	8.7	8.1	7.5 2/	8	10.6	12.2	11.4	13.6	12.5	15.1	9.8	11.3	10.1	..
Mendoza	4.1	4.4	5.9	4.8	4.4	4.4	2.9	3.5	2.8	3.4	1.4	3.1	4.2	5.3	4.8	3.3	4.5	4.5	3.3	3.7	3.6	3.7	4.9	3.3	3.6	3.1	4.7	4.0	..
TOTAL UNDEREMPLOYMENT c/	..	..	..	5.3	4.1	3.8	5.5	3.8	3.9	3.6	4.5	5.8	5.0	6.0	6.7	6.4	5.9	5.9	5.4	5.9	7.5	7.2	..	7.3	8.2	8.1	7.8	7.9	..

Source: INDEC.

a/ Includes 27 urban centers in 1979 to 1981, and 26 since 1982.

b/ Includes Taji-Viejo.

c/ Persons employed less than 30 hours per week.

August 1989

Table 1.4: ARGENTINA - POPULATION AGE GROUPS AND SETTLEMENT STRUCTURE, 1980 a/
(Thousands)

Age Group	Total	Male	Female	Urban				Rural			
				Total	%	Male	Female	Total	%	Male	Female
Total	27947	13756	14191	23193	83.0	11215	11978	4754	17.0	2541	2213
0 - 4	3241	1640	1601	2591	79.9	1311	1280	650	20.1	329	321
5 - 9	2784	1407	1377	2213	79.5	1117	1096	571	20.5	290	281
10 - 14	2456	1240	1216	1945	79.2	973	972	511	20.8	267	244
15 - 19	2342	1174	1168	1918	81.9	941	977	424	18.1	233	191
20 - 24	2224	1100	1124	1865	83.9	909	956	359	16.1	191	168
25 - 29	2124	1050	1074	1791	84.3	870	921	333	15.7	180	153
30 - 34	1975	980	995	1672	84.7	815	857	303	15.3	165	138
35 - 39	1725	856	869	1455	84.3	709	746	270	15.7	147	123
40 - 44	1549	773	776	1301	84.0	636	665	248	16.0	137	111
45 - 49	1497	748	749	1267	84.6	619	648	230	15.4	129	101
50 - 54	1458	709	749	1243	85.3	589	654	215	14.7	120	95
55 - 59	1281	621	660	1098	85.7	518	580	183	14.3	103	80
60 - 64	1002	470	532	854	85.2	387	467	148	14.8	83	65
65 - 69	874	398	476	751	85.9	329	422	123	14.1	69	54
70 - 74	634	279	355	549	86.6	232	317	85	13.4	47	38
75 - 79	426	181	245	370	86.9	151	219	56	13.1	30	26
80 - 84	224	86	138	196	87.5	72	124	28	12.5	14	14
85 and more	131	44	87	114	87.0	37	77	17	13.0	7	10

Source: INDEC.
a/ 1980 census data. Next census will be held in 1990.

May 1987

Table 1.5: ARGENTINA - ENROLLMENT BY LEVEL AND TYPE OF EDUCATION, 1985
(Thousands)

Level	Total		National		Provincial		Municipal		Private	
	No.	%	No.	%	No.	%	No.	%	No.	%
Pre-Primary and Primary	5505	100	143	2.6	4038	73.4	242	4.4	1082	19.7
Secondary and Post-Secondary	1684	100	756	44.9	433	25.7	2	0.1	493	29.3
Non-University	182	100	58	31.9	63	34.6	1	0.5	60	33.0
University	664	100	586	88.3	2	0.3	..	0.0	76	11.4

Source: Ministry of Education and Justice, Argentina.

May 1987

Table 1.6: ARGENTINA - SCHOOL ENROLLMENT AND LITERACY BY AGE GROUP, 1980 a/
(Thousands)

Age and Sex	Population: 5 years and more	Total	Currently Enrolled				Graduated b/	Not Attended	Illiterates	% of Attendance		% of Literacy
			Pre-school	Elementary	Secondary	Higher				Current c/	Total d/	
TOTAL	24706	6374	462	4132	1333	447	16943	1389	1264	25.8	94.4	94.9
5 - 9	2784	2376	462	1914	0	0	56	352	0	85.3	87.4	100.0
10 - 14	2456	2205	0	1829	376	0	224	27	80	89.8	98.9	96.7
15 - 19	2342	994	0	132	778	84	1309	38	70	42.4	98.3	97.0
20 - 24	2224	338	0	31	83	224	1839	47	72	15.2	97.9	96.8
25 - 29	2124	145	0	29	30	86	1919	59	83	6.8	97.2	96.1
30 - 34	1975	74	0	29	18	27	1834	68	93	3.7	96.6	95.3
35 - 39	1725	51	0	27	12	12	1606	68	91	3.0	96.1	94.7
40 - 44	1549	40	0	25	9	6	1438	70	87	2.6	95.4	94.4
45 - 49	1497	28	0	18	7	3	1393	75	88	1.9	94.9	94.1
50 and more	6030	123	0	98	20	5	5325	585	600	2.0	90.3	90.0
Males	12116	3201	232	2102	646	221	8280	635	589	26.4	94.8	95.1
Females	12590	3173	230	2030	687	226	8663	754	675	25.2	94.0	94.6

Source: INDEC.
a/ 1980 census data. Next census will be held in 1990.
b/ Includes those who have not completed.
c/ Current enrollment/age group population.
d/ Current enrollment plus graduates/age group population.

April 1987

Table 1.7: ARGENTINA - EDUCATION LEVEL OF POPULATION BY AGE GROUPS, 1980 a/
(Percent)

Education Level	14 + Population Attending / Attended School	AGE GROUPS									
		14-19	20-24	25-29	30-34	35-39	40-44	45-49	50-54	55-59	60 & more
TOTAL	100.0	100.0	100.0	100.0	100.0	100.0	100.0	100.0	100.0	100.0	100.0
Elementary	64.7	46.2	48.9	54.8	59.9	65.0	69.6	74.9	78.2	80.3	84.2
Incomplete	31.5	22.0	17.3	20.9	25.4	29.6	33.0	37.3	41.2	44.2	50.8
Complete	33.2	24.2	31.5	33.9	34.5	35.4	36.6	37.6	37.0	36.1	33.4
Secondary	27.6	50.5	36.1	30.9	28.8	26.2	23.6	19.7	17.1	15.7	12.6
Incomplete	17.3	46.8	20.8	16.7	14.7	13.5	11.8	9.7	8.1	7.3	5.9
Complete	10.3	3.7	15.3	14.3	14.1	12.7	11.8	10.0	9.0	8.4	6.8
Superior	1.7	1.1	4.0	3.0	2.3	1.6	1.1	1.1	1.0	0.9	0.7
Incomplete	0.7	1.1	2.6	1.0	0.6	0.4	0.3	0.2	0.1	0.2	0.1
Complete	1.0	0.0	1.4	1.9	1.7	1.2	0.9	0.8	0.9	0.8	0.6
University	6.0	2.2	11.0	11.3	9.0	7.2	5.6	4.4	3.6	3.0	2.4
Incomplete	3.5	2.2	10.2	7.1	4.0	2.9	2.3	1.7	1.3	1.0	0.7
Complete	2.5	0.0	0.8	4.2	5.0	4.3	3.3	2.7	2.3	2.0	1.7

Source: INDEC.
a/ 1980 census data. Next census will be held in 1990.

April 1987

Table 1.8: ARGENTINA - ECONOMICALLY ACTIVE POPULATION, 1980 a/
(Thousands)

Age	14 + Population	Economically Active	Economically not Active					Activity Ratio
			Total	Retired	Students	House-care	Other	
Total	19936	10034	9901	1944.3	1518	5449	990	50.3
14	470	67	403	0.1	338	27	38	14.3
15	481	122	359	0.1	280	39	40	25.4
16	473	165	307	0.2	224	46	37	34.9
17	470	192	277	0.2	190	54	33	40.9
18	473	229	244	0.1	133	65	46	48.4
19	446	222	223	0.1	86	72	65	49.8
20	447	268	178	0.2	65	83	30	60.0
21	432	275	158	0.3	48	87	23	63.7
22	447	287	159	0.3	38	100	21	64.2
23	456	297	158	0.3	29	110	19	65.1
24	443	289	154	0.4	22	116	16	65.2
25 - 29	2124	1388	736	3	45	622	66	65.3
30 - 34	1975	1287	687	5	9	620	53	65.2
35 - 39	1725	1117	607	10	4	549	44	64.8
40 - 44	1549	990	560	19	2	496	43	63.9
45 - 49	1497	917	580	43	2	489	46	61.3
50 - 54	1458	812	648	95	2	496	55	55.7
55 - 59	1281	598	683	192	1	433	57	46.7
60 - 64	1002	296	706	324	0	328	54	29.5
65 - 69	873	136	738	419	0	267	52	15.6
70 - 74	634	53	581	362	0	171	48	8.4
75 +	780	27	753	470	0	179	104	3.5

Source: INDEC.
a/ 1980 census data. Next census will be held in 1990.

May 1987

Table 2.1: ARGENTINA - GROSS DOMESTIC PRODUCT BY SECTORAL ORIGIN, 1970-1988
(1970 Australes)

	1970	1971	1972	1973	1974	1975	1976	1977	1978	1979	1980	1981	1982	1983	1984	1985	1986	1987	1988
GDP AT MARKET PRICES	8775	9105	9294	9642	10163	10103	10102	10747	10400	11130	11295	10543	10021	10321	10585	10105	10656	10863	10531
NET INDIRECT TAXES	1000	1038	1061	1100	1159	1152	1152	1226	1186	1269	1288	1203	1142	1177	1207	1152	1215	1238	1201
GDP AT FACTOR COST	7774	8067	8233	8542	9004	8951	8950	9522	9214	9861	10008	9340	8879	9144	9378	8953	9441	9624	9330
Agriculture	1023	1039	1059	1173	1205	1172	1227	1257	1292	1329	1256	1280	1370	1403	1446	1421	1375	1416	1421
Mining	178	192	197	192	196	193	198	215	219	233	246	248	249	250	248	242	232	233	255
Manufacturing	2099	2228	2317	2409	2550	2485	2410	2598	2325	2556	2465	2076	1970	2170	2253	2020	2280	2267	2109
Construction	503	534	529	467	504	527	606	680	647	644	652	567	437	380	304	284	311	357	305
Electricity, Gas and Water	181	198	217	234	248	263	272	285	295	326	351	347	358	387	412	418	449	476	499
Commerce, Restaurants and Hotels	1183	1221	1242	1273	1349	1342	1297	1401	1314	1491	1619	1464	1261	1315	1374	1260	1370	1390	1305
Transport and Communication	881	895	893	942	967	959	953	1004	982	1054	1061	1019	993	1034	1090	1049	1102	1131	1070
Banking	592	598	585	611	692	634	607	692	738	797	895	847	750	698	707	698	747	768	762
Public and Private Services a/	1135	1163	1193	1242	1293	1376	1380	1390	1403	1431	1462	1491	1490	1508	1545	1562	1574	1587	1605

Source: Central Bank of the Republic of Argentina (BCRA).
a/ 1970-1975 figures for public and private services are estimates.

August 1989

Table 2.2: ARGENTINA - GROSS DOMESTIC PRODUCT BY SECTORAL ORIGIN, 1970-1988
(Growth Rates)

	1970	1971	1972	1973	1974	1975	1976	1977	1978	1979	1980	1981	1982	1983	1984	1985	1986	1987	1988
GDP AT MARKET PRICES		3.8	2.1	3.7	5.4	-0.6	0.0	6.4	-3.2	7.0	1.5	-6.7	-5.0	3.0	2.6	-4.5	5.5	1.9	-3.1
NET INDIRECT TAXES		3.7	2.2	3.6	5.4	-0.7	0.0	6.4	-3.3	7.0	1.5	-6.6	-5.0	3.0	2.5	-4.5	5.5	1.9	-3.0
GDP AT FACTOR COST		3.8	2.1	3.8	5.4	-0.6	0.0	6.4	-3.2	7.0	1.5	-6.7	-4.9	3.0	2.6	-4.5	5.5	1.9	-3.1
Agriculture		1.6	1.9	10.7	2.7	-2.7	4.7	2.4	2.8	2.9	-5.5	1.9	7.0	2.4	3.1	-1.7	-3.2	3.0	0.4
Mining		7.8	2.9	-2.8	2.4	-1.5	2.4	8.5	1.9	6.3	5.8	0.6	0.5	0.2	-0.6	-2.6	-3.8	0.2	9.5
Manufacturing		6.1	4.0	4.0	5.9	-2.6	-3.0	7.8	-10.5	10.0	-3.6	-15.8	-5.1	10.2	3.8	-10.3	12.9	-0.6	-7.0
Construction		6.3	-1.0	-11.7	7.9	4.6	14.9	12.2	-4.8	-0.5	1.1	-13.0	-22.9	-13.1	-20.0	-6.7	9.7	14.7	-14.5
Electricity, Gas and Water		9.3	9.9	7.5	5.9	6.1	3.7	4.6	3.3	10.7	7.8	-1.1	3.1	8.0	6.5	1.4	7.4	6.1	4.9
Commerce, Restaurants and Hotels		3.2	1.7	2.5	6.0	-0.5	-3.4	8.0	-6.2	13.5	8.6	-9.6	-13.9	4.3	4.5	-8.3	8.7	1.4	-6.1
Transport and Communication		1.6	-0.2	5.5	2.7	-0.8	-0.7	5.4	-2.2	7.3	0.7	-4.0	-2.6	4.1	5.5	-3.8	5.1	2.7	-5.4
Banking		1.1	-2.2	4.5	13.2	-8.4	-4.2	13.8	6.7	8.0	12.3	-5.3	-11.5	-7.0	1.2	-1.2	7.0	2.8	39.3
Public and Private Services a/		2.4	2.6	4.1	4.1	6.4	0.3	0.8	0.9	2.0	2.2	2.0	-0.1	1.2	2.5	1.1	0.8	0.8	-52.0

Source: Table 2.1
a/ 1970-1975 figures for public and private services are estimates.

August 1989

Table 2.3: ARGENTINA - GROSS DOMESTIC PRODUCT BY SECTORAL ORIGIN, 1970-1988
(Percent of GDP at Factor Cost, 1970 Prices)

	1970	1971	1972	1973	1974	1975	1976	1977	1978	1979	1980	1981	1982	1983	1984	1985	1986	1987	1988
GDP AT MARKET PRICES	112.9	112.9	112.9	112.9	112.9	112.9	112.9	112.9	112.9	112.9	112.9	112.9	112.9	112.9	112.9	112.9	112.9	112.9	112.9
NET INDIRECT TAXES	12.9	12.9	12.9	12.9	12.9	12.9	12.9	12.9	12.9	12.9	12.9	12.9	12.9	12.9	12.9	12.9	12.9	12.9	12.9
GDP AT FACTOR COST	100.0	100.0	100.0	100.0	100.0	100.0	100.0	100.0	100.0	100.0	100.0	100.0	100.0	100.0	100.0	100.0	100.0	100.0	100.0
Agriculture	13.2	12.9	12.9	13.7	13.4	13.1	13.7	13.2	14.0	13.5	12.5	13.7	15.4	15.3	15.4	15.9	14.6	14.7	15.2
Mining	2.3	2.4	2.4	2.2	2.2	2.2	2.2	2.3	2.4	2.4	2.5	2.7	2.8	2.7	2.6	2.7	2.5	2.4	2.7
Manufacturing	27.0	27.6	28.1	28.2	28.3	27.8	26.9	27.3	25.2	25.9	24.6	22.2	22.2	23.7	24.0	22.6	24.2	23.6	22.6
Construction	6.5	6.6	6.4	5.5	5.6	5.9	6.8	7.1	7.0	6.5	6.5	6.1	4.9	4.2	3.2	3.2	3.3	3.7	3.3
Electricity, Gas and Water	2.3	2.5	2.6	2.7	2.8	2.9	3.0	3.0	3.2	3.3	3.5	3.7	4.0	4.2	4.4	4.7	4.8	4.9	5.3
Commerce, Restaurants and Hotels	15.2	15.1	15.1	14.9	15.0	15.0	14.5	14.7	14.3	15.1	16.2	15.7	14.2	14.4	14.6	14.1	14.5	14.4	14.0
Transport and Communication	11.3	11.1	10.8	11.0	10.7	10.7	10.6	10.5	10.7	10.7	10.6	10.9	11.2	11.3	11.6	11.7	11.7	11.8	11.5
Banking	7.6	7.4	7.1	7.2	7.7	7.1	6.8	7.3	8.0	8.1	8.9	9.1	8.4	7.6	7.5	7.8	7.9	8.0	11.5
Public and Private Services a/	14.6	14.4	14.5	14.5	14.4	15.4	15.4	14.6	15.2	14.5	14.6	16.0	16.8	16.5	16.5	17.4	16.7	16.5	8.2

Source: Table 2.1.

a/ 1970-1975 figures for public and private services are estimates.

August 1989

Table 2.4: ARGENTINA - GROSS DOMESTIC PRODUCT BY SECTORAL ORIGIN, 1970-1988 a/
(Thousands of Australes)

	1970	1971	1972	1973	1974	1975	1976	1977	1978	1979	1980	1981	1982	1983	1984	1985	1986	1987	1988 b/
GROSS DOMESTIC PRODUCT	8.775	12.518	20.690	35.485	48.647	143.0	758.7	2093.4	5234	14251	28336	54752	147613	682652	5281000	39592600	74309000	173109400	821468000
Agriculture	1.068	1.797	3.070	5.610	6.604	12.7	83.3	229.3	531	1504	2436	4939	17383	86377	668214	5010694	9404274	21908089	103961967
Mining	0.176	0.246	0.368	0.721	1.028	2.8	14.8	25.8	101	343	669	1590	4514	26955	208857	1565842	2938836	6846278	32488115
Manufacturing	2.641	3.781	6.349	10.119	13.929	46.4	250.5	650.1	1513	3951	7082	13147	41918	210154	1625994	12190368	22879379	53299540	252925990
Construction	0.507	0.749	1.183	1.794	2.743	11.2	54.8	147.6	381	991	2000	3306	7206	42344	327210	2453152	4604176	10725835	50898047
Electricity	0.204	0.266	0.429	0.743	1.035	2.6	19.1	55.3	145	304	718	1632	2513	17216	133050	997499	1872147	4361332	20696132
Commmerce	1.464	2.002	3.474	5.299	7.440	22.4	132.0	351.5	884	2522	5106	9716	24499	112520	870239	6524341	12245148	28526157	135367145
Transport	0.840	1.064	1.733	2.783	3.808	11.2	60.0	175.3	443	1143	2171	4414	9244	38661	299362	2244373	4212331	9812998	46566298
Banking	0.706	0.990	1.590	3.492	4.549	11.8	58.5	234.1	562	1659	3885	7583	21383	43522	336493	2522745	4734791	11030114	52341963
Government	1.169	1.623	2.494	4.924	7.511	22.0	85.7	224.4	674	1835	4270	8426	18952	104903	811450	6083585	11417920	26599057	126222342

Source: Central Bank of the Republic of Argentina (BCRA).
a/ GDP distribution by sector for 1984-1988 are estimates.

August 1989

Table 2.5: ARGENTINA - GROSS DOMESTIC PRODUCT BY SECTORAL ORIGIN, 1970-1988 a/
(Percent)

	1970	1971	1972	1973	1974	1975	1976	1977	1978	1979	1980	1981	1982	1983	1984	1985	1986	1987	1988
GROSS DOMESTIC PRODUCT	100.0	100.0	100.0	100.0	100.0	100.0	100.0	100.0	100.0	100.0	100.0	100.0	100.0	98.6	100.0	100.0	100.0	100.0	100.0
Agriculture	12.2	14.4	14.8	15.8	13.6	8.9	11.0	11.0	10.1	10.6	8.6	9.0	11.8	12.7	12.7	12.7	12.7	12.7	12.7
Mining	2.0	2.0	1.8	2.0	2.1	2.0	2.0	1.2	1.9	2.4	2.4	2.9	3.1	3.9	4.0	4.0	4.0	4.0	4.0
Manufacturing	30.1	30.2	30.7	28.5	28.6	32.4	33.0	31.1	28.9	27.7	25.0	24.0	28.4	30.8	30.8	30.8	30.8	30.8	30.8
Construction	5.8	6.0	5.7	5.1	5.6	7.8	7.2	7.0	7.3	7.0	7.1	6.0	4.9	6.2	6.2	6.2	6.2	6.2	6.2
Electricity	2.3	2.1	2.1	2.1	2.1	1.8	2.5	2.6	2.8	2.1	2.5	3.0	1.7	2.5	2.5	2.5	2.5	2.5	2.5
Commmerce	16.7	16.0	16.8	14.9	15.3	15.7	17.4	16.8	16.9	17.7	18.0	17.7	16.6	16.5	16.5	16.5	16.5	16.5	16.5
Transport	9.6	8.5	8.4	7.8	7.8	7.8	7.9	8.4	8.5	8.0	7.7	8.1	6.3	5.7	5.7	5.7	5.7	5.7	5.7
Banking	8.0	7.9	7.7	9.8	9.4	8.2	7.7	11.2	10.7	11.6	13.7	13.8	14.5	5.0	6.4	6.4	6.4	6.4	6.4
Government	13.3	13.0	12.1	13.9	15.4	15.4	11.3	10.7	12.9	12.9	15.1	15.4	12.8	15.4	15.4	15.4	15.4	15.4	15.4

Source: Table 2.4.
a/ GDP distribution by sector for 1984-1988 are estimates.

August 1989

Table 2.6: ARGENTINA - GROSS DOMESTIC PRODUCT BY EXPENDITURE, 1970-1988
(1970 Australes)

	1970	1971	1972	1973	1974	1975	1976	1977	1978	1979	1980	1981	1982	1983	1984	1985	1986	1987	1988
Gross Domestic Product	8775	9105	9294	9642	10163	10103	10102	10747	10400	11130	11300	10542	10018	10311	10565	10102	10648	10870	10531
Terms of Trade Effect	0	107	134	302	138	-26	-118	2	-84	58	258	328	33	22	145	-11	-117	-479	-404
Gross Domestic Income	8775	9212	9428	9944	10301	10077	9984	10749	10316	11188	11558	10870	10051	10333	10710	10091	10531	10391	10127
Imports of Goods and NFS a/	789	883	840	829	874	890	703	964	914	1413	2068	1869	1077	1026	1089	932	1101	1161	1024
Exports of Goods and NFS a/	810	729	744	848	850	771	1014	1291	1406	1361	1291	1362	1410	1519	1508	1698	1560	1563	1752
Exports Adjusted by Terms of Trade	810	836	878	1150	988	745	896	1293	1322	1419	1549	1690	1443	1541	1653	1687	1443	1084	1348
Resource Gap b/	-21	47	-38	-321	-114	145	-193	-329	-408	-6	519	179	-366	-515	-564	-755	-342	77	-324
Total Expenditures	8754	9259	9390	9623	10187	10222	9791	10420	9908	11182	12077	11049	9685	9818	10146	9336	10189	10468	9802
Consumption	6843	7196	7305	7636	8179	8177	7605	7776	7685	8733	9403	9000	8041	8348	8839	8291	8954	9038	8486
Public	908	949	932	992	1069	1070	1119	1167	1204	1271	1288	1308	1215	1248	1274	1252	1318	..	..
Private	5935	6247	6373	6644	7110	7107	6486	6609	6481	7462	8115	7692	6826	7100	7565	7039	7636	..	..
Gross Domestic Investment	1860	2063	2085	1967	2009	2045	2186	2644	2223	2450	2675	2049	1645	1470	1306	1046	1235	1430	1316
Changes in Inventories	-1	51	50	100	46	78	14	18	-67	5	99	-73	107	12	-20	-126	-28	-21	59
Gross Domestic Fixed Investment	1861	2012	2035	1867	1963	1967	2172	2626	2290	2445	2576	2122	1538	1458	1326	1172	1263	1451	1258
Public	708	780	795	661	684	778	989	1219	1068	1017	1010	907	712	671	491	410	515	..	..
Private	1153	1232	1240	1226	1279	1189	1183	1407	1222	1428	1566	1215	826	787	835	762	748	..	..
Gross Domestic Savings c/	1882	2016	2123	2308	2123	1900	2379	2973	2631	2456	2156	1670	2011	1985	1870	1801	1577	1354	1641
Net Factor Income Payments	-96	-39	-120	-145	-117	-124	-127	-143	-174	-209	-308	-620	-766	-854	-857	-712	-574	-578	-669
Net Transfers	-1	-1	-1	3	0	1	4	7	10	6	4	-3	4	2	0	0	0	-1	0
Gross National Savings	1784	1976	2002	2166	2006	1778	2256	2836	2466	2253	1851	1247	1249	1133	1013	1089	1003	775	972
Gross National Product	8677	9065	9173	9500	10046	9980	9979	10611	10236	10927	10995	9918	9256	9459	9708	9390	10074	10291	9862
Gross National Income	8678	9172	9307	9802	10184	9954	9861	10613	10152	10985	11253	10246	9289	9481	9853	9379	9957	9812	9458

Source: Central Bank of the Republic of Argentina (BCRA) and IBRD staff estimates.
a/ Balance of Payments figures deflated by respective price indices and converted at 1970 exchange rate (3.8 pesos per US$)
b/ - (Exports adjusted by terms of trade - imports).
c/ Gross domestic investment - resource gap.

August 1989

Table 2.7: ARGENTINA - GROSS DOMESTIC PRODUCT BY EXPENDITURE, 1970-1988
(Growth Rates)

	1971	1972	1973	1974	1975	1976	1977	1978	1979	1980	1981	1982	1983	1984	1985	1986	1987	1988
Gross Domestic Product	3.8	2.1	3.7	5.4	-0.6	0.0	6.4	-3.2	7.0	1.5	-6.7	-5.0	2.9	2.5	-4.4	5.4	2.1	-3.1
Terms of Trade Effect		25.5	124.8	-54.3	-118.8	-353.8	101.7	-4300.0	169.0	344.8	27.1	-89.9	-33.3	559.1	-107.6	-963.6	-309.1	15.6
Gross Domestic Income	5.0	2.3	5.5	3.6	-2.2	-0.9	7.7	-4.0	8.5	3.3	-6.0	-7.5	2.8	3.7	-5.8	4.4	-1.3	-2.5
Imports of Goods and NFS	11.9	-4.9	-1.3	5.4	1.8	-21.0	37.1	-5.2	54.6	46.4	-9.6	-42.4	-4.7	6.1	-14.4	18.1	5.4	-11.8
Exports of Goods and NFS	-10.0	2.1	14.0	0.2	-9.3	31.5	27.3	8.9	-3.2	-5.1	5.5	3.5	7.7	-0.7	12.6	-8.1	0.2	12.1
Exports Adjusted by Terms of Trade	3.2	5.1	30.9	-14.1	-24.6	20.3	44.3	2.2	7.3	9.2	9.1	-14.6	6.8	7.3	2.1	-14.5	-24.9	24.4
Resource Gap	322.3	-181.6	-737.4	64.5	227.2	-233.1	-70.5	-24.0	98.5	8750.0	-65.5	-304.5	-40.7	-9.5	-33.9	54.7	122.4	522.7
Total Expenditures	5.8	1.4	2.5	5.9	0.3	-4.2	6.4	-4.9	12.9	8.0	-8.5	-12.3	1.4	3.3	-8.0	9.1	2.7	-6.4
Consumption	5.2	1.5	4.5	7.1	0.0	-7.0	2.2	-1.2	13.6	7.7	-4.3	-10.7	3.8	5.9	-6.2	8.0	0.9	-6.1
Public	4.5	-1.8	6.4	7.8	0.1	4.6	4.3	3.2	5.6	1.3	1.6	-7.1	2.7	2.1	-1.7	5.3	..	..
Private	5.3	2.0	4.3	7.0	0.0	-8.7	1.9	-1.9	15.1	8.8	-5.2	-11.3	4.0	6.5	-7.0	8.5	..	..
Gross Domestic Investment	10.9	1.1	-4.7	1.1	1.8	6.9	20.9	-15.9	10.2	9.2	-23.4	-19.7	-10.6	-11.1	-19.9	18.1	15.8	-8.0
Changes in Inventories	8616.7	-2.2	99.2	-54.3	72.3	-81.9	25.4	-478.7	107.3	1914.3	-173.8	246.4	-89.1	-272.4	-530.0	77.8	26.1	382.6
Gross Domestic Fixed Investment	8.1	1.1	-7.3	4.0	0.2	10.4	20.9	-12.8	6.8	5.4	-17.6	-27.5	-5.2	-9.1	-11.6	7.8	14.9	-13.3
Public	10.2	1.9	-16.9	3.5	13.7	27.1	23.3	-12.4	-4.8	-0.7	-10.2	-21.5	-5.8	-26.8	-16.5	25.6	..	..
Private	6.9	0.6	-1.1	4.3	-7.0	-0.5	18.9	-13.1	16.9	9.7	-22.4	-32.0	-4.7	6.1	-8.7	-1.8	..	..
Gross Domestic Savings	7.2	5.3	8.7	-8.0	-10.5	25.2	24.9	-11.5	-6.6	-12.2	-13.2	7.5	-1.3	-5.8	-3.7	-12.4	-14.2	21.2
Net Factor Income Payments	59.4	-207.7	-20.8	19.3	-6.0	-2.4	-12.6	-21.7	-20.1	-47.4	-101.3	-23.5	-11.5	-0.4	16.9	19.4	-0.7	-15.7
Net Transfers	3.2	-27.7	343.2	-100.0	0.0	244.4	62.2	43.6	-35.2	-42.4	-187.7	242.4	-50.7	-81.8	-100.0	0.0	-489.5	-100.0
Gross National Savings	10.7	1.3	8.2	-7.4	-11.4	26.9	25.7	-13.1	-8.6	-17.8	-32.6	0.2	-9.3	-10.5	7.5	-7.9	-22.8	25.5
Gross National Product	4.5	1.2	3.6	5.7	-0.7	0.0	6.3	-3.5	6.8	0.6	-9.8	-6.7	2.2	2.6	-3.3	7.3	2.2	-4.2
Gross National Income	5.7	1.5	5.3	3.9	-2.3	-0.9	7.6	-4.3	8.2	2.4	-8.9	-9.3	2.1	3.9	-4.8	6.2	-1.5	-3.6

Source: Table 2.6.

August 1989

Table 2.8: ARGENTINA - GROSS DOMESTIC PRODUCT BY EXPENDITURE, 1970-1988
(Percent of GDP, 1970 Prices)

	1970	1971	1972	1973	1974	1975	1976	1977	1978	1979	1980	1981	1982	1983	1984	1985	1986	1987	1988
Gross Domestic Product	100.0	100.0	100.0	100.0	100.0	100.0	100.0	100.0	100.0	100.0	100.0	100.0	100.0	100.0	100.0	100.0	100.0	100.0	100.0
Terms of Trade Effect	0.0	1.2	1.4	3.1	1.4	-0.3	-1.2	0.0	-0.8	0.5	2.3	3.1	0.3	0.2	1.4	-0.1	-1.1	-4.4	-3.8
Gross Domestic Income	100.0	101.2	101.4	103.1	101.4	99.7	98.8	100.0	99.2	100.5	102.3	103.1	100.3	100.2	101.4	99.9	98.9	95.6	96.2
Imports of Goods and NFS	9.0	9.7	9.0	8.6	8.6	8.8	7.0	9.0	8.8	12.7	18.3	17.7	10.8	10.0	10.3	9.2	10.3	10.7	9.7
Exports of Goods and NFS	9.2	8.0	8.0	8.8	8.4	7.6	10.0	12.0	13.5	12.2	11.4	12.9	14.1	14.7	14.3	16.8	14.7	14.4	16.6
Exports Adjusted by Terms of Trade	9.2	9.2	9.5	11.9	9.7	7.4	8.9	12.0	12.7	12.7	13.7	16.0	14.4	14.9	15.6	16.7	13.6	10.0	12.8
Resource Gap	-0.2	0.5	-0.4	-3.3	-1.1	1.4	-1.9	-3.1	-3.9	-0.1	4.6	1.7	-3.7	-5.0	-5.3	-7.5	-3.2	0.7	-3.1
Total Expenditures	99.8	101.7	101.0	99.8	100.2	101.2	96.9	97.0	95.3	100.5	106.9	104.8	96.7	95.2	96.0	92.4	95.7	96.3	93.1
Consumption	78.0	79.0	78.6	79.2	80.5	80.9	75.3	72.4	73.9	78.5	83.2	85.4	80.3	81.0	83.7	82.1	84.1	83.1	80.6
Public	10.3	10.4	10.0	10.3	10.5	10.6	11.1	10.9	11.6	11.4	11.4	12.4	12.1	12.1	12.1	12.4	12.4	..	..
Private	67.6	68.6	68.6	68.9	70.0	70.3	64.2	61.5	62.3	67.0	71.8	73.0	68.1	68.9	71.6	69.7	71.7	..	..
Gross Domestic Investment	21.2	22.7	22.4	20.6	19.8	20.2	21.6	24.6	21.4	22.0	23.7	19.4	16.4	14.3	12.4	10.4	11.6	13.2	12.5
Changes in Inventories	-0.01	0.6	0.5	1.0	0.4	0.8	0.1	0.2	-0.6	0.0	0.9	-0.7	1.1	0.1	-0.2	-1.2	-0.3	-0.2	0.6
Gross Domestic Fixed Investment	21.2	22.1	21.9	19.6	19.3	19.5	21.5	24.4	22.0	22.0	22.8	20.1	15.4	14.1	12.6	11.6	11.9	13.3	11.9
Public	8.1	8.6	8.6	6.9	6.7	7.7	9.8	11.3	10.3	9.1	8.9	8.6	7.1	6.5	4.6	4.1	4.8	..	..
Private	13.1	13.5	13.3	12.7	12.6	11.8	11.7	13.1	11.8	12.8	13.9	11.5	8.2	7.6	7.9	7.5	7.0	..	..
Gross Domestic Savings	21.4	22.1	22.8	23.9	20.9	18.8	23.6	27.7	25.3	22.1	19.1	17.7	20.1	19.2	17.7	17.8	14.8	12.5	15.6
Net Factor Income Payments	-1.1	-0.4	-1.3	-1.5	-1.2	-1.2	-1.3	-1.3	-1.7	-1.9	-2.7	-5.9	-7.6	-8.3	-8.1	-7.0	-5.4	-5.3	-6.3
Net Transfers	-0.01	-0.01	-0.02	0.04	0.00	0.01	0.04	0.06	0.09	0.06	0.03	-0.03	0.04	0.02	0.0	0.0	0.0	0.0	0.0
Gross National Savings	20.3	21.7	21.5	22.5	19.7	17.6	22.3	26.4	23.7	20.2	16.4	11.8	12.5	11.0	9.6	10.8	9.4	7.1	9.2
Gross National Product	98.9	99.6	98.7	98.5	98.8	98.8	98.8	98.7	98.4	98.2	97.3	94.1	92.4	91.7	91.9	93.0	94.6	94.7	93.7
Gross National Income	98.9	100.7	100.1	101.7	100.2	98.5	97.6	98.7	97.6	98.7	99.6	97.2	92.7	92.0	93.3	92.8	93.5	90.3	89.8

Source: Table 2.6.

August 1989

Table 2.9: ARGENTINA - GROSS FIXED INVESTMENT, 1970-1988
(1970 Australes)

	1970	1971	1972	1973	1974	1975	1976	1977	1978	1979	1980	1981	1982	1983	1984	1985	1986	1987	1988
Gross Fixed Investment, by sector	1861	2012	2035	1887	1963	1967	2172	2626	2290	2445	2576	2122	1538	1458	1326	1171	1263	1451	1258
Public	708	780	795	661	684	778	989	1214	1068	1017	1010	907	712	671	491	410	515	331	274
Private	1153	1233	1240	1226	1279	1189	1184	1412	1222	1428	1566	1215	826	787	835	761	748	1120	984
Gross Fixed Investment, by sector	1861	2012	2035	1887	1963	1967	2172	2626	2290	2445	2576	2122	1538	1458	1326	1172	1264	1451	1258
Construction	1161	1244	1221	1089	1172	1226	1393	1531	1452	1449	1457	1279	998	885	738	670	716	818	698
Public	463	516	529	440	469	412	595	790	709	623	603	511	399	361	219	197	280	331	274
Private	698	729	692	649	703	814	798	741	743	826	854	768	599	524	519	472	436	488	424
Machinery and Equipment	700	768	814	798	791	741	779	1096	838	996	1119	843	540	573	589	502	548	633	560
Public	245	264	266	221	215	366	394	424	359	394	407	396	313	310	272	213	235	..	..
Private	455	504	548	577	576	375	385	672	479	602	712	447	227	263	316	289	313	..	..
of which:																			
Machinery, tools and furnitu	482	527	541	519	543	521	573	795	594	680	803	634	411	417	425	363	392	455	407
Domestic	274	305	322	356	388	344	416	500	335	358	303	204	187	254	281	211	260	273	227
Imported	208	222	219	163	155	177	157	295	259	322	500	430	224	164	144	153	131	182	181
Transport Equipment	219	240	272	279	249	121	207	300	245	317	315	209	128	155	163	139	156	178	153
Domestic	215	230	257	269	240	109	196	244	209	250	252	147	115	138	144	128	154	173	147
Imported	4	10	15	10	9	12	11	56	36	67	63	62	14	18	19	11	3	5	6
Memo item:																			
Residential Construction	493	486	457	426	476	596	567	502	502	545	558	488	391	339	337	315	..	..	..
Non-residential Construction	668	758	764	663	696	630	826	1029	950	904	899	791	607	546	401	355	..	..	..

Source: Central Bank of the Republic of Argentina (BCRA) and IBRD estimates.

August 1989

Table 2.10: ARGENTINA - GROSS FIXED INVESTMENT, 1970-1988
(Growth Rates)

	1971	1972	1973	1974	1975	1976	1977	1978	1979	1980	1981	1982	1983	1984	1985	1986	1987	1988
Gross Fixed Investment, by sector	8.1	1.1	-7.3	4.0	0.2	10.4	20.9	-12.8	6.8	5.3	-17.6	-27.5	-5.2	-9.0	-11.7	7.9	14.9	-13.3
Public	10.2	2.0	-16.9	3.5	13.8	27.0	22.8	-12.0	-4.8	-0.7	-10.2	-21.5	-5.7	-26.8	-16.5	25.6	-35.8	-17.3
Private	6.9	0.6	-1.1	4.3	-7.0	-0.4	19.3	-13.5	16.9	9.6	-22.4	-32.0	-4.8	6.2	-8.9	-1.7	49.7	-12.1
Gross Fixed Investment, by sector	8.1	1.1	-7.3	4.0	0.2	10.4	20.9	-12.8	6.8	5.3	-17.6	-27.5	-5.2	-9.0	-11.6	7.8	14.8	-13.3
Construction	7.2	-1.9	-10.8	7.6	4.6	13.6	9.9	-5.1	-0.2	0.6	-12.2	-22.0	-11.3	-16.6	-9.3	6.9	14.3	-14.7
Public	11.4	2.7	-16.9	6.6	-12.1	44.2	32.9	-10.3	-12.1	-3.2	-15.2	-21.9	-9.5	-39.3	-10.0	42.1	18.1	-17.3
Private	4.4	-5.1	-6.2	8.2	15.8	-1.9	-7.2	0.3	11.2	3.4	-10.1	-22.0	-12.5	-1.0	-8.9	-7.8	11.9	-13.0
Machinery and Equipment	9.7	6.0	-2.0	-0.9	-6.3	5.1	40.7	-23.5	18.9	12.3	-24.7	-35.9	6.1	2.8	-14.6	9.0	15.5	-11.5
Public	7.8	0.8	-16.9	-2.7	70.2	7.7	7.6	-15.3	9.7	3.3	-2.7	-21.0	-1.0	-12.3	-21.7	10.3	..	..
Private	10.8	8.7	5.3	-0.2	-34.9	2.7	74.5	-28.7	25.7	18.3	-37.2	-49.2	15.9	20.2	-8.5	8.3	..	..
of which:																		
Machinery, tools and furniture	9.4	2.7	-4.1	4.6	-4.1	10.0	38.7	-25.3	14.5	18.1	-21.0	-35.2	1.5	1.9	-14.6	7.8	16.2	-10.5
Domestic	11.4	5.6	10.6	9.0	-11.3	20.9	20.2	-33.0	6.9	-15.4	-32.7	-8.2	35.6	10.7	-25.0	23.6	4.8	-16.9
Imported	6.7	-1.4	-25.6	-4.9	14.2	-11.3	87.9	-12.2	24.3	55.4	-14.0	-47.9	-27.0	-11.9	5.8	-14.0	38.8	-0.9
Transport Equipment	9.6	13.6	2.4	-10.7	-51.4	71.1	44.9	-18.3	29.4	-0.5	-33.8	-38.6	21.1	5.2	-14.8	12.4	13.7	-14.0
Domestic	7.0	12.0	4.5	-10.7	-54.6	79.8	24.5	-14.3	19.5	1.0	-41.8	-21.9	19.9	4.7	-11.0	19.8	12.8	-15.4
Imported	150.0	50.0	-33.3	-10.0	33.3	-8.3	409.1	-35.7	86.9	-6.4	-1.9	-78.2	31.9	8.4	-43.5	-75.2	66.7	40.0
Memo item:																		
Residential Construction	-1.4	-6.0	-6.8	11.7	25.2	-4.9	-11.5	0.0	8.6	2.4	-12.5	-19.9	-13.3	-0.6	-6.5	..	..	..
Non-residential Construction	13.6	0.8	-13.2	4.9	-9.4	31.0	24.5	-7.7	-4.9	-0.5	-12.0	-23.2	-10.0	-26.6	-11.6	..	..	..

Source: Table 2.9

August 1989

Table 2.11: ARGENTINA - GROSS FIXED INVESTMENT, 1970-1988
(Percent of GDP, 1970 Prices)

	1970	1971	1972	1973	1974	1975	1976	1977	1978	1979	1980	1981	1982	1983	1984	1985	1986	1987	1988
Gross Fixed Investment, by sector	21.2	22.1	21.9	19.6	19.3	19.5	21.5	24.4	22.0	22.0	22.8	20.1	15.4	14.1	12.6	11.6	11.9	13.3	11.9
Public	8.1	8.6	8.6	6.9	6.7	7.7	9.8	11.3	10.3	9.1	8.9	8.6	7.1	6.5	4.6	4.1	4.8	3.0	2.6
Private	13.1	13.5	13.3	12.7	12.6	11.8	11.7	13.1	11.8	12.8	13.9	11.5	8.2	7.6	7.9	7.5	7.0	10.3	9.3
Gross Fixed Investment, by sector	21.2	22.1	21.9	19.6	19.3	19.5	21.5	24.4	22.0	22.0	22.8	20.1	15.4	14.1	12.6	11.6	11.9	13.3	11.9
Construction	13.2	13.7	13.1	11.3	11.5	12.1	13.8	14.2	14.0	13.0	12.9	12.1	10.0	8.6	7.0	6.6	6.7	7.5	6.6
Public	5.3	5.7	5.7	4.6	4.6	4.1	5.9	7.3	6.8	5.6	5.3	4.8	4.0	3.5	2.1	2.0	2.6	3.0	2.6
Private	8.0	8.0	7.4	6.7	6.9	8.1	7.9	6.9	7.1	7.4	7.6	7.3	6.0	5.1	4.9	4.7	4.1	4.5	4.0
Machinery and Equipment	8.0	8.4	8.8	8.3	7.8	7.3	7.7	10.2	8.1	8.9	9.9	8.0	5.4	5.6	5.6	5.0	5.1	5.8	5.3
Public	2.8	2.9	2.9	2.3	2.1	3.6	3.9	3.9	3.5	3.5	3.6	3.8	3.1	3.0	2.6	2.1	2.2	..	..
Private	5.2	5.5	5.9	6.0	5.7	3.7	3.8	6.3	4.6	5.4	6.3	4.2	2.3	2.6	3.0	2.9	2.9	..	..
of which:																			
Machinery, tools and furniture	5.5	5.8	5.8	5.4	5.3	5.2	5.7	7.4	5.7	6.1	7.1	6.0	4.1	4.0	4.0	3.6	3.7	4.2	3.9
Domestic	3.1	3.3	3.5	3.7	3.8	3.4	4.1	4.7	3.2	3.2	2.7	1.9	1.9	2.5	2.7	2.1	2.4	2.5	2.2
Imported	2.4	2.4	2.4	1.7	1.5	1.8	1.6	2.7	2.5	2.9	4.4	4.1	2.2	1.6	1.4	1.5	1.2	1.7	1.7
Transport Equipment	2.5	2.6	2.9	2.9	2.5	1.2	2.0	2.8	2.4	2.8	2.8	2.0	1.3	1.5	1.5	1.4	1.5	1.6	1.5
Domestic	2.4	2.5	2.8	2.8	2.4	1.1	1.9	2.3	2.0	2.2	2.2	1.4	1.1	1.3	1.4	1.3	1.4	1.6	1.4
Imported	0.0	0.1	0.2	0.1	0.1	0.1	0.1	0.5	0.3	0.6	0.6	0.6	0.1	0.2	0.2	0.1	0.0	0.0	0.1
Memo item:																			
Residential Construction	5.6	5.3	4.9	4.4	4.7	5.9	5.6	4.7	4.8	4.9	4.9	4.6	3.9	3.3	3.2	3.1	..	..	..
Non-residential Construction	7.6	8.3	8.2	6.9	6.8	6.2	8.2	9.6	9.1	8.1	8.0	7.5	6.1	5.3	3.8	3.5	..	..	..

Source: Table 2.9.

August 1989

Table 2.12: ARGENTINA - GROSS DOMESTIC PRODUCT BY EXPENDITURE, 1970-1988
(1970 Australes Per Capita)

	1970	1971	1972	1973	1974	1975	1976	1977	1978	1979	1980	1981	1982	1983	1984	1985	1986	1987	1988
Gross Domestic Product	366	374	375	383	397	388	381	399	380	401	400	367	344	348	351	331	343	345	329
Terms of Trade Effect	0	4	5	12	5	-1	-4	0	-3	2	9	11	1	1	5	-0	-4	-15	-13
Gross Domestic Income	366	378	381	395	402	387	377	399	377	403	409	379	345	349	356	330	339	330	317
Imports of Goods and NFS	33	36	34	33	34	34	27	36	33	51	73	65	37	35	36	30	35	37	32
Exports of Goods and NFS	34	30	30	34	33	30	38	48	51	49	46	47	48	51	50	56	50	50	55
Exports Adjusted by Terms of Trade	34	34	35	46	39	29	34	48	48	51	55	59	49	52	55	55	47	34	42
Resource Gap	-1	2	-2	-13	-4	6	-7	-12	-15	-0	18	6	-13	-17	-19	-25	-11	2	-10
Total Expenditures	365	380	379	382	398	392	370	387	362	402	428	385	332	331	337	305	328	332	307
Consumption	286	295	295	303	319	314	287	289	281	314	333	314	276	282	294	271	289	287	265
Public	38	39	38	39	42	41	42	43	44	46	46	46	42	42	42	41	42	..	..
Private	248	257	257	264	278	273	245	246	237	269	287	268	234	240	251	230	246	..	..
Gross Domestic Investment	78	85	84	79	78	79	83	98	81	88	95	71	56	50	43	34	40	45	41
Changes in Inventories	-0	2	2	4	2	3	1	1	-2	0	3	-3	4	0	-1	-4	-1	-1	2
Gross Domestic Fixed Investment	78	83	82	75	77	76	82	98	84	88	91	74	53	49	44	38	41	46	39
Public	30	32	32	26	27	30	37	45	39	37	36	32	24	23	16	13	17	..	..
Private	48	51	50	49	50	46	45	52	45	51	55	42	28	27	28	25	24	..	..
Gross Domestic Savings	79	83	86	92	83	73	90	110	96	88	76	65	69	67	62	59	51	43	51
Net Factor Income Payments	-4	-2	-5	-6	-5	-5	-5	-5	-6	-8	-11	-22	-26	-29	-28	-23	-18	-18	-21
Net Transfers	-0	-0	-0	0	0	0	0	0	0	0	0	-0	0	0	0	0	0	-0	0
Gross National Savings	74	81	81	86	78	68	85	105	90	81	66	43	43	38	34	36	32	25	30
Gross National Product	362	372	370	377	392	383	377	394	374	393	389	346	317	319	323	307	325	327	309
Gross National Income	362	377	376	389	397	382	372	394	371	395	399	357	319	320	327	307	321	312	296

Source: Tables 1.1 and 2.6.

August 1989

Table 3.1: ARGENTINA - NOMINAL EXCHANGE RATES, 1970-1988 a/
(Annual Averages; Australes/US$)

	Exchange Rate
1970	0.00000038
1971	0.00000045
1972	0.00000050
1973	0.00000050
1974	0.00000050
1975	0.00000235
1976	0.00001400
1977	0.00004080
1978	0.00007960
1979	0.00013170
1980	0.00018560
1981	0.00044170
1982	0.00259000
1983	0.01053000
1984	0.06765000
1985	0.60406000
1986	0.94303000
1987	2.14509083
1988	8.77030000

Source: 1970-87 data are from IMF, International Financial Statistics (IFS) and correspond to lines wf (rf), representing exchange rates of members maintaining a multiple exchange rate system. 1988 data is from Central Bank of the Republic of Argentina (BCRA) and corresponds to the official exchange market.

August 1989

Table 3.2: ARGENTINA - NOMINAL EXCHANGE RATES, 1980-1989 a/
(Monthly Averages; Australes/US$)

	1980	1981	1982	1983	1984	1985	1986	1987	1988	1989
Annual Average	0.0001858	0.0004419	0.0021712	0.0105300	0.0676	0.6018	0.9439	2.15	8.77	
January	0.0001670	0.0002020	0.0009970	0.0051400	0.0249	0.1999	0.8005	1.29	3.90	13.67
February	0.0001710	0.0002250	0.0010050	0.0057200	0.0278	0.2424	0.8005	1.39	4.33	14.49
March	0.0001750	0.0002340	0.0010870	0.0064000	0.0309	0.3063	0.8005	1.54	4.92	15.29
April	0.0001790	0.0003100	0.0011820	0.0071700	0.0350	0.3963	0.8280	1.54	5.72	51.07
May	0.0001830	0.0003280	0.0014020	0.0077700	0.0409	0.5247	0.8491	1.59	6.74	121.97
June	0.0001860	0.0004400	0.0015160	0.0085200	0.0476	0.7492	0.8731	1.70	8.07	211.14
July	0.0001890	0.0004710	0.0020670	0.0094400	0.0563	0.8005	0.9027	1.89	9.53	571.64
August	0.0001910	0.0005120	0.0023260	0.0106400	0.0684	0.8005	0.9650	2.11	12.00	
September	0.0001940	0.0005580	0.0026820	0.0121100	0.0834	0.8005	1.0500	2.44	12.00	
October	0.0001960	0.0006040	0.0030730	0.0145100	0.1042	0.8005	1.0935	3.23	12.22	
November	0.0001980	0.0006510	0.0041150	0.0175700	0.1318	0.8005	1.1500	3.50	12.67	
December	0.0002000	0.0007680	0.0046020	0.0213700	0.1606	0.8005	1.2140	3.51	13.13	

Source: 1980-87 data are from IMF, International Financial Statistics (IFS) and correspond to lines wf (rf) in IFS, representing exchange rates for members maintaining a dual exchange rate system. 1988-89 data are from Central Bank of the Republic of Argentina and correspond to the official exchange market.

August 1989

Table 3.3: ARGENTINA - REAL EFFECTIVE EXCHANGE RATE INDEX, 1970-1989
(1987=100; increase = depreciation of Australes)

(continues...)

	Commercial Rate Index a/	Free Rate Index b/	Free Rate Premium	Average Rate Index c/
1970 I	78.1	78.1	0.0%	
II	78.9	79.0	0.1%	
III	83.2	83.2	0.0%	
IV	76.4	81.6	6.8%	
1971 I	71.5	77.4	8.3%	
II	71.0	81.4	14.6%	
III	75.8	95.5	25.9%	
IV	73.1	129.1	76.6%	
1972 I	61.6	121.9	98.0%	
II	55.1	119.4	116.6%	
III	50.1	128.5	156.6%	
IV	45.8	115.5	152.2%	
1973 I	41.4	100.1	141.6%	
II	38.3	93.7	144.6%	
III	39.9	87.0	118.0%	
IV	38.9	84.3	116.6%	
1974 I	39.5	95.3	141.0%	
II	39.5	108.0	173.4%	
III	36.8	128.6	249.0%	
IV	33.4	141.5	323.0%	
1975 I	67.5	161.4	139.1%	
II	86.6	201.5	132.8%	
III	97.7	183.2	87.4%	
IV	122.1	231.2	89.4%	
1976 I	95.0	245.0	157.8%	
II	72.9	129.4	77.5%	
III	63.5	113.2	78.1%	
IV	68.1	93.9	37.9%	
1977 I	89.5	92.3	3.1%	
II	92.3	93.4	1.1%	
III	86.7	86.2	-0.6%	
IV	84.1	84.4	0.3%	
1978 I	84.8	84.6	-0.3%	
II	79.1	78.6	-0.7%	
III	73.0	72.5	-0.7%	
IV	67.3	67.6	0.4%	
1979 I	61.5	60.9	-1.0%	
II	56.7	56.3	-0.6%	
III	51.4	51.1	-0.7%	
IV	50.1	51.3	2.4%	

Table 3.3: ARGENTINA - REAL EFFECTIVE EXCHANGE RATE INDEX, 1970-1989
(1987=100; increase = depreciation of Australes)

		Commercial Rate Index a/	Free Rate Index b/	Free Rate Premium	Average Rate Index c/
1980	I	48.0	48.1	0.2%	
	II	45.3	45.4	0.2%	
	III	43.8	43.9	0.2%	
	IV	40.0	40.1	0.2%	
1981	I	39.8	39.9	0.2%	
	II	52.2	53.3	2.1%	
	III	52.2	75.9	45.4%	
	IV	54.9	85.7	56.1%	
1982	I	65.9	65.0	-1.4%	
	II	73.3	114.4	56.0%	
	III	125.3	163.2	30.2%	
	IV	94.9	134.0	41.3%	
1983	I	90.3	121.8	34.9%	
	II	88.3	109.7	24.2%	
	III	78.6	125.4	59.6%	
	IV	78.6	108.8	38.4%	
1984	I	79.8	116.0	45.3%	
	II	71.7	109.6	52.8%	
	III	68.7	94.8	38.0%	
	IV	75.8	89.8	18.5%	
1985	I	77.7	97.5	25.4%	
	II	81.5	93.0	14.1%	
	III	89.4	105.9	18.5%	
	IV	90.9	101.5	11.6%	
1986	I	91.4	101.9	11.5%	
	II	89.3	95.3	6.7%	
	III	88.8	97.5	9.8%	
	IV	89.7	106.5	18.8%	
1987	I	96.2	121.0	25.9%	
	II	97.2	124.5	28.1%	
	III	95.8	129.0	34.7%	
	IV	107.3	131.2	22.2%	
1988	I	107.9	144.4	33.8%	
	II	104.7	129.9	24.1%	
	III	90.4	109.2	20.8%	94.9
	IV	82.4	100.1	21.5%	91.3
1989	I	76.2	135.6	78.0%	88.6

Source: IBRD estimates based on data from IMF, International Financial Statistics.
Nominal free exchange rates from FIEL.

a/ Trade weighted geometric average of bilateral real exchange rates with Argentina's leading trading partners, using commercial nominal exchange rates and combined price indices.

b/ Trade weighted exchange rate based on the free market rate.

c/ A trade-weighted average of the real exchange rates applying to different types of merchandise trade.

August 1989

Table 3.4: ARGENTINA - BALANCE OF PAYMENTS, 1970-1988 a/
(Millions of US dollars)

	1970	1971	1972	1973	1974	1975	1976	1977	1978	1979	1980	1981	1982	1983	1984	1985	1986	1987	1988
Exports (FOB)	1773	1740	1941	3266	3930	2961	3918	5651	6401	7810	8021	9143	7624	7836	8107	8396	6852	6360	9134
Imports (CIF)	1694	1868	1905	2230	3635	3947	3033	4162	3834	6700	10541	9430	5337	4505	4584	3814	4724	5820	5324
Trade Balance	79	-128	36	1036	295	-986	885	1489	2567	1110	-2520	-287	2287	3331	3523	4582	2128	540	3810
Non-Factor Services (net)	-12	-2	78	68	164	172	272	387	-100	-762	-739	-705	43	-341	-205	-231	-573	-285	-298
Receipts	424	457	458	557	861	743	836	1117	1314	1791	2744	2402	1901	1676	1921	1846	1865	2112	2167
Payments	437	459	380	489	696	571	564	730	1414	2553	3483	3107	1858	2017	2125	2077	2438	2397	2465
Balance of Goods and NFS	67	-130	114	1104	459	-814	1157	1876	2467	348	-3259	-992	2330	2990	3318	4351	1555	255	3512
Net Factor Service Income	-223	-256	-334	-394	-333	-475	-508	-618	-681	-920	-1531	-3700	-4719	-5408	-5712	-5304	-4416	-4485	-5127
Net Interest Payments	..	..	-273	-317	-298	-460	-465	-370	-405	-493	-947	-2965	-4403	-4983	-5273	-4879	-3934	-3927	-4467
Interest Receipts	..	..	..	..	..	..	..	..	769	907	1228	885	523	440	264	253	357	218	211
Interest Payments	..	..	..	..	..	..	..	..	1174	1400	2175	3850	4926	5423	5537	5132	4291	4145	4678
Direct Investment Income	..	..	-60	-78	-36	-16	-27	-208	-276	-427	-584	-735	-316	-425	-439	-425	-482	-558	-660
Other Factor Services	..	..	..	..	..	1	-16	-40	..	..	..	..	..	..	..	..	..	..	..
Current Transfers (net)	-3	-3	-4	11	0	5	18	31	48	35	23	-22	32	16	3	0	2	-8	0
Balance on Current Account	-159	-389	-223	721	126	-1284	667	1289	1834	-537	-4767	-4714	-2357	-2402	-2391	-953	-2859	-4238	-1615
Total M< Loans (net) b/	229	208	-1	-136	8	-12	1230	875	907	2648	3400	8557	7401	2610	-756	2786	5763	2653	446
Disbursement	..	..	..	..	..	1018	2311	2123	4265	4064	5809	11969	10981	10490	3595	7564	7437	..	..
Amortization	..	..	..	..	..	1030	1081	1248	3358	1416	2409	3412	3580	7880	4351	4778	1674	..	..
Public M< Loans (net)	83	128	-87	-98	106	-83	1351	356	-332	823	2511	7088	5182	1927	2652	673	1363	2676	565
Disbursement	..	..	..	..	..	437	1956	1064	2509	1633	3425	8370	7202	7628	3518	980	1808	..	..
Amortization	..	..	..	..	..	520	605	708	2841	810	914	1282	2020	5701	866	307	445	..	..
Private M< Loans (net)	146	80	86	-38	-97	71	-121	519	1239	1825	889	1469	2219	683	-3408	2113	4400	-23	-119
Disbursment	..	..	..	..	..	581	355	1059	1756	2431	2384	3599	3779	2862	77	6584	5629	..	..
Amortization	..	..	..	..	..	510	476	540	517	606	1495	2130	1560	2179	3485	4471	1229	..	..
Total Short-term Loans (net)	185	-398	-74	157	-62	226	-923	31	-1215	1635	-1780	-8093	-5446	-935	2340	-2384	-4186	-720	-1899
Disbursement	..	..	..	..	..	..	..	..	3503	8690	19063	24833	8647	3976	3680	..	..	..	..
Amortization	..	..	..	..	..	..	..	..	4718	7055	20843	32926	14093	4911	1340	..	..	..	..
Public ST Loans (net)	..	..	..	..	..	..	..	..	341	187	329	432	-782	758	822	..	..	-238	-648
Disbursement	..	..	..	..	..	..	..	..	691	813	824	2512	1027	1282	1495	..	..	..	..
Amortization	..	..	..	..	..	..	..	..	350	626	495	2080	1809	524	673	..	..	..	..
Private ST Loans (net)	..	..	..	..	..	..	..	..	-1556	1448	-2109	-8525	-4664	-1693	1518	..	..	-482	-1251
Disbursement	..	..	..	..	..	..	..	..	2812	7877	18239	22321	7620	2694	2185	..	..	..	..
Amortization	..	..	..	..	..	..	..	..	4368	6429	20348	30846	12284	4387	667	..	..	..	..
Direct Investment	..	..	..	..	..	..	..	136	274	265	788	927	257	183	268	919	574	-19	1147
Net Use of IMF Resources	..	..	..	..	..	216	235	-115	0	0	0	0	0	1178	0	1007	145	614	18
Capital Transactions n.e.i. c/	..	..	..	..	..	-253	-1244	-218	198	431	-437	-135	-610	811	746	496	0	609	409
Changes in Gross Reserves (- = increase)	-259	420	110	-864	45	1107	35	-1998	-1998	-4442	2796	3458	755	-1445	-207	-1871	563	1111	-1785
Memo Items:																			
Total Net Public Borrowing	83	128	-87	-98	106	97	113	-23	9	1010	2840	2349	5489	6371	4052	3481	1782	3537	2718
Total Net Private Borrowing	331	-318	12	119	-159	95	-220	1310	-317	3273	-1220	-1538	-1773	-537	-9	-762	-386	-505	-1370

Source: Central Bank of the Republic of Argentina (BCRA); IBRD estimates.

a/ For 1970-1974, private sector transactions include "banking" sector.

b/ 1985-1987 data on public and private disbursements and amortizations are IBRD estimates, based on data provided by the Central Bank.

c/ Valuation adjustments, SDRs, changes in arrears and errors and omissions.

August 1989

Table 3.5: ARGENTINA - EXPORTS BY COMMODITY GROUPS, 1970-1988
(Millions of US dollars)

	1970	1971	1972	1973	1974	1975	1976	1977	1978	1979	1980	1981	1982	1983	1984	1985	1986	1987	1988
TOTAL EXPORTS (FOB)	1773	1740	1941	3266	3931	2961	3916	5652	6400	7810	8021	9143	7624	7836	8107	8396	6852	6360	9136
I- Livestock and animal products	349	340	624	712	350	240	478	621	798	1154	891	889	882	680	460	425	561	655	744
II- Cereals, oilseeds and other agricultural products	598	634	447	986	1545	1285	1465	2071	2200	2775	2688	3799	2612	3540	3475	3306	2205	1373	1868
III- Animal and vegetable fats and oils	104	82	56	139	203	91	176	371	391	540	524	395	428	538	930	993	656	546	918
IV- Food, beverages, and tobacco	282	290	282	468	710	443	574	857	846	1014	1174	1084	885	1148	1115	855	1169	1337	1953
V- Mineral products	14	16	13	13	26	22	28	38	74	69	315	658	581	372	365	657	184	125	183
VI- Chemical products	54	57	72	85	120	112	133	156	200	242	367	376	348	298	270	330	298	344	526
VII- Plastic products and rubber manufactures	11	9	12	20	34	10	11	23	31	27	26	34	57	80	87	92	74	122	223
VIII- Leather, fur and their manufactures	111	77	130	153	141	91	201	307	433	670	555	490	364	300	331	318	381	419	443
IX- Wood products	0	0	0	0	0	0	0	0	1	1	1	1	1	1	4	4	6	10	41
X- Paper and paper products	17	17	21	39	61	29	32	130	54	59	69	58	47	39	43	52	53	79	136
XI- Textiles and clothing	121	79	95	213	132	152	239	387	473	363	474	398	338	224	307	323	246	314	512
XII- Footwear, etc.	0	1	3	21	30	5	7	23	22	9	3	3	7	6	2	2	9	38	37
XIII- Nonmetallic mineral products	2	3	4	9	13	8	12	24	35	34	28	23	21	9	10	12	20	38	45
XIV- Precious metals and products	0	1	2	1	0	0	5	11	6	6	24	27	6	0	0	0	0	0	1
XV- Basic metal manufactures	39	47	56	138	179	61	131	129	302	305	325	450	514	318	316	509	474	532	913
XVI- Machinery and electrical equipment	55	63	82	148	208	222	202	244	286	304	345	307	296	181	211	268	280	270	384
XVII- Transport equipment	12	19	35	105	161	175	201	231	218	182	174	126	216	92	168	236	212	135	171
XVIII- Optic and scientific instruments	3	4	5	10	12	11	15	19	22	47	27	18	14	6	9	10	14	14	23
XIX- Arms and ammunition	0	0	0	2	2	0	2	4	1	1	2	2	1	1	0	1	1	1	2
XX- Other products NES	0	0	1	3	3	3	3	5	6	6	7	3	4	2	3	2	5	6	11
XXI- Antiques and art objects	0	0	0	0	0	0	0	0	0	0	0	0	0	0	0	0	0	0	0
XXII- Samples and small packages	1	1	1	1	1	1	1	1	1	2	2	2	2	1	1	1	4	2	2

Source: INDEC.

August 1989

Table 3.6: ARGENTINA - EXPORTS BY COMMODITY GROUPS, 1970-1988
(Percent)

	1970	1971	1972	1973	1974	1975	1976	1977	1978	1979	1980	1981	1982	1983	1984	1985	1986	1987	1988
TOTAL EXPORTS (FOB)	100.0	100.0	100.0	100.0	100.0	100.0	100.0	100.0	100.0	100.0	100.0	100.0	100.0	100.0	100.0	100.0	100.0	100.0	100.0
I- Livestock and animal products	19.7	19.5	32.1	21.8	8.9	8.1	12.2	11.0	12.5	14.8	11.1	9.7	11.6	8.7	5.7	5.1	8.2	10.3	8.1
II- Cereals, oilseeds and other agricultural products	33.7	36.4	23.0	30.2	39.3	43.4	37.4	36.6	34.4	35.5	33.5	41.6	34.3	45.2	42.9	39.4	32.2	21.6	20.4
III- Animal and vegetable fats and oils	5.9	4.7	2.9	4.3	5.2	3.1	4.5	6.6	6.1	6.9	6.5	4.3	5.6	6.9	11.5	11.8	9.6	8.6	10.0
IV- Food, beverages, and tobacco	15.9	16.7	14.5	14.3	18.1	15.0	14.7	15.2	13.2	13.0	14.6	11.9	11.6	14.7	13.8	10.2	17.1	21.0	21.4
V- Mineral products	0.8	0.9	0.7	0.4	0.7	0.7	0.7	0.7	1.2	0.9	3.9	7.2	7.6	4.7	4.5	7.8	2.7	2.0	2.0
VI- Chemical products	3.0	3.3	3.7	2.6	3.1	3.8	3.4	2.8	3.1	3.1	4.6	4.1	4.6	3.8	3.3	3.9	4.3	5.4	5.8
VII- Plastic products and rubber manufactures	0.6	0.5	0.6	0.6	0.9	0.3	0.3	0.4	0.5	0.3	0.3	0.4	0.7	1.0	1.1	1.1	1.1	1.9	2.4
VIII- Leather, fur and their manufactures	6.3	4.4	6.7	4.7	3.6	3.1	5.1	5.4	6.8	8.6	6.9	5.4	4.8	3.8	4.1	3.8	5.6	6.6	4.8
IX- Wood products	0.0	0.0	0.0	0.0	0.0	0.0	0.0	0.0	0.0	0.0	0.0	0.0	0.0	0.0	0.0	0.0	0.1	0.2	0.4
X- Paper and paper products	1.0	1.0	1.1	1.2	1.6	1.0	0.8	2.3	0.8	0.8	0.9	0.6	0.6	0.5	0.5	0.6	0.8	1.2	1.5
XI- Textiles and clothing	6.8	4.5	4.9	6.5	3.4	5.1	6.1	6.8	7.4	4.6	5.9	4.4	4.4	2.9	3.8	3.8	3.6	4.9	5.6
XII- Footwear, etc.	0.0	0.1	0.2	0.6	0.8	0.2	0.2	0.4	0.3	0.1	0.0	0.0	0.1	0.1	0.0	0.0	0.1	0.6	0.4
XIII- Nonmetallic mineral products	0.1	0.2	0.2	0.3	0.3	0.3	0.3	0.4	0.5	0.4	0.3	0.3	0.3	0.1	0.1	0.1	0.3	0.6	0.5
XIV- Precious metals and products	0.0	0.1	0.1	0.0	0.0	0.0	0.1	0.2	0.1	0.1	0.3	0.3	0.1	0.0	0.0	0.0	0.0	0.0	0.0
XV- Basic metal manufactures	2.2	2.7	2.9	4.2	4.6	2.1	3.3	2.3	4.7	3.9	4.1	4.9	6.7	4.1	3.9	6.1	6.9	8.4	10.0
XVI- Machinery and electrical equipment	3.1	3.6	4.2	4.5	5.3	7.5	5.2	4.3	4.5	3.9	4.3	3.4	3.9	2.3	2.6	3.2	4.1	4.2	4.2
XVII- Transport equipment	0.7	1.1	1.8	3.2	4.1	5.9	5.1	4.1	3.4	2.3	2.2	1.4	2.8	1.2	2.1	2.8	3.1	2.1	1.9
XVIII- Optic and scientific instruments	0.2	0.2	0.3	0.3	0.3	0.4	0.4	0.3	0.3	0.6	0.3	0.2	0.2	0.1	0.1	0.1	0.2	0.2	0.3
XIX- Arms and ammunition	0.0	0.0	0.0	0.1	0.1	0.0	0.1	0.1	0.0	0.0	0.0	0.0	0.0	0.0	0.0	0.0	0.0	0.0	0.0
XX- Other products NES	0.0	0.0	0.1	0.1	0.1	0.1	0.1	0.1	0.1	0.0	0.0	0.0	0.1	0.0	0.0	0.0	0.1	0.1	0.1
XXI- Antiques and art objects	0.0	0.0	0.0	0.0	0.0	0.0	0.0	0.0	0.0	0.0	0.0	0.0	0.0	0.0	0.0	0.0	0.0	0.0	0.0
XXII- Samples and small packages	0.1	0.1	0.1	0.0	0.0	0.0	0.0	0.0	0.0	0.0	0.0	0.0	0.0	0.0	0.0	0.0	0.1	0.0	0.0

Source: INDEC.

August 1989

Table 3.7: ARGENTINA - IMPORTS BY COMMODITY GROUPS, 1970-1988
(Millions of US dollars)

	1970	1971	1972	1973	1974	1975	1976	1977	1978	1979	1980	1981	1982	1983	1984	1985	1986	1987	1988
TOTAL IMPORTS (CIF)	1694	1868	1905	2230	3635	3947	3033	4162	3834	6700	10541	9430	5337	4564	4584	3814	4724	5817	5324
I- Livestock and animal products	12	9	6	5	14	17	4	25	30	69	108	76	29	31	21	20	40	55	27
II- Cereals, oilseeds and other agricultural products	69	62	104	183	128	133	119	160	142	257	316	229	153	120	136	124	245	176	160
III- Animal and vegetable fats and oils	2	6	3	4	8	9	6	7	6	11	14	10	9	9	10	6	7	6	8
IV- Food, beverages, and tobacco	24	26	21	23	39	40	26	43	53	164	206	184	67	45	67	53	68	69	48
V- Mineral products	127	168	109	211	607	637	651	809	583	1292	1257	1174	816	600	614	570	571	823	708
VI- Chemical products	214	249	286	299	644	638	516	590	528	848	1125	1016	920	942	1001	799	1036	1040	1033
VII- Plastic products and rubber manufactures	54	57	65	75	168	166	130	144	169	352	457	404	271	299	275	182	285	313	309
VIII- Leather, fur and their manufactures	1	0	0	0	0	1	0	1	2	5	9	17	7	3	3	2	3	4	5
IX- Wood products	72	72	51	57	99	102	36	53	54	123	193	134	65	63	62	33	55	52	43
X- Paper and paper products	111	105	94	116	201	199	139	159	181	246	398	370	198	150	91	75	123	142	106
XI- Textiles and clothing	43	46	43	39	79	65	47	43	49	193	430	433	151	130	145	71	110	117	80
XII- Footwear, etc.	2	1	0	0	0	1	0	1	2	19	57	41	6	2	2	1	1	2	1
XIII- Nonmetallic mineral products	22	24	23	22	27	36	34	34	41	88	170	123	46	50	42	35	47	56	53
XIV- Precious metals and products	8	24	3	1	10	10	7	6	6	12	23	16	8	7	10	5	5	5	1
XV- Basic metal manufactures	368	363	395	549	802	997	505	516	380	608	971	681	538	492	514	325	396	567	655
XVI- Machinery and electrical equipment	418	481	524	469	551	641	607	998	1097	1417	2994	2913	1478	1120	1061	1041	1197	1725	1528
XVII- Transport equipment	89	110	126	111	145	149	144	455	335	728	1114	991	341	304	349	294	283	351	256
XVIII- Optic and scientific instruments	51	58	48	62	105	96	55	104	154	199	518	457	196	172	157	157	224	288	261
XIX- Arms and ammunition	0	0	0	0	0	0	0	0	0	1	6	5	2	1	0	0	1	1	1
XX- Other products nes.	7	7	4	4	8	10	6	14	22	67	170	138	35	22	22	20	24	25	43
XXI- Antiques and art objects	0	0	0	0	0	0	0	0	0	0	1	14	0	1	1	1	0	..	..
XXII- Samples and small packages	0	0	0	0	0	0	1	0	0	1	4	4	1	1	1	0	3	..	..

Source: INDEC.

August 1989

Table 3.8: ARGENTINA - IMPORTS BY COMMODITY GROUPS, 1970-1988
(Percent)

	1970	1971	1972	1973	1974	1975	1976	1977	1978	1979	1980	1981	1982	1983	1984	1985	1986	1987	1988
TOTAL IMPORTS (CIF)	100.0	100.0	100.0	100.0	100.0	100.0	100.0	100.0	100.0	100.0	100.0	100.0	100.0	100.0	100.0	100.0	100.0	100.0	100.0
I- Livestock and animal products	0.7	0.5	0.3	0.2	0.4	0.4	0.1	0.6	0.8	1.0	1.0	0.8	0.5	0.7	0.5	0.5	0.8	0.9	0.5
II- Cereals, oilseeds and other agricultural products	4.1	3.3	5.5	8.2	3.5	3.4	3.9	3.8	3.7	3.8	3.0	2.4	2.9	2.6	3.0	3.3	5.2	3.0	3.0
III- Animal and vegetable fats and oils	0.1	0.3	0.2	0.2	0.2	0.2	0.2	0.2	0.2	0.2	0.1	0.1	0.2	0.2	0.2	0.2	0.2	0.1	0.2
IV- Food, beverages, and tobacco	1.4	1.4	1.1	1.0	1.1	1.0	0.9	1.0	1.4	2.4	2.0	2.0	1.3	1.0	1.5	1.4	1.4	1.2	0.9
V- Mineral products	7.5	9.0	5.7	9.5	16.7	16.1	21.5	19.4	15.2	19.3	11.9	12.4	15.3	13.1	13.4	14.9	12.1	14.1	13.3
VI- Chemical products	12.6	13.3	15.0	13.4	17.7	16.2	17.0	14.2	13.8	12.7	10.7	10.8	17.2	20.6	21.8	20.9	21.9	17.9	19.4
VII- Plastic products and rubber manufactures	3.2	3.1	3.4	3.4	4.6	4.2	4.3	3.5	4.4	5.3	4.3	4.3	5.1	6.6	6.0	4.8	6.0	5.4	5.8
VIII- Leather, fur and their manufactures	0.1	0.0	0.0	0.0	0.0	0.0	0.0	0.0	0.1	0.1	0.1	0.2	0.1	0.1	0.1	0.1	0.1	0.1	0.1
IX- Wood products	4.3	3.9	2.7	2.6	2.7	2.6	1.2	1.3	1.4	1.8	1.8	1.4	1.2	1.4	1.4	0.9	1.2	0.9	0.8
X- Paper and paper products	6.6	5.6	4.9	5.2	5.5	5.0	4.6	3.8	4.7	3.7	3.8	3.9	3.7	3.3	2.0	2.0	2.6	2.4	2.0
XI- Textiles and clothing	2.5	2.5	2.3	1.7	2.2	1.6	1.5	1.0	1.3	2.9	4.1	4.6	2.8	2.8	3.2	1.9	2.3	2.0	1.5
XII- Footwear, etc.	0.1	0.1	0.0	0.0	0.0	0.0	0.0	0.0	0.1	0.3	0.5	0.4	0.1	0.0	0.0	0.0	0.0	0.0	0.0
XIII- Nonmetallic mineral products	1.3	1.3	1.2	1.0	0.7	0.9	1.1	0.8	1.1	1.3	1.6	1.3	0.9	1.1	0.9	0.9	1.0	1.0	1.0
XIV- Precious metals and products	0.5	1.3	0.2	0.0	0.3	0.3	0.2	0.1	0.2	0.2	0.2	0.2	0.1	0.2	0.2	0.1	0.1	0.1	0.0
XV- Basic metal manufactures	21.7	19.4	20.7	24.6	22.1	25.3	16.7	12.4	9.9	9.1	9.2	7.2	10.1	10.8	11.2	8.5	8.4	9.7	12.3
XVI- Machinery and electrical equipment	24.7	25.7	27.5	21.0	15.2	16.2	20.0	24.0	28.6	21.1	28.4	30.9	27.7	24.5	23.1	27.3	25.3	29.7	28.7
XVII- Transport equipment	5.3	5.9	6.6	5.0	4.0	3.8	4.7	10.9	8.7	10.9	10.6	10.5	6.4	6.7	7.6	7.7	6.0	6.0	4.8
XVIII- Optic and scientific instruments	3.0	3.1	2.5	2.8	2.9	2.4	1.8	2.5	4.0	3.0	4.9	4.8	3.7	3.8	3.4	4.1	4.8	5.0	4.9
XIX- Arms and ammunition	0.0	0.0	0.0	0.0	0.0	0.0	0.0	0.0	0.0	0.0	0.1	0.1	0.0	0.0	0.0	0.0	0.0	0.0	0.0
XX- Other products nes.	0.4	0.4	0.2	0.2	0.2	0.3	0.2	0.3	0.6	1.0	1.6	1.5	0.7	0.5	0.5	0.5	0.5	0.4	0.8
XXI- Antiques and art objects	0.0	0.0	0.0	0.0	0.0	0.0	0.0	0.0	0.0	0.0	0.0	0.1	0.0	0.0	0.0	0.0	0.0	0.0	0.0
XXII- Samples and small packages	0.0	0.0	0.0	0.0	0.0	0.0	0.0	0.0	0.0	0.0	0.0	0.0	0.0	0.0	0.0	0.0	0.1	0.0	0.0

Source: INDEC.

August 1989

Table 3.9: ARGENTINA - DIRECTION OF TRADE, 1970-1988

	1970	1971	1972	1973	1974	1975	1976	1977	1978	1979	1980	1981	1982	1983	1984	1985	1986	1987	1988
EXPORTS								(Percent of Total Exports)											
LAFTA	20.6	21.0	24.9	24.4	23.7	25.5	26.3	24.3	23.6	25.8	21.6	19.9	19.9	13.2	17.1	17.7	22.6	20.7	18.7
EEC	46.7	45.8	48.3	40.2	33.7	28.9	32.4	31.4	33.5	32.7	27.1	21.4	21.0	19.9	24.1	20.9	28.7	27.4	30.2
U.S.	8.7	9.1	9.5	7.8	8.1	6.4	6.9	6.8	8.4	7.3	8.7	9.2	13.2	9.7	10.5	12.0	9.9	14.1	12.9
Japan	6.2	5.1	3.0	4.1	4.5	4.6	5.3	5.4	6.0	5.0	2.6	1.8	3.7	4.8	3.3	4.2	5.7	3.5	3.6
Rest of the world	17.8	19.0	14.3	23.5	30.0	34.6	29.1	32.1	28.5	29.2	40.0	47.7	42.2	52.4	45.0	45.2	33.1	34.3	34.6
IMPORTS								(Percent of Total Imports)											
LAFTA	22.0	20.9	19.6	19.5	21.7	22.9	26.8	23.6	21.7	21.8	20.3	20.0	28.8	32.1	35.8	34.0	33.8	29.6	33.3
EEC	30.8	30.6	35.8	30.0	27.3	27.4	27.3	26.3	31.1	26.6	25.9	25.8	22.1	23.8	22.6	26.5	28.8	31.6	29.0
U.S.	24.7	22.2	20.2	21.4	16.8	16.2	17.7	18.5	18.4	21.1	22.4	22.0	21.7	21.6	18.3	18.0	17.5	16.1	17.1
Japan	5.0	8.4	7.5	11.5	10.8	12.5	8.2	8.7	7.0	5.3	9.3	10.2	8.0	6.8	8.2	7.0	7.1	7.6	6.6
Rest of the world	17.5	17.9	16.9	17.6	23.4	21.0	20.0	22.9	21.8	25.2	22.1	22.0	19.4	15.7	15.1	14.5	12.8	15.0	14.0

Source: Central Bank of the Republic of Argentina.

August 1989

Table 3.10: ARGENTINA - INTERNATIONAL RESERVES, 1971-1988
(Millions of US dollars, end of period)

	1971	1972	1973	1974	1975	1976	1977	1978	1979	1980	1981	1982	1983	1984	1985	1986	1987	1988
TOTAL NET INTERNATIONAL RESERVES	71.7	-84.5	536.7	587.2	-520.1	-556.2	1659.4	4905.3	9347.2	6724.3	3165	-3001	-5573	-7317	-7873	-9948	-14049	-15429
Central Bank	194.1	135.6	801.3	858.4	-285.9	-86.1	2134.5	5284.0	9585.5	6737.3	2735	-1953	-4293	-5796	-6327	-8444	-13393	-15069
A. ASSETS	305.2	500.2	1364.4	1319.5	464.2	1623.9	3424.2	5305.4	9671.9	6743.3	3222	2507	2671	2632	4801	4287	3018	4979
Gold a/	89.7	151.7	168.7	168.7	168.7	168.7	176.7	180.6	183.6	185.0	185	185	1421	1421	1421	1421	1421	1421
SDRs	2.8	19.3	80.7	101.8	40.9	90.7	88.9	209.0	315.9	325.3	401	..	23	26	13	6	..	38
Reserve Position in the IMF	110.0	..	..	13.7	..	..	..	173.0	202.5	339.0	276	100	..	..	..	..	..	..
Foreign Exchange	69.8	293.7	1068.4	1027.7	246.8	1353.4	3063.4	4573.3	8878.7	6055.0	2584	2406	1172	1238	3273	2717	1410	3363
Bilateral (LAFTA) b/	32.9	35.5	46.6	7.6	7.8	11.1	95.2	169.5	91.2	-161.0	-224	-184	55	-53	94	143	187	157
B. LIABILITIES	233.5	584.7	827.7	732.3	984.3	2180.1	1764.8	400.1	324.7	100.0	502	6058	9054	10904	13386	14857	18637	21988
Central Bank Liabilities	111.1	364.6	563.1	461.1	750.1	1710.0	1289.7	21.4	86.4	6.0	487	4460	6964	8428	11128	12731	16411	20048
IMF	..	188.9	209.9	77.8	293.4	528.7	414.2	..	..	..	..	..	1173	1139	2289	2719	3825	3678
Private Banks	99.0	167.1	337.3	267.8	387.0	1008.1	856.4	0	0	0	0	0	1250	1850	3596	5059	6790	7479
US banks	..	62.0	137.3	105.9	69.6	533.3	464.3	..	..	..	..	..	..	..	..	..	..	..
European banks	..	80.1	200.0	161.9	104.7	454.8	392.1	..	..	..	..	..	..	..	..	..	..	..
Other banks	99.0	25.0	..	..	212.7	20.0	..	..	..	..	..	..	..	..	..	..	..	..
Other	2.3	2.7	3.1	3.6	3.9	1.2	0.6	0.2	..	6.0	3	540	91	339	1178	1256	1744	2680
Export letters discounted abroad	..	..	..	6.7	5.5	..	..	..	..	..	..	..	..	..	..	..	..	..
Bilateral (LAFTA)	9.8	5.9	12.8	105.2	60.3	172.0	18.5	21.2	86.4	..	..	..	..	..	..	..	..	..
Foreign Currency Swaps	..	..	..	..	..	..	..	..	..	..	484	1380	1228	937	730	639	571	235
Arrears	..	..	..	..	..	..	..	..	..	..	..	2540	3222	4163	1718	544	583	2691
Paris Club	..	..	..	..	..	..	..	..	..	..	..	..	..	..	1617	2514	2898	3285
Treasury Liabilities	122.4	220.1	264.6	271.2	234.2	470.1	475.1	378.7	238.3	94.0	15	1598	2090	2476	2258	2126	2226	1940
External Bonds	79.8	166.3	244.6	255.1	222.0	461.8	470.5	376.1	236.6	94.0	15	1598	2090	2476	2258	2126	2226	1940
Other	42.6	53.8	20.0	16.1	12.2	8.3	4.6	2.6	1.7	0.0	..	..	..	..	..	..	..	..
Valuation and other adjustments c/	..	..	..	..	..	..	..	..	..	81.0	445	550	810	955	712	622	1570	1580

Source: Central Bank of the Republic of Argentina (BCRA), and IMF.

a/ Before 1982, valued at US$42 per fine troy ounce; after 1982, valued at US$ 325 per troy once.

b/ Latin American Free Trade Association (LAFTA). For 1984: net bilateral.

c/ Currency revaluation and statistical discrepancies.

August 1989

Table 3.11: ARGENTINA - INTERNATIONAL PRICES, TRADE VOLUME INDICES AND THE TERMS OF TRADE, 1970-1988

	Export Volume Index a/	Import Volume Index a/	Export Price Index a/	Import Price Index a/	Terms of Trade	Fuel Index b/	MUV Index c/
1970	100.0	100.0	100.0	100.0	100.0	100	100.0
1971	86.1	109.6	114.0	100.6	113.3	120	105.4
1972	83.8	107.9	130.6	104.2	125.3	140	115.1
1973	101.9	103.4	180.7	127.3	141.9	200	132.8
1974	97.7	107.3	226.8	200.0	113.4	800	162.2
1975	81.3	108.0	205.4	215.7	95.2	780	180.3
1976	117.3	85.1	188.3	210.5	89.5	840	183.0
1977	164.6	118.7	193.7	207.0	93.6	920	200.8
1978	182.1	105.0	198.2	215.6	91.9	920	230.9
1979	173.7	156.2	253.6	253.2	100.2	1340	261.8
1980	149.6	242.2	302.3	256.9	117.7	2200	287.3
1981	165.8	223.0	311.0	249.6	124.6	2460	288.8
1982	166.6	128.4	258.2	245.4	105.2	2240	284.6
1983	189.9	116.4	232.7	228.4	101.9	2020	277.2
1984	185.9	121.4	246.0	223.0	110.3	1980	272.6
1985	215.8	97.5	219.4	231.0	95.0	1920	275.3
1986	199.1	116.1	194.0	240.1	80.8	980	325.9
1987	188.2	127.8	186.4	268.7	69.4	1240	357.5
1988	215.3	109.3	221.2	287.5	76.9	..	386.1

Source: Central Bank of the Republic of Argentina (BCRA), and IBRD estimates.
a/ Indices refer to merchandise trade.
b/ Based on current US$ petroleum index.
c/ Manufacturing Unit Value index of manufactured exports from five industrial market economies to developing countries on a cif basis.

August 1989

Table 4.1: ARGENTINA - EXTERNAL DEBT BY BORROWER, 1975-1988
(Millions of US dollars; end of period)

	1975	1976	1977	1978	1979	1980	1981	1982	1983 a/	1984	1985	1986	1987	1988 c/
TOTAL EXTERNAL DEBT b/	7875	8280	9678	12496	19034	27162	35671	43634	45087	46903	48312	51422	58299	58810
PUBLIC SECTOR	4021	5189	6044	8357	9960	14459	20024	28616	33175	36139	39868	44722	..	..
A. General Government	1168	1425	1791	3759	4418	6344	9545	15951	17762	..	..	..	..	..
National Government	1117	1381	1755	3688	4024	5471	8361	14869	17156	..	..	..	..	..
Central Administration	733	991	1298	3024	2592	3100	5134	11417	13345	..	..	..	..	..
Binational Entities	..	..	..	..	748	1244	1373	1509	1529	..	..	..	..	..
Decentralized Agencies	384	390	457	664	684	1127	1854	1943	2282	..	..	..	..	..
Provincial Governments	45	36	30	58	367	829	1053	987	598	..	..	..	..	..
Municipal Authorities	6	8	6	13	27	44	131	95	8	..	..	..	..	..
B. Public Enterprises	921	1047	848	944	965	1370	1772	2682	11007	..	..	..	..	..
C. Mixed Enterprises	1170	1117	1972	3274	3983	5704	7324	7864	1457	..	..	..	..	..
D. Banks	762	1600	1433	380	594	1041	1383	2119	2949	..	..	..	..	..
Central Bank	705	1547	1271	..	..	..	..	299	2427	..	..	..	..	..
Other Official Banks	57	53	162	380	594	1041	1383	1820	522	..	..	..	..	..
PRIVATE SECTOR	3854	3091	3634	4139	9074	12703	15647	15018	11912	10764	8444	6700	..	..
A. Commercial Debt	1441	1182	1393	1796	3279	3791	2759	1919	2160	..	..	..	..	..
B. Financial Debt	2413	1909	2241	2343	5795	8912	12888	13099	9752	..	..	..	..	..

Source: Central Bank of the Republic of Argentina (BCRA) and IBRD estimates.
a/ 1983 figures are IBRD estimates.
b/ Includes IMF.
c/ Preliminary estimate.
August 1989

Table 4.2: ARGENTINA - EXTERNAL DEBT BY CREDITOR, 1983-1988 a/ b/
(Millions of US dollars; end of period)

	1983	1984	1985	1986	1987	1988 d/
Total Debt	45087	46903	48312	51422	58299	58810
Medium and Long Term	36853	40333	44126	51422	54724	58810
Short Term	8234	6570	4186	NA	3575	NA e/
MLT Debt Outstanding and Disbursed (DOD)	36853	40333	44126	51422	58299	58810
Commercial Banks	28412	30053	32681	37607	40774	42089 f/
Bilateral	1654	3290	3621	5132	5809	5585
Bonds	4208	4332	3817	3638	3528	2935
IBRD	533	549	622	920	2146	2265
IDB	873	936	1205	1405	2185	2263
IMF	1173	1173	2180	2720	3857	3673
MLT Disbursements (Gross)	8593	14951	8571	7984	5027	4298
Commercial Banks	6106	12698	6584	5446	2202	2562
Bilateral c/	704	1814	510	1238	571	477
Bonds	451	196	0	183	83	0
IBRD	70	96	144	408	795	487
IDB	89	147	326	162	123	231
IMF purchases	1173	0	1007	547	1253	541
MLT Amortiztions	4209	11471	4778	2076	2490	2463
Commercial Banks	3214	11057	3956	583	1258	943
Bilateral	510	178	179	258	187	0
Bonds	400	72	515	646	179	656
IBRD	40	80	71	110	133	188
IDB	45	84	57	77	94	153
IMF repurchases	0	0	0	402	639	523
MLT Disbursements (Net)	4384	3480	3793	5908	2537	1835
Commercial Banks	2892	1641	2628	4863	944	1619
Bilateral	194	1636	331	980	384	477
Bonds	51	124	-515	-463	-96	-656
IBRD	30	16	73	298	662	299
IDB	44	63	269	85	29	78
IMF	1173	0	1007	145	614	18
Interest Payments	5423	5537	5132	4291	4362	5198
Commercial Banks	..	4427	3802	3098	2998	3457
Bilateral	..	189	297	122	482	513
Bonds	..	437	449	320	490	323
IBRD	36	56	46	75	91	154
IDB	56	80	72	108	180	158
IMF charges	74	90	140	159	121	282
Short Term	..	258	326	409	0	311

Source: IBRD staff estimates based on data from the Central Bank of the Republic of Argentina, the IMF and the IDB.

a/ Disbursments and amortizations may not explain changes in debt stock from year-to-year due to valuation changes.
b/ Figures adjusted by the changes in cross-rates between US dollar and other currencies.
c/ Bilaterals for 1986 includes new disbursements from bilaterals (US$ mill. 283) plus US$ mill. 180 capitalized interest plus US$ mill. 302 in interest arrears.
d/ Preliminary estimate.
e/ Short term debt is included in medium and long term debt for 1986 and 1988.
f/ For 1988-1989 all arrears are included in commercial banks DOD.

August 1989

Table 5.1: ARGENTINA - PUBLIC SECTOR REVENUES, EXPENDITURES AND FINANCING, 1970-1987
(Thousands of Australes)

	1970	1971	1972	1973	1974	1975	1976	1977	1978	1979	1980	1981	1982	1983	1984	1985	1986	1987 a/
Current Account																		
I- Current Revenues	2.77	3.51	5.44	9.62	14.93	34.6	210	685	1896	4722	10065	18954	47622	233246	1738773	16030000	27677400	54660900
Tax Revenues	1.71	2.13	3.16	5.85	9.60	19.5	120	402	1093	2931	6590	11139	27644	126655	958400	8714100	16298800	31218900
Non-tax Revenues	1.06	1.38	2.28	3.77	5.32	15.2	90	283	803	1791	3475	7815	19978	106591	780373	7315900	11378600	23442000
II- Current Expenditures	2.20	3.09	4.88	9.62	14.85	43.7	202	514	1612	4256	9745	21540	58512	280173	2015951	16040300	26268900	59994600
Personnel Expenditures	1.04	1.46	2.28	4.94	7.48	22.7	79	197	609	1640	3786	6866	14555	87050	728478	5055200	8942700	12832700
Goods and Services	0.55	0.71	1.23	2.11	2.65	8.8	63	146	416	1023	1998	4129	14139	74172	487050	4820300	7098800	14732900
Interest on Debt	0.06	0.11	0.21	0.33	0.56	1.8	15	42	159	444	971	4052	15311	40705	261946	2157500	2833900	7452800
Domestic	0.03	0.05	0.11	0.22	0.36	1.0	12	29	129	372	747	2830	8438	6404	42191	309700	238500	1475800
Foreign	0.03	0.06	0.10	0.11	0.20	0.8	4	13	31	73	223	1222	6872	34301	219755	1847800	2595400	5977000
Other Current Expenditures	0.01	0.01	0.04	0.00	0.38	0.4	2	4	37	45	358	632	1852	7227	20124	228100	489400	1918600
Current & Capital Transfers	0.54	0.81	1.12	2.24	3.77	10.0	43	125	391	1104	2632	5861	12656	71019	518353	3779200	6904100	24047600
Economies	0.00	0.00	0.00	0.00	0.00	0.0	0	0	0	0	0	0	0	0	0	0	0	-1040000
III- Savings	0.57	0.42	0.57	0.00	0.08	-9.1	8	171	284	466	320	-2586	-10890	-46927	-277178	-10300	1408500	-5333700
Capital Account																		
IV- Capital Revenues	0.03	0.04	0.06	0.03	0.24	0.2	1	8	33	44	89	136	726	1555	12542	99800	117800	456300
V- Capital Expenditures	0.75	1.01	1.80	2.63	4.28	12.8	99	277	656	1497	2694	5284	12638	66116	413056	2796900	5283600	9102200
Fixed Investments	0.70	0.96	1.71	2.48	3.74	12.1	87	246	609	1409	2514	4515	11716	59837	377055	2580000	4865000	8169500
Machinery and Equipment	0.25	0.31	0.59	0.85	1.19	3.9	28	80	211	489	675	1356	3390	16405	122900	663800	1146600	3145500
Construction	0.45	0.64	1.13	1.64	2.56	8.2	59	166	399	921	1839	3160	8326	43432	254155	1916200	3718400	5024000
Changes in Inventories	0.00	0.00	0.00	0.00	0.00	0.0	0	0	0	0	50	130	182	733	4388	22200	38300	65000
Financial Investments	0.05	0.05	0.09	0.15	0.54	0.6	12	32	46	88	130	638	740	5546	31613	194700	380300	1077700
Economies	0.00	0.00	0.00	0.00	0.00	0.0	0	0	0	0	0	0	0	0	0	0	0	-210000
VI-Financing Due to Economic Emergency	0.00	0.00	0.00	0.00	0.00	0.0	0	0	0	0	0	0	0	0	0	261000	430500	379400
VII-Capital from Previous Period	0.00	0.00	0.00	0.00	0.00	0.0	0	0	0	60	166	474	494	1713	10357	34600	130200	520900
VIII-Contributions from:	0.31	0.48	0.88	2.11	3.65	15.8	72	115	288	959	1688	4077	10574	113857	566242	5135400	9398700	13459100
Central Administration	0.23	0.40	0.75	1.75	2.83	13.6	62	79	154	457	971	2663	7174	96484	461760	4062800	7468500	10920100
Special Accounts	0.05	0.06	0.11	0.34	0.77	2.1	10	35	125	323	669	1324	3183	16446	103682	1064600	1881700	2493800
Decentralized Agencies	0.03	0.02	0.01	0.02	0.05	0.1	0	2	9	16	48	90	218	927	800	8000	48500	45200
Provinces & MCBA	0.00	0.00	0.00	0.00	0.00	0.0	0	0	0	51	0	0	0	0	0	0	0	0
Public Enterprises	0.00	0.00	0.00	0.00	0.00	0.0	0	0	0	111	0	0	0	0	0	0	0	0
Social Security System	0.00	0.00	0.00	0.00	0.00	0.0	0	0	0	0	0	0	0	0	0	0	0	0
IX- Contributions to:	0.30	0.48	0.89	2.11	3.65	15.8	72	116	288	960	1688	4078	10573	113857	566242	5135400	9398700	13459100
Central Administration	0.00	0.00	0.00	0.00	0.00	0.0	0	0	0	158	0	0	0	0	0	0	0	88700
Special Accounts	0.01	0.07	0.14	0.08	0.23	1.3	7	14	55	201	376	904	2418	10735	62608	443600	826800	2142500
Decentralized Agencies	0.11	0.15	0.27	0.63	0.97	2.5	13	29	77	175	388	860	2351	20260	102669	714300	1163500	3183500
Provinces & MCBA	0.07	0.11	0.28	0.84	1.48	8.8	26	33	87	224	488	1465	2987	34831	202171	2240500	4646500	0
Public Enterprises	0.12	0.15	0.19	0.53	0.95	3.2	25	40	68	197	426	850	2817	48031	160236	1166500	1688200	6558400
Social Security System	0.00	0.00	0.00	0.04	0.01	0.1	0	0	1	5	10	0	0	0	38558	570500	1073700	1486000
X- Revenues (I+IV+VI+VII+VIII)	3.11	4.03	6.38	11.76	18.81	50.7	284	809	2217	5786	12008	23641	59417	350371	2327914	21560800	37754600	69476600
XI- Expenditures (II+V+IX)	3.25	4.59	7.57	14.37	22.77	72.3	373	907	2556	6713	14127	30902	81723	460146	2995249	23972600	40951200	82555900
XII- Financing Needs (XI-X)	0.15	0.56	1.19	2.60	3.96	21.6	89	98	339	927	2119	7260	22307	109775	667335	2411800	3196600	13079300
XIII-Net Financing	0.07	0.24	0.39	0.34	0.90	2.7	31	59	287	789	847	4399	9305	-11184	-52068	231400	2809600	8851400
a. Net Use of Credit	0.07	0.24	0.39	0.34	0.90	2.7	31	59	287	789	965	4532	9441	-9199	-51438	250100	2854800	9044400
Domestic	0.01	0.09	0.18	0.29	0.47	2.5	23	32	185	538	433	2175	7485	-12923	-30804	-117300	-100800	1633600
Credit	0.07	0.17	0.32	0.68	1.00	3.7	28	56	223	777	753	3646	13494	884	3341	55000	775000	3812900
Amortization	0.06	0.08	0.14	0.39	0.53	1.2	4	23	37	239	320	1471	6009	13807	34145	172300	875800	2179300
Foreign	0.06	0.15	0.22	0.05	0.43	0.2	8	26	102	250	532	2357	1955	3724	-20634	367400	2955600	7410800
Credit	0.13	0.23	0.46	0.52	0.77	1.3	15	43	149	551	835	5132	23883	61025	462899	6053100	6438100	11242400
Amortization	0.07	0.08	0.24	0.47	0.34	1.2	7	17	47	301	303	2774	21928	57301	483533	5685700	3482500	3831600
b. Net Use of Advances	0.00	0.00	0.00	0.00	0.00	0.0	0	0	0	0	-118	-133	-135	-1985	-630	-18700	-45200	-193000
Credit	0.00	0.00	0.00	0.00	0.00	0.0	0	0	0	0	0	146	105	92	719	100	900	900
Debit	0.00	0.00	0.00	0.00	0.00	0.0	0	0	0	0	118	280	240	2077	1349	18800	46100	193900
XIV- Overall Balance (XIII-XII)	-0.07	-0.33	-0.80	-2.26	-3.06	-18.9	-58	-39	-52	-138	-1272	-2861	-13001	-120959	-719403	-2180400	-387000	-4227900
Central Bank	0.30	0.30	0.37	1.58	2.72	13.9	26	51	45	-43	1017	2911	10767	113321	326936	920700	0	0
Net Variation in Passive Financing	-0.22	0.03	0.42	0.68	0.34	5.0	32	-12	7	181	256	-50	2234	7638	392467	1259700	387000	4227900

Source: Ministry of Economy.

a/ Excludes provincial governments after 1986.

June 1989

Table 5.2: ARGENTINA - PUBLIC SECTOR REVENUES, EXPENDITURES AND FINANCING, 1970-1987
(Percent of GDP)

	1970	1971	1972	1973	1974	1975	1976	1977	1978	1979	1980	1981	1982	1983	1984	1985	1986	1987 a/
Current Account																		
I- Current Revenues	31.5	28.0	26.3	27.1	30.7	24.2	27.8	32.7	36.2	33.1	35.5	34.6	32.3	34.2	32.9	40.5	37.2	30.8
Tax Revenues	19.5	17.0	15.3	16.5	19.7	13.6	15.9	19.2	20.9	20.6	23.3	20.3	18.7	18.6	18.1	22.0	21.9	17.6
Non-tax Revenues	12.0	11.0	11.0	10.6	10.9	10.6	11.9	13.5	15.3	12.6	12.3	14.3	13.5	15.6	14.8	18.5	15.3	13.2
II- Current Expenditures	25.1	24.7	23.6	27.1	30.5	30.6	26.7	24.6	30.8	29.9	34.4	39.3	39.6	41.0	38.2	40.5	35.4	33.9
Personnel Expenditures	11.8	11.7	11.0	13.9	15.4	15.9	10.4	9.4	11.6	11.5	13.4	12.5	9.9	12.8	13.8	12.8	12.0	7.2
Goods and Services	6.3	5.6	6.0	5.9	5.4	6.2	8.3	7.0	8.0	7.2	7.1	7.5	9.6	10.9	9.2	12.2	9.6	8.3
Interest on Debt	0.7	0.9	1.0	0.9	1.2	1.3	2.0	2.0	3.0	3.1	3.4	7.4	10.4	6.0	5.0	5.4	3.8	4.2
Domestic	0.4	0.4	0.5	0.6	0.7	0.7	1.5	1.4	2.5	2.6	2.6	5.2	5.7	0.9	0.8	0.8	0.3	0.8
Foreign	0.3	0.5	0.5	0.3	0.4	0.5	0.5	0.6	0.6	0.5	0.8	2.2	4.7	5.0	4.2	4.7	3.5	3.4
Other Current Expenditures	0.1	0.1	0.2	0.0	0.8	0.3	0.3	0.2	0.7	0.3	1.3	1.2	1.3	1.1	0.4	0.6	0.7	1.1
Current & Capital Transfers	6.1	6.4	5.4	6.3	7.8	7.0	5.6	6.0	7.5	7.7	9.3	10.7	8.6	10.4	9.8	9.5	9.3	13.6
Economies	0.0	0.0	0.0	0.0	0.0	0.0	0.0	0.0	0.0	0.0	0.0	0.0	0.0	0.0	0.0	0.0	0.0	-0.6
III- Savings	6.5	3.3	2.7	0.0	0.2	-6.3	1.1	8.2	5.4	3.3	1.1	-4.7	-7.4	-6.9	-5.2	-0.0	1.9	-3.0
Capital Account																		
IV- Capital Revenues	0.3	0.3	0.3	0.1	0.5	0.2	0.2	0.4	0.6	0.3	0.3	0.2	0.5	0.2	0.2	0.3	0.2	0.3
V- Capital Expenditures	8.6	8.1	8.7	7.4	8.8	8.9	13.0	13.2	12.5	10.5	9.5	9.7	8.6	9.7	7.8	7.1	7.1	5.1
Fixed Investments	8.0	7.7	8.3	7.0	7.7	8.5	11.5	11.7	11.6	9.9	8.9	8.2	7.9	8.8	7.1	6.5	6.5	4.6
Machinery and Equipment	2.8	2.5	2.8	2.4	2.4	2.8	3.7	3.8	4.0	3.4	2.4	2.5	2.3	2.4	2.3	1.7	1.5	1.8
Construction	5.1	5.1	5.4	4.6	5.3	5.7	7.8	7.9	7.6	6.5	6.5	5.8	5.6	6.4	4.8	4.8	5.0	2.8
Changes in Inventories	0.0	0.0	0.0	0.0	0.0	0.0	0.0	0.0	0.0	0.0	0.2	0.2	0.1	0.1	0.1	0.1	0.1	0.0
Financial Investments	0.6	0.4	0.4	0.4	1.1	0.4	1.5	1.5	0.9	0.6	0.5	1.2	0.5	0.8	0.6	0.5	0.5	0.6
Economies	0.0	0.0	0.0	0.0	0.0	0.0	0.0	0.0	0.0	0.0	0.0	0.0	0.0	0.0	0.0	0.0	0.0	-0.1
VI-Financing Due to Economic Emergency	0.0	0.0	0.0	0.0	0.0	0.0	0.0	0.0	0.0	0.0	0.0	0.0	0.0	0.0	0.0	0.7	0.6	0.2
VII-Capital from Previous Period	0.0	0.0	0.0	0.0	0.0	0.0	0.0	0.0	0.0	0.4	0.6	0.9	0.3	0.3	0.2	0.1	0.2	0.3
VIII-Contributions from:	3.6	3.8	4.2	5.9	7.5	11.1	9.5	5.5	5.5	6.7	6.0	7.4	7.2	16.7	10.7	13.0	12.6	7.6
Central Administration	2.7	3.2	3.6	4.9	5.8	9.5	8.1	3.8	2.9	3.2	3.4	4.9	4.9	14.1	8.7	10.3	10.1	6.2
Special Accounts	0.6	0.4	0.5	0.9	1.6	1.5	1.4	1.7	2.4	2.3	2.4	2.4	2.2	2.4	2.0	2.7	2.5	1.4
Decentralized Agencies	0.3	0.2	0.0	0.1	0.1	0.1	0.0	0.1	0.2	0.1	0.2	0.2	0.1	0.1	0.0	0.0	0.1	0.0
Provinces & MCBA	0.0	0.0	0.0	0.0	0.0	0.0	0.0	0.0	0.0	0.4	0.0	0.0	0.0	0.0	0.0	0.0	0.0	0.0
Public Enterprises	0.0	0.0	0.0	0.0	0.0	0.0	0.0	0.0	0.0	0.8	0.0	0.0	0.0	0.0	0.0	0.0	0.0	0.0
Social Security System	0.0	0.0	0.0	0.0	0.0	0.0	0.0	0.0	0.0	0.0	0.0	0.0	0.0	0.0	0.0	0.0	0.0	0.0
IX- Contributions to:	3.5	3.9	4.3	6.0	7.5	11.1	9.5	5.5	5.5	6.7	6.0	7.4	7.2	16.7	10.7	13.0	12.6	7.6
Central Administration	0.0	0.0	0.0	0.0	0.0	0.0	0.0	0.0	0.0	1.1	0.0	0.0	0.0	0.0	0.0	0.0	0.0	0.1
Special Accounts	0.1	0.6	0.7	0.2	0.5	0.9	1.0	0.7	1.0	1.4	1.3	1.7	1.6	1.6	1.2	1.1	1.1	1.2
Decentralized Agencies	1.3	1.2	1.3	1.8	2.0	1.8	1.8	1.4	1.5	1.2	1.4	1.6	1.6	3.0	1.9	1.8	1.6	1.8
Provinces & MCBA	0.8	0.9	1.4	2.4	3.0	6.2	3.4	1.6	1.7	1.6	1.7	2.7	2.0	5.1	3.8	5.7	6.3	0.0
Public Enterprises	1.3	1.2	0.9	1.5	2.0	2.2	3.3	1.9	1.3	1.4	1.5	1.6	1.9	7.0	3.0	2.9	2.3	3.7
Social Security System	0.0	0.0	0.0	0.1	0.0	0.1	0.0	0.0	0.0	0.0	0.0	0.0	0.0	0.0	0.7	1.4	1.4	0.8
X- Revenues (I+IV+VI+VII+VIII)	35.4	32.2	30.8	33.1	38.7	35.5	37.4	38.6	42.4	40.6	42.4	43.2	40.3	51.3	44.1	54.5	50.8	39.2
XI- Expenditures (II+V+IX)	**37.1**	**36.7**	**36.6**	**40.5**	**46.8**	**50.6**	**49.2**	**43.3**	**48.8**	**47.1**	**49.9**	**56.4**	**55.4**	**67.4**	**56.7**	**60.5**	**55.1**	**46.6**
XII- Financing Needs (XI-X)	**1.7**	**4.5**	**5.8**	**7.3**	**8.1**	**15.1**	**11.8**	**4.7**	**6.5**	**6.5**	**7.5**	**13.3**	**15.1**	**16.1**	**12.6**	**6.1**	**4.3**	**7.4**
XIII-Net Financing	0.8	1.9	1.9	1.0	1.8	1.9	4.2	2.8	5.5	5.5	3.0	8.0	6.3	-1.6	[illegible]	0.6	3.8	5.0
a. Net Use of Credit	0.8	1.9	1.9	1.0	1.8	1.9	4.2	2.8	5.5	5.5	3.4	8.3	6.4	-1.3	[illegible]	0.6	3.8	5.1
Domestic	0.1	0.7	0.9	0.8	1.0	1.8	3.0	1.5	3.5	3.8	1.5	4.0	5.1	-1.9	[illegible]	-0.3	-0.1	0.9
Credit	0.8	1.3	1.6	1.9	2.1	2.6	3.6	2.7	4.3	5.5	2.7	6.7	9.1	0.1	0.1	0.1	1.0	2.2
Amortization	0.7	0.6	0.7	1.1	1.1	0.8	0.6	1.1	0.7	1.7	1.1	2.7	4.1	2.0	0.6	0.4	1.2	1.2
Foreign	0.7	1.2	1.0	0.1	0.9	0.1	1.1	1.3	1.9	1.8	1.9	4.3	1.3	0.5	-0.4	0.9	4.0	4.2
Credit	1.5	1.9	2.2	1.5	1.6	0.9	2.0	2.1	2.8	3.9	2.9	9.4	16.2	8.9	8.8	15.3	8.7	6.3
Amortization	0.8	0.6	1.2	1.3	0.7	0.8	0.9	0.8	0.9	2.1	1.1	5.1	14.9	8.4	9.2	14.4	4.7	2.2
b. Net Use of Advances	0.0	0.0	0.0	0.0	0.0	0.0	0.0	0.0	0.0	0.0	-0.4	-0.2	-0.1	-0.3	-0.0	-0.0	-0.1	-0.1
Credit	0.0	0.0	0.0	0.0	0.0	0.0	0.0	0.0	0.0	0.0	0.0	0.3	0.1	0.0	0.0	0.0	0.0	0.0
Debit	0.0	0.0	0.0	0.0	0.0	0.0	0.0	0.0	0.0	0.0	0.4	0.5	0.2	0.3	0.0	0.0	0.1	0.1
XIV- Overall Balance (XIII-XII)	**-0.9**	**-2.6**	**-3.9**	**-6.4**	**-6.3**	**-13.2**	**-7.6**	**-1.9**	**-1.0**	**-1.0**	**-4.5**	**-5.2**	**-8.8**	**-17.7**	**-13.6**	**-5.5**	**-0.5**	**-2.4**
Central Bank	**3.4**	**2.4**	**1.8**	**4.5**	**5.6**	**9.7**	**3.4**	**2.4**	**0.9**	**-0.3**	**3.6**	**5.3**	**7.3**	**16.6**	**6.2**	**2.3**	**0.0**	**0.0**
Net Variation in Passive Financing	**-2.5**	**0.2**	**2.0**	**1.9**	**0.7**	**3.5**	**4.2**	**-0.6**	**0.1**	**1.3**	**0.9**	**-0.1**	**1.5**	**1.1**	**7.4**	**3.2**	**0.5**	**2.4**

Source: Ministry of Economy.

a/ Excludes provincial governments after 1986.

June 1989

Table 5.3: ARGENTINA - CENTRAL ADMINISTRATION: REVENUES, EXPENDITURES AND FINANCING, 1970-1987
(Thousands of Australes)

	1970	1971	1972	1973	1974	1975	1976	1977	1978	1979	1980	1981	1982	1983	1984	1985	1986	1987 a/
Current Account																		
I- Current Revenues	0.70	0.87	1.37	2.00	2.97	6.30	44.9	138	363	786	1743	3440	9193	40549	329159	4661300	8180900	18587700
Tax Revenues	0.63	0.75	1.14	1.74	2.75	5.18	38.2	119	278	663	1539	2975	8037	37167	281677	4302100	7925600	18120300
Non-tax Revenues	0.07	0.12	0.23	0.26	0.22	1.12	6.7	19	85	124	204	465	1157	3382	47482	359200	255300	467400
II- Current Expenditures	0.48	0.67	1.03	2.15	3.05	9.07	48.3	116	374	944	2000	4830	14202	56429	375554	2701000	4796700	20526900
Personnel Expenditures	0.31	0.40	0.61	1.35	1.79	5.24	19.7	52	160	383	924	1799	3663	20073	147391	930100	1746900	4805300
Goods and Services	0.07	0.10	0.17	0.30	0.39	1.28	10.7	23	65	159	288	560	1988	9787	37740	371000	748900	1739900
Interest on Debt	0.03	0.06	0.10	0.15	0.33	0.89	10.4	21	87	247	436	1658	7006	18204	129229	984600	1515400	4238300
Domestic	0.02	0.03	0.04	0.09	0.27	0.68	9.7	19	80	236	362	1162	4351	1746	22836	36700	164200	1286300
Foreign	0.01	0.03	0.06	0.06	0.06	0.21	0.8	2	7	11	74	496	2655	16458	106393	947900	1351200	2952000
Other Current Expenditures	0.00	0.00	0.00	0.00	0.00	0.00	0.0	0	0	0	0	0	0	0	0	0	0	0
Current & Capital Transfers	0.07	0.11	0.15	0.35	0.54	1.66	7.4	20	62	155	352	813	1546	8365	61194	415300	785500	10518400
Economies	0.00	0.00	0.00	0.00	0.00	0.00	0.0	0	0	0	0	0	0	0	0	0	0	-775000
III- Savings	0.22	0.20	0.34	-0.15	-0.08	-2.77	-3.4	22	-11	-158	-257	-1390	-5008	-15880	-46395	1960300	3384200	-1939200
Capital Account																		
IV- Capital Revenues	0.00	0.00	0.01	0.00	0.02	0.02	0.1	0	0	1	0	0	2	10	10	100	300	220000
V- Capital Expenditures	0.07	0.09	0.15	0.24	0.38	1.31	9.4	21	47	102	143	377	768	6076	16925	129400	193400	551400
Fixed Investments	0.07	0.09	0.15	0.23	0.37	1.26	7.6	19	46	101	131	290	607	3786	12524	74400	162600	545300
Machinery and Equipment	0.02	0.02	0.03	0.06	0.19	0.83	5.2	14	32	71	73	180	349	1308	3975	34800	108500	327800
Construction	0.05	0.07	0.12	0.17	0.18	0.43	2.4	5	14	30	58	110	257	2478	8549	39600	54100	217500
Changes in Inventories	0.00	0.00	0.00	0.00	0.00	0.00	0.0	0	0	0	10	40	45	448	2434	5500	6900	25600
Financial Investments	0.00	0.00	0.00	0.01	0.01	0.05	1.8	2	1	1	2	47	116	1842	1967	49500	23900	50500
Economies	0.00	0.00	0.00	0.00	0.00	0.00	0.0	0	0	0	0	0	0	0	0	0	0	-70000
VI-Financing Due to Economic Emergency	0.00	0.00	0.00	0.00	0.00	0.00	0.0	0	0	0	0	0	0	0	0	261000	430500	379400
VII-Capital from Previous Period	0.00	0.00	0.00	0.00	0.00	0.00	0.0	0	0	0	5	26	63	808	5432	11200	0	0
VIII-Contributions from:	0.00	0.00	0.00	0.00	0.00	0.00	0.0	0	0	158	0	0	0	0	0	0	0	88700
Central Administration	0.00	0.00	0.00	0.00	0.00	0.00	0.0	0	0	0	0	0	0	0	0	0	0	0
Special Accounts	0.00	0.00	0.00	0.00	0.00	0.00	0.0	0	0	0	0	0	0	0	0	0	0	71600
Decentralized Agencies	0.00	0.00	0.00	0.00	0.00	0.00	0.0	0	0	0	0	0	0	0	0	0	0	17100
Provinces & MCBA	0.00	0.00	0.00	0.00	0.00	0.00	0.0	0	0	46	0	0	0	0	0	0	0	0
Public Enterprises	0.00	0.00	0.00	0.00	0.00	0.00	0.0	0	0	111	0	0	0	0	0	0	0	0
Social Security System	0.00	0.00	0.00	0.00	0.00	0.00	0.0	0	0	0	0	0	0	0	0	0	0	0
IX- Contributions to:	0.23	0.40	0.76	1.75	2.82	13.62	61.7	79	154	458	971	2664	7174	96484	461760	4062800	7468500	10920100
Central Administration	0.00	0.00	0.00	0.00	0.00	0.00	0.0	0	0	0	0	0	0	0	0	0	0	0
Special Accounts	0.01	0.01	0.02	0.05	0.13	1.05	6.5	9	37	164	292	724	1896	8423	50747	292000	554900	1435800
Decentralized Agencies	0.09	0.13	0.26	0.57	0.78	2.21	12.6	26	63	138	363	807	2272	20242	102584	713800	1136700	3169700
Provinces & MCBA	0.06	0.11	0.28	0.74	1.24	7.84	20.6	14	23	48	101	710	1173	25509	144961	1748700	3566800	0
Public Enterprises	0.08	0.15	0.19	0.35	0.67	2.43	21.7	29	30	102	205	424	1833	42310	124910	737800	1136400	4828600
Social Security System	0.00	0.00	0.00	0.04	0.01	0.08	0.2	0	1	5	10	0	0	0	38558	570500	1073700	1486000
X- Revenues (I+IV+VI+VII+VIII)	0.70	0.87	1.38	2.00	2.99	6.32	45.0	138	363	944	1748	3466	9258	41367	334601	4933600	8611700	19275800
XI- Expenditures (II+V+IX)	0.78	1.16	1.94	4.14	6.25	24.00	119.4	215	575	1504	3114	7871	22143	158989	854239	6893200	12458600	31998400
XII- Financing Needs (XI-X)	0.08	0.29	0.56	2.14	3.26	17.68	74.4	77	212	559	1366	4405	12885	117622	519638	1959600	3846900	12722600
XIII-Net Financing	0.02	0.09	0.03	-0.00	0.26	2.03	24.5	23	136	334	-55	956	679	-6186	-5948	148200	2854600	8393900
a. Net Use of Credit	0.02	0.09	0.03	-0.00	0.26	2.03	24.5	23	136	334	-14	999	754	-6070	-5948	148200	2854600	8393900
Domestic	0.01	0.07	0.06	0.15	0.24	2.19	21.7	21	117	292	-104	191	154	-9538	-29008	71900	-7200	1733700
Credit	0.04	0.12	0.13	0.35	0.51	2.98	24.5	40	133	337	74	591	4306	-2914	1500	7100	701000	3755000
Amortization	0.03	0.05	0.07	0.20	0.27	0.79	2.9	19	16	45	178	400	4153	6624	30508	79000	708200	2021300
Foreign	0.01	0.02	-0.03	-0.15	0.02	-0.16	2.9	2	19	42	90	808	600	3468	23060	220100	2861800	6660200
Credit	0.04	0.05	0.10	0.10	0.22	0.27	6.0	7	33	70	162	2283	4311	15772	32667	2359600	5124700	8280800
Amortization	0.03	0.03	0.13	0.25	0.20	0.43	3.2	5	14	28	72	1475	3710	12304	9607	2139500	2262900	1620600
b. Net Use of Advances	0.00	0.00	0.00	0.00	0.00	0.00	0.0	0	0	0	-41	-43	-75	-116	0	0	0	0
Credit	0.00	0.00	0.00	0.00	0.00	0.00	0.0	0	0	0	0	0	0	20	0	0	0	0
Debit	0.00	0.00	0.00	0.00	0.00	0.00	0.0	0	0	0	41	43	75	136	0	0	0	0
XIV- Overall Balance (XIII-XII)	-0.06	-0.20	-0.53	-2.14	-3.00	-15.65	-49.9	-54	-76	-225	-1421	-3449	-12206	-123808	-525586	-1811400	-992300	-4328700
Central Bank	0.27	0.26	0.33	1.51	2.42	11.80	24.0	51	45	-43	1017	2911	10767	113321	290804	920700	0	0
Net Variation in Passive Financing	-0.21	-0.06	0.20	0.63	0.58	3.86	25.9	3	31	268	404	538	1439	10487	234782	890700	992300	4328700

Source: Ministry of Economy.

a/ Excludes provincial governments.

June 1989

Table 5.4: ARGENTINA - CENTRAL ADMINISTRATION: REVENUES, EXPENDITURES AND FINANCING, 1970-1987
(Percent of GDP)

	1970	1971	1972	1973	1974	1975	1976	1977	1978	1979	1980	1981	1982	1983	1984	1985	1986	1987 a/
Current Account																		
I- Current Revenues	7.98	6.95	6.62	5.64	6.11	4.41	5.92	6.59	6.94	5.52	6.15	6.28	6.23	5.94	6.23	11.77	11.01	10.49
Tax Revenues	7.18	5.99	5.51	4.90	5.65	3.62	5.04	5.69	5.31	4.65	5.43	5.43	5.44	5.44	5.33	10.87	10.67	10.23
Non-tax Revenues	0.80	0.96	1.11	0.73	0.45	0.78	0.88	0.91	1.62	0.87	0.72	0.85	0.78	0.50	0.90	0.91	0.34	0.26
II- Current Expenditures	5.47	5.35	4.98	6.06	6.27	6.34	6.37	5.52	7.15	6.62	7.06	8.82	9.62	8.27	7.11	6.82	6.46	11.58
Personnel Expenditures	3.53	3.20	2.95	3.80	3.68	3.66	2.60	2.48	3.06	2.69	3.26	3.29	2.48	2.94	2.79	2.35	2.35	2.71
Goods and Services	0.80	0.80	0.82	0.85	0.80	0.89	1.42	1.10	1.24	1.12	1.02	1.02	1.35	1.43	0.71	0.94	1.01	0.98
Interest on Debt	0.34	0.48	0.48	0.42	0.68	0.62	1.38	1.00	1.66	1.73	1.54	3.03	4.75	2.67	2.45	2.49	2.04	2.39
Domestic	0.23	0.24	0.19	0.25	0.56	0.48	1.28	0.91	1.53	1.66	1.28	2.12	2.95	0.26	0.43	0.09	0.22	0.73
Foreign	0.11	0.24	0.29	0.17	0.12	0.15	0.10	0.10	0.13	0.08	0.26	0.91	1.80	2.41	2.01	2.39	1.82	1.67
Other Current Expenditures	0.00	0.00	0.00	0.00	0.00	0.00	0.00	0.00	0.00	0.00	0.00	0.00	0.00	0.00	0.00	0.00	0.00	0.00
Current & Capital Transfers	0.80	0.88	0.73	0.99	1.11	1.16	0.98	0.93	1.18	1.09	1.24	1.48	1.05	1.23	1.16	1.05	1.06	5.94
Economies	0.00	0.00	0.00	0.00	0.00	0.00	0.00	0.00	0.00	0.00	0.00	0.00	0.00	0.00	0.00	0.00	0.00	-0.44
III- Savings	2.51	1.60	1.64	-0.42	-0.16	-1.94	-0.45	1.07	-0.21	-1.11	-0.91	-2.54	-3.39	-2.33	-0.88	4.95	4.55	-1.09
Capital Account																		
IV- Capital Revenues	0.00	0.00	0.05	0.00	0.04	0.01	0.01	0.00	0.00	0.00	0.00	0.00	0.00	0.00	0.00	0.00	0.00	0.12
V- Capital Expenditures	0.80	0.72	0.73	0.68	0.78	0.92	1.24	1.00	0.90	0.72	0.50	0.69	0.52	0.89	0.32	0.33	0.26	0.31
Fixed Investments	0.80	0.72	0.73	0.65	0.76	0.88	1.00	0.91	0.88	0.71	0.46	0.53	0.41	0.55	0.24	0.19	0.22	0.31
Machinery and Equipment	0.23	0.16	0.15	0.17	0.39	0.58	0.68	0.67	0.61	0.50	0.26	0.33	0.24	0.19	0.08	0.09	0.15	0.18
Construction	0.57	0.56	0.58	0.48	0.37	0.30	0.32	0.24	0.27	0.21	0.20	0.20	0.17	0.36	0.16	0.10	0.07	0.12
Changes in Inventories	0.00	0.00	0.00	0.00	0.00	0.00	0.00	0.00	0.00	0.00	0.04	0.07	0.03	0.07	0.05	0.01	0.01	0.01
Financial Investments	0.00	0.00	0.00	0.03	0.02	0.03	0.24	0.10	0.02	0.01	0.01	0.09	0.08	0.27	0.04	0.13	0.03	0.03
Economies	0.00	0.00	0.00	0.00	0.00	0.00	0.00	0.00	0.00	0.00	0.00	0.00	0.00	0.00	0.00	0.00	0.00	-0.04
VI-Financing Due to Economic Emergency	0.00	0.00	0.00	0.00	0.00	0.00	0.00	0.00	0.00	0.00	0.00	0.00	0.00	0.00	0.00	0.66	0.58	0.21
VII-Capital from Previous Period	0.00	0.00	0.00	0.00	0.00	0.00	0.00	0.00	0.00	0.00	0.02	0.05	0.04	0.12	0.10	0.03	0.00	0.00
VIII-Contributions from:	0.00	0.00	0.00	0.00	0.00	0.00	0.00	0.00	0.00	1.11	0.00	0.00	0.00	0.00	0.00	0.00	0.00	0.05
Central Administration	0.00	0.00	0.00	0.00	0.00	0.00	0.00	0.00	0.00	0.00	0.00	0.00	0.00	0.00	0.00	0.00	0.00	0.00
Special Accounts	0.00	0.00	0.00	0.00	0.00	0.00	0.00	0.00	0.00	0.00	0.00	0.00	0.00	0.00	0.00	0.00	0.00	0.04
Decentralized Agencies	0.00	0.00	0.00	0.00	0.00	0.00	0.00	0.00	0.00	0.00	0.00	0.00	0.00	0.00	0.00	0.00	0.00	0.01
Provinces & MCBA	0.00	0.00	0.00	0.00	0.00	0.00	0.00	0.00	0.00	0.33	0.00	0.00	0.00	0.00	0.00	0.00	0.00	0.00
Public Enterprises	0.00	0.00	0.00	0.00	0.00	0.00	0.00	0.00	0.00	0.78	0.00	0.00	0.00	0.00	0.00	0.00	0.00	0.00
Social Security System	0.00	0.00	0.00	0.00	0.00	0.00	0.00	0.00	0.00	0.00	0.00	0.00	0.00	0.00	0.00	0.00	0.00	0.00
IX- Contributions to:	2.61	3.22	3.68	4.94	5.80	9.52	8.14	3.76	2.94	3.21	3.43	4.87	4.86	14.13	8.74	10.26	10.05	6.16
Central Administration	0.00	0.00	0.00	0.00	0.00	0.00	0.00	0.00	0.00	0.00	0.00	0.00	0.00	0.00	0.00	0.00	0.00	0.00
Special Accounts	0.11	0.08	0.11	0.15	0.26	0.73	0.86	0.45	0.71	1.15	1.03	1.32	1.28	1.23	0.96	0.74	0.75	0.81
Decentralized Agencies	0.99	1.02	1.27	1.61	1.60	1.54	1.66	1.26	1.20	0.97	1.28	1.47	1.54	2.97	1.94	1.80	1.53	1.79
Provinces & MCBA	0.63	0.91	1.36	2.09	2.56	5.48	2.72	0.67	0.44	0.34	0.36	1.30	0.79	3.74	2.74	4.42	4.80	0.00
Public Enterprises	0.89	1.21	0.93	0.98	1.37	1.70	2.87	1.37	0.56	0.72	0.72	0.77	1.24	6.20	2.37	1.86	1.53	2.73
Social Security System	0.00	0.00	0.00	0.11	0.02	0.06	0.03	0.01	0.03	0.04	0.04	0.00	0.00	0.00	0.73	1.44	1.44	0.84
X- Revenues (I+IV+VI+VII+VIII)	7.98	6.95	6.67	5.64	6.15	4.42	5.94	6.59	6.94	6.63	6.17	6.33	6.27	6.06	6.34	12.46	11.59	10.88
XI- Expenditures (II+V+IX)	8.88	9.29	9.38	11.67	12.85	16.78	15.76	10.28	10.98	10.55	10.99	14.38	15.00	23.29	16.18	17.41	16.77	18.06
XII- Financing Needs (XI-X)	0.90	2.34	2.71	6.03	6.71	12.36	9.82	3.69	4.05	3.92	4.82	8.04	8.73	17.23	9.84	4.95	5.18	7.18
XIII-Net Financing	0.23	0.72	0.15	-0.00	0.53	1.42	3.24	1.10	2.60	2.34	-0.19	1.75	0.46	-0.91	-0.11	0.37	3.84	4.74
a. Net Use of Credit	0.23	0.72	0.15	-0.00	0.53	1.42	3.24	1.10	2.60	2.34	-0.05	1.82	0.51	-0.89	-0.11	0.37	3.84	4.74
Domestic	0.11	0.56	0.29	0.42	0.49	1.53	2.86	1.00	2.24	2.05	-0.37	0.35	0.10	-1.40	-0.55	-0.18	-0.01	0.98
Credit	0.46	0.96	0.63	0.99	1.05	2.08	3.24	1.91	2.54	2.36	0.26	1.08	2.92	-0.43	-0.03	0.02	0.94	2.12
Amortization	0.34	0.40	0.34	0.56	0.56	0.55	0.38	0.91	0.31	0.32	0.63	0.73	2.81	0.97	0.58	0.20	0.95	1.14
Foreign	0.11	0.16	-0.15	-0.42	0.04	-0.11	0.38	0.10	0.36	0.29	0.32	1.48	0.41	0.51	0.44	0.56	3.85	3.76
Credit	0.46	0.40	0.48	0.28	0.45	0.19	0.80	0.33	0.63	0.49	0.57	4.17	2.92	2.31	0.62	5.96	6.90	4.67
Amortization	0.34	0.24	0.63	0.70	0.41	0.30	0.42	0.24	0.27	0.20	0.25	2.69	2.51	1.80	0.18	5.40	3.05	0.91
b. Net Use of Advances	0.00	0.00	0.00	0.00	0.00	0.00	0.00	0.00	0.00	0.00	-0.14	-0.08	-0.05	-0.02	0.00	0.00	0.00	0.00
Credit	0.00	0.00	0.00	0.00	0.00	0.00	0.00	0.00	0.00	0.00	0.00	0.00	0.00	0.00	0.00	0.00	0.00	0.00
Debit	0.00	0.00	0.00	0.00	0.00	0.00	0.00	0.00	0.00	0.00	0.14	0.08	0.05	0.02	0.00	0.00	0.00	0.00
XIV- Overall Balance (XIII-XII)	-0.68	-1.62	-2.56	-6.03	-6.17	-10.94	-6.58	-2.59	-1.45	-1.58	-5.01	-6.30	-8.27	-18.14	-9.95	-4.58	-1.34	-2.44
Central Bank	3.05	2.08	1.62	4.27	4.98	8.25	3.16	2.44	0.86	-0.30	3.59	5.32	7.29	16.60	5.51	2.33	0.00	0.00
Net Variation in Passive Financing	-2.38	-0.46	0.95	1.77	1.19	2.70	3.42	0.15	0.59	1.88	1.43	0.98	0.97	1.54	4.45	2.25	1.34	2.44

Source: Ministry of Economy.
a/ Excludes provincial governments.
June 1989

Table 5.5: ARGENTINA - DECENTRALIZED AGENCIES: REVENUES, EXPENDITURES AND FINANCING, 1970-1987
(Thousands of Australes)

	1970	1971	1972	1973	1974	1975	1976	1977	1978	1979	1980	1981	1982	1983	1984	1985	1986	1987 a/
Current Account																		
I- Current Revenues	0.22	0.27	0.45	0.71	0.95	1.90	11.66	35.8	107.3	278	443	592	1819	15314	117829	886600	1096900	3239100
Tax Revenues	0.14	0.16	0.31	0.47	0.75	1.54	7.97	24.0	66.8	180	382	353	1102	5644	43284	383600	655500	2068400
Non-tax Revenues	0.08	0.11	0.14	0.24	0.20	0.36	3.69	11.8	40.5	98	61	239	717	9670	74545	503000	441400	1170700
II- Current Expenditures	0.19	0.25	0.44	0.88	1.31	3.21	12.13	32.9	113.5	270	626	1325	4752	23170	166904	1193000	1514400	4728200
Personnel Expenditures	0.12	0.16	0.29	0.63	0.87	2.21	6.99	17.8	59.1	157	366	646	1608	10924	91108	596400	716600	2124800
Goods and Services	0.06	0.06	0.11	0.18	0.29	0.70	4.00	11.2	34.2	60	152	267	1068	7415	45122	337700	402700	1103200
Interest on Debt	0.00	0.00	0.01	0.02	0.03	0.05	0.54	2.0	12.9	32	59	322	1794	3488	17669	147900	187900	470100
Domestic	0.00	0.00	0.01	0.02	0.01	0.03	0.24	1.3	10.8	28	49	199	1142	775	2617	35700	21300	83900
Foreign	0.00	0.00	0.00	0.00	0.02	0.02	0.30	0.7	2.1	3	10	123	652	2713	15052	112200	166600	386200
Other Current Expenditures	0.00	0.00	0.00	0.00	0.00	0.00	0.00	0.0	0.0	0	0	0	0	0	0	0	0	0
Current & Capital Transfers	0.01	0.03	0.03	0.05	0.12	0.25	0.60	2.0	7.4	22	49	90	283	1343	13005	111000	207200	1161100
Economies	0.00	0.00	0.00	0.00	0.00	0.00	0.00	0.0	0.0	0	0	0	0	0	0	0	0	-131000
III- Savings	0.03	0.02	0.01	-0.17	-0.36	-1.31	-0.47	2.9	-6.3	8	-183	-733	-2933	-7856	-49075	-306400	-417500	-1489100
Capital Account																		
IV- Capital Revenues	0.00	0.00	0.00	0.00	0.01	0.01	0.03	0.2	0.4	1	3	5	22	79	433	1700	2700	3500
V- Capital Expenditures	0.11	0.16	0.33	0.49	0.53	1.15	13.10	37.4	92.6	186	332	664	2301	10491	56030	404400	728400	1981900
Fixed Investments	0.11	0.16	0.30	0.45	0.49	1.01	12.57	34.7	77.3	159	287	609	2179	10052	53216	385400	699800	1962900
Machinery and Equipment	0.01	0.01	0.02	0.03	0.05	0.15	0.94	2.3	5.3	13	21	44	99	715	5396	54100	76700	221300
Construction	0.10	0.15	0.28	0.42	0.44	0.86	11.63	32.4	71.9	146	266	565	2080	9337	47820	331300	623100	1741600
Changes in Inventories	0.00	0.00	0.00	0.00	0.00	0.00	0.00	0.0	0.0	0	2	1	2	33	7	400	1400	8800
Financial Investments	0.00	0.00	0.03	0.04	0.04	0.14	0.53	2.7	15.3	27	43	54	120	406	2807	18600	27200	74200
Economies	0.00	0.00	0.00	0.00	0.00	0.00	0.00	0.0	0.0	0	0	0	0	0	0	0	0	-64000
VI-Financing Due to Economic Emergency	0.00	0.00	0.00	0.00	0.00	0.00	0.00	0.0	0.0	0	0	0	0	0	0	0	0	0
VII-Capital from Previous Period	0.00	0.00	0.00	0.00	0.00	0.00	0.00	0.0	0.0	8	34	101	40	324	2628	10600	47000	175000
VIII-Contributions from:	0.10	0.14	0.27	0.61	0.94	2.49	13.43	29.1	77.0	175	388	859	2351	20260	102669	714300	1163500	3183500
Central Administration	0.09	0.13	0.26	0.57	0.78	2.21	12.61	26.3	62.9	138	363	806	2272	20242	102584	713800	1136700	3169700
Special Accounts	0.00	0.00	0.01	0.03	0.14	0.23	0.81	2.1	13.3	36	24	51	75	14	43	100	500	13800
Decentralized Agencies	0.01	0.01	0.00	0.01	0.02	0.05	0.01	0.7	0.8	0	1	2	3	4	42	400	26300	0
Provinces & MCBA	0.00	0.00	0.00	0.00	0.00	0.00	0.00	0.0	0.0	0	0	0	0	0	0	0	0	0
Public Enterprises	0.00	0.00	0.00	0.00	0.00	0.00	0.00	0.0	0.0	0	0	0	0	0	0	0	0	0
Social Security System	0.00	0.00	0.00	0.00	0.00	0.00	0.00	0.0	0.0	0	0	0	0	0	0	0	0	0
IX- Contributions to:	0.02	0.02	0.01	0.03	0.06	0.09	0.01	2.2	8.9	16	48	90	217	927	800	8000	48500	45200
Central Administration	0.00	0.00	0.00	0.00	0.00	0.00	0.00	0.0	0.0	0	0	0	0	0	0	0	0	17100
Special Accounts	0.00	0.00	0.00	0.00	0.00	0.00	0.00	1.1	7.3	14	32	72	189	817	554	3700	17000	28100
Decentralized Agencies	0.02	0.02	0.01	0.03	0.06	0.09	0.01	0.7	0.8	0	1	2	3	4	42	400	26300	0
Provinces & MCBA	0.00	0.00	0.00	0.00	0.00	0.00	0.00	0.3	0.9	2	15	14	25	101	204	3900	5200	0
Public Enterprises	0.00	0.00	0.00	0.00	0.00	0.00	0.00	0.0	0.0	0	1	1	1	5	0	0	0	0
Social Security System	0.00	0.00	0.00	0.00	0.00	0.00	0.00	0.0	0.0	0	0	0	0	0	0	0	0	0
X- Revenues (I+IV+VI+VII+VIII)	0.32	0.41	0.72	1.32	1.90	4.40	25.12	65.0	184.6	462	868	1557	4232	35977	223559	1613200	2310100	6601100
XI- Expenditures (II+V+IX)	**0.32**	**0.43**	**0.78**	**1.40**	**1.90**	**4.45**	**25.24**	**72.6**	**215.1**	**473**	**1006**	**2079**	**7270**	**34588**	**223734**	**1605400**	**2291300**	**6755300**
XII- Financing Needs (XI-X)	0.00	0.02	0.06	0.08	0.00	0.05	0.12	7.5	30.4	11	138	522	3039	-1389	175	-7800	-18800	154200
XIII-Net Financing	0.01	0.02	0.06	0.08	-0.01	0.06	0.12	7.5	27.1	22	73	476	1845	-43	-34151	-2300	11500	349100
a. Net Use of Credit	0.01	0.02	0.06	0.08	-0.01	0.06	0.12	7.5	27.1	22	103	513	1862	-66	-33789	1400	13400	369100
Domestic	0.01	0.00	0.01	0.06	0.00	0.08	0.29	3.1	22.5	21	33	306	1866	-487	450	10500	9100	29400
Credit	0.01	0.00	0.01	0.07	0.06	0.14	0.69	4.4	30.9	54	70	310	2012	1299	733	13100	16600	32200
Amortization	0.00	0.00	0.00	0.01	0.06	0.06	0.40	1.3	8.4	34	37	3	146	1786	283	2600	7500	2800
Foreign	0.00	0.02	0.05	0.02	-0.01	-0.02	-0.17	4.4	4.6	1	70	207	-3	421	-34239	-9100	4300	339700
Credit	0.00	0.02	0.05	0.02	0.02	0.04	0.39	5.5	9.3	14	108	255	344	950	7820	33300	73400	513200
Amortization	0.00	0.00	0.00	0.00	0.03	0.06	0.56	1.1	4.7	14	38	48	347	529	42059	42400	69100	173500
b. Net Use of Advances	0.00	0.00	0.00	0.00	0.00	0.00	0.00	0.0	0.0	0	-30	-37	-17	23	-362	-3700	-1900	-20000
Credit	0.00	0.00	0.00	0.00	0.00	0.00	0.00	0.0	0.0	0	0	3	27	111	37	0	900	900
Debit	0.00	0.00	0.00	0.00	0.00	0.00	0.00	0.0	0.0	0	30	40	44	88	399	3700	2800	20900
XIV- Overall Balance (XIII-XII)	**0.01**	**0.00**	**0.00**	**0.00**	**-0.01**	**0.01**	**0.00**	**-0.0**	**-3.4**	**11**	**-64**	**-46**	**-1193**	**1346**	**-34326**	**5500**	**30300**	**194900**
Central Bank	**0.00**	**0.00**	**0.00**	**0.00**	**0.00**	**0.00**	**0.00**	0.0	0.0	0	0	0	0	0	0	0	0	0
Net Variation in Passive Financing	**-0.01**	**-0.00**	**-0.00**	**-0.00**	**0.01**	**-0.01**	**-0.00**	0.0	3.4	-11	64	46	1193	-1346	34326	-5500	-30300	-194900

Source: Ministry of Economy.
a/ Excludes provincial governments.
June 1989

Table 5.6: ARGENTINA - DECENTRALIZED AGENCIES REVENUES, EXPENDITURES AND FINANCING, 1970-1987
(Percent of GDP)

	1970	1971	1972	1973	1974	1975	1976	1977	1978	1979	1980	1981	1982	1983	1984	1985	1986	1987 a/
Current Account																		
I- Current Revenues	2.51	2.16	2.17	2.00	1.95	1.33	1.54	1.71	2.05	1.95	1.56	1.08	1.23	2.24	2.23	2.24	1.48	1.83
Tax Revenues	1.60	1.28	1.50	1.32	1.54	1.08	1.05	1.15	1.28	1.27	1.35	0.64	0.75	0.83	0.82	0.97	0.88	1.17
Non-tax Revenues	0.91	0.88	0.68	0.68	0.41	0.25	0.49	0.56	0.77	0.68	0.22	0.44	0.49	1.42	1.41	1.27	0.59	0.66
II- Current Expenditures	2.17	2.00	2.13	2.48	2.69	2.24	1.60	1.57	2.17	1.90	2.21	2.42	3.22	3.39	3.16	3.01	2.04	2.67
Personnel Expenditures	1.37	1.28	1.40	1.78	1.79	1.55	0.92	0.85	1.13	1.10	1.29	1.18	1.09	1.60	1.73	1.51	0.96	1.20
Goods and Services	0.68	0.48	0.53	0.51	0.60	0.49	0.53	0.53	0.65	0.42	0.54	0.49	0.72	1.09	0.85	0.85	0.54	0.62
Interest on Debt	0.00	0.00	0.05	0.06	0.06	0.03	0.07	0.09	0.25	0.22	0.21	0.59	1.22	0.51	0.33	0.37	0.25	0.27
Domestic	0.00	0.00	0.05	0.06	0.02	0.02	0.03	0.06	0.21	0.20	0.17	0.36	0.77	0.11	0.05	0.09	0.03	0.05
Foreign	0.00	0.00	0.00	0.00	0.04	0.01	0.04	0.03	0.04	0.02	0.04	0.22	0.44	0.40	0.29	0.28	0.22	0.22
Other Current Expenditures	0.00	0.00	0.00	0.00	0.00	0.00	0.00	0.00	0.00	0.00	0.00	0.00	0.00	0.00	0.00	0.00	0.00	0.00
Current & Capital Transfers	0.11	0.24	0.14	0.14	0.25	0.17	0.08	0.10	0.14	0.16	0.17	0.16	0.19	0.20	0.25	0.28	0.28	0.66
Economies	0.00	0.00	0.00	0.00	0.00	0.00	0.00	0.00	0.00	0.00	0.00	0.00	0.00	0.00	0.00	0.00	0.00	-0.07
III- Savings	0.34	0.16	0.05	-0.48	-0.74	-0.92	-0.06	0.14	-0.12	0.05	-0.65	-1.34	-1.99	-1.15	-0.93	-0.77	-0.56	-0.84
Capital Account																		
IV- Capital Revenues	0.00	0.00	0.00	0.00	0.02	0.00	0.00	0.01	0.01	0.01	0.01	0.01	0.01	0.01	0.01	0.00	0.00	0.00
V- Capital Expenditures	1.25	1.28	1.59	1.38	1.09	0.80	1.73	1.79	1.77	1.31	1.17	1.21	1.56	1.54	1.06	1.02	0.98	1.12
Fixed Investments	1.25	1.28	1.45	1.27	1.01	0.71	1.66	1.66	1.48	1.12	1.01	1.11	1.48	1.47	1.01	0.97	0.94	1.11
Machinery and Equipment	0.11	0.08	0.10	0.08	0.10	0.10	0.12	0.11	0.10	0.09	0.07	0.08	0.07	0.10	0.10	0.14	0.10	0.12
Construction	1.14	1.20	1.35	1.18	0.90	0.60	1.53	1.55	1.37	1.02	0.94	1.03	1.41	1.37	0.91	0.84	0.84	0.98
Changes in Inventories	0.00	0.00	0.00	0.00	0.00	0.00	0.00	0.00	0.00	0.00	0.01	0.00	0.00	0.00	0.00	0.00	0.00	0.00
Financial Investments	0.00	0.00	0.14	0.11	0.08	0.10	0.07	0.13	0.29	0.19	0.15	0.10	0.08	0.06	0.05	0.05	0.04	0.04
Economies	0.00	0.00	0.00	0.00	0.00	0.00	0.00	0.00	0.00	0.00	0.00	0.00	0.00	0.00	0.00	0.00	0.00	-0.04
VI-Financing Due to Economic Emergency	0.00	0.00	0.00	0.00	0.00	0.00	0.00	0.00	0.00	0.00	0.00	0.00	0.00	0.00	0.00	0.00	0.00	0.00
VII-Capital from Previous Period	0.00	0.00	0.00	0.00	0.00	0.00	0.00	0.00	0.00	0.06	0.12	0.18	0.03	0.05	0.05	0.03	0.06	0.10
VIII-Contributions from:	1.14	1.12	1.30	1.72	1.93	1.74	1.77	1.39	1.47	1.23	1.37	1.57	1.59	2.97	1.94	1.80	1.57	1.80
Central Administration	1.03	1.04	1.26	1.61	1.60	1.55	1.66	1.26	1.20	0.97	1.28	1.47	1.54	2.97	1.94	1.80	1.53	1.79
Special Accounts	0.00	0.00	0.05	0.08	0.29	0.16	0.11	0.10	0.25	0.25	0.08	0.09	0.05	0.00	0.00	0.00	0.00	0.01
Decentralized Agencies	0.11	0.08	0.00	0.03	0.04	0.03	0.00	0.03	0.02	0.00	0.00	0.00	0.00	0.00	0.00	0.00	0.04	0.00
Provinces & MCBA	0.00	0.00	0.00	0.00	0.00	0.00	0.00	0.00	0.00	0.00	0.00	0.00	0.00	0.00	0.00	0.00	0.00	0.00
Public Enterprises	0.00	0.00	0.00	0.00	0.00	0.00	0.00	0.00	0.00	0.00	0.00	0.00	0.00	0.00	0.00	0.00	0.00	0.00
Social Security System	0.00	0.00	0.00	0.00	0.00	0.00	0.00	0.00	0.00	0.00	0.00	0.00	0.00	0.00	0.00	0.00	0.00	0.00
IX- Contributions to:	0.23	0.16	0.05	0.08	0.12	0.06	0.00	0.11	0.17	0.12	0.17	0.16	0.15	0.14	0.02	0.02	0.07	0.03
Central Administration	0.00	0.00	0.00	0.00	0.00	0.00	0.00	0.00	0.00	0.00	0.00	0.00	0.00	0.00	0.00	0.00	0.00	0.01
Special Accounts	0.00	0.00	0.00	0.00	0.00	0.00	0.00	0.05	0.14	0.10	0.11	0.13	0.13	0.12	0.01	0.01	0.02	0.02
Decentralized Agencies	0.23	0.16	0.05	0.08	0.12	0.06	0.00	0.03	0.02	0.00	0.00	0.00	0.00	0.00	0.00	0.00	0.04	0.00
Provinces & MCBA	0.00	0.00	0.00	0.00	0.00	0.00	0.00	0.02	0.02	0.01	0.05	0.03	0.02	0.01	0.00	0.01	0.01	0.00
Public Enterprises	0.00	0.00	0.00	0.00	0.00	0.00	0.00	0.00	0.00	0.00	0.00	0.00	0.00	0.00	0.00	0.00	0.00	0.00
Social Security System	0.00	0.00	0.00	0.00	0.00	0.00	0.00	0.00	0.00	0.00	0.00	0.00	0.00	0.00	0.00	0.00	0.00	0.00
X- Revenues (I+IV+VI+VII+VIII)	3.65	3.28	3.48	3.72	3.91	3.08	3.31	3.11	3.53	3.24	3.06	2.84	2.87	5.27	4.23	4.07	3.11	3.73
XI Expenditures (II+V+IX)	3.65	3.44	3.77	3.95	3.91	3.11	3.33	3.47	4.11	3.32	3.55	3.80	4.93	5.07	4.24	4.05	3.08	3.81
XII- Financing Needs (XI-X)	0.00	0.16	0.29	0.23	0.00	0.04	0.02	0.36	0.58	0.08	0.49	0.95	2.06	-0.20	0.00	-0.02	-0.03	0.09
XIII-Net Financing	0.11	0.16	0.29	0.23	-0.02	0.04	0.02	0.36	0.52	0.15	0.26	0.87	1.25	-0.01	-0.65	-0.01	0.02	0.20
a. Net Use of Credit	0.11	0.16	0.29	0.23	-0.02	0.04	0.02	0.36	0.52	0.15	0.36	0.94	1.26	-0.01	-0.64	0.00	0.02	0.21
Domestic	0.11	0.00	0.05	0.17	0.00	0.06	0.04	0.15	0.43	0.15	0.12	0.56	1.26	-0.07	0.01	0.03	0.01	0.02
Credit	0.11	0.00	0.05	0.20	0.12	0.10	0.09	0.21	0.59	0.38	0.25	0.57	1.36	0.19	0.01	0.03	0.02	0.02
Amortization	0.00	0.00	0.00	0.03	0.12	0.04	0.05	0.06	0.16	0.24	0.13	0.01	0.10	0.26	0.01	0.01	0.01	0.00
Foreign	0.00	0.16	0.24	0.06	-0.02	-0.01	-0.02	0.21	0.09	0.01	0.25	0.38	-0.00	0.06	-0.65	-0.02	0.01	0.19
Credit	0.00	0.16	0.24	0.06	0.04	0.03	0.05	0.26	0.18	0.10	0.38	0.47	0.23	0.14	0.15	0.08	0.10	0.29
Amortization	0.00	0.00	0.00	0.00	0.06	0.04	0.07	0.05	0.09	0.09	0.13	0.09	0.24	0.08	0.80	0.11	0.09	0.10
b. Net Use of Advances	0.00	0.00	0.00	0.00	0.00	0.00	0.00	0.00	0.00	0.00	-0.11	-0.07	-0.01	0.00	-0.01	-0.01	-0.00	-0.01
Credit	0.00	0.00	0.00	0.00	0.00	0.00	0.00	0.00	0.00	0.00	0.00	0.01	0.02	0.02	0.00	0.00	0.00	0.00
Debit	0.00	0.00	0.00	0.00	0.00	0.00	0.00	0.00	0.00	0.00	0.11	0.07	0.03	0.01	0.01	0.01	0.00	0.01
XIV- Overall Balance (XIII-XII)	0.11	0.00	0.00	0.00	-0.02	0.01	0.00	-0.00	-0.06	0.08	-0.23	-0.08	-0.81	0.20	-0.65	0.01	0.04	0.11
Central Bank	0.00	0.00	0.00	0.00	0.00	0.00	0.00	0.00	0.00	0.00	0.00	0.00	0.00	0.00	0.00	0.00	0.00	0.00
Net Variation in Passive Financing	-0.11	-0.00	-0.00	-0.00	0.02	-0.01	-0.00	0.00	0.06	-0.08	0.23	0.08	0.81	-0.20	0.65	-0.01	-0.04	-0.11

Source: Ministry of Economy.
a/ Excludes provincial governments.
June 1989

Table 5.7: ARGENTINA - SPECIAL ACCOUNTS: REVENUES, EXPENDITURES AND FINANCING, 1970-1987
(Thousands of Australes)

	1970	1971	1972	1973	1974	1975	1976	1977	1978	1979	1980	1981	1982	1983	1984	1985	1986	1987 a/
Current Account																		
I- Current Revenues	0.11	0.13	0.24	0.57	1.34	2.76	12.0	51.9	159.6	484	1086	1586	4270	22168	135592	1335700	2614500	5847000
Tax Revenues	0.05	0.07	0.12	0.44	1.05	2.20	10.2	36.7	135.6	351	736	1091	2938	16237	104259	1102500	1951100	4343000
Non-tax Revenues	0.05	0.06	0.12	0.13	0.29	0.56	1.8	15.2	24.0	133	350	495	1332	5931	31333	233200	663400	1504000
II- Current Expenditures	0.04	0.05	0.09	0.20	0.43	1.15	4.8	13.9	45.1	174	382	719	1934	8176	61648	485600	948900	5062700
Personnel Expenditures	0.02	0.02	0.03	0.09	0.17	0.49	2.2	5.2	13.3	39	92	167	339	1857	14600	99400	200600	510000
Goods and Services	0.02	0.02	0.04	0.08	0.14	0.35	2.3	7.6	27.8	98	193	325	809	2888	18140	178100	353600	1040100
Interest on Debt	0.00	0.00	0.00	0.00	0.00	0.00	0.0	0.0	0.1	4	0	2	250	980	6987	8400	15000	62100
Domestic	0.00	0.00	0.00	0.00	0.00	0.00	0.0	0.0	0.0	4	0	0	0	375	1	0	4100	13800
Foreign	0.00	0.00	0.00	0.00	0.00	0.00	0.0	0.0	0.0	0	0	2	250	605	6986	8400	10900	48300
Other Current Expenditures	0.00	0.00	0.00	0.00	0.00	0.00	0.0	0.0	0.0	0	0	0	0	0	0	0	0	0
Current & Capital Transfers	0.01	0.01	0.02	0.03	0.11	0.31	0.3	1.0	3.9	33	97	225	536	2451	21921	199700	379700	3584500
Economies	0.00	0.00	0.00	0.00	0.00	0.00	0.0	0.0	0.0	0	0	0	0	0	0	0	0	-134000
III- Savings	0.07	0.08	0.15	0.37	0.91	1.61	7.2	38.0	114.5	310	704	867	2336	13992	73944	850100	1665600	784300
Capital Account																		
IV- Capital Revenues	0.00	0.00	0.01	0.00	0.05	0.01	0.1	0.1	0.7	1	7	15	42	220	1072	7800	16800	64100
V- Capital Expenditures	0.03	0.03	0.06	0.13	0.46	0.79	4.5	18.6	45.5	132	193	433	765	3316	9273	72000	298100	599800
Fixed Investments	0.02	0.03	0.05	0.08	0.15	0.74	4.4	18.5	43.7	122	167	361	626	2553	6637	45500	121500	462300
Machinery and Equipment	0.01	0.01	0.02	0.03	0.07	0.39	3.3	15.1	31.4	89	119	256	451	1837	4159	16600	47000	256800
Construction	0.02	0.02	0.03	0.06	0.08	0.35	1.0	3.4	12.2	32	48	105	175	716	2478	28900	74500	205500
Changes in Inventories	0.00	0.00	0.00	0.00	0.00	0.00	0.0	0.0	0.0	0	8	12	10	22	86	1700	14800	30600
Financial Investments	0.00	0.00	0.01	0.05	0.31	0.05	0.1	0.1	1.8	11	18	60	129	741	2550	24800	161800	182900
Economies	0.00	0.00	0.00	0.00	0.00	0.00	0.0	0.0	0.0	0	0	0	0	0	0	0	0	-76000
VI-Financing Due to Economic Emergency	0.00	0.00	0.00	0.00	0.00	0.00	0.0	0.0	0.0	0	0	0	0	0	0	0	0	0
VII-Capital from Previous Period	0.00	0.00	0.00	0.00	0.00	0.00	0.0	0.0	0.0	46	97	310	354	500	1999	9900	83200	345900
VIII-Contributions from:	0.02	0.01	0.03	0.09	0.26	1.31	7.4	13.6	54.6	201	376	904	2419	10735	62608	443600	826800	2142500
Central Administration	0.01	0.01	0.02	0.05	0.13	1.05	6.5	9.4	37.1	164	292	724	1896	8423	50747	292000	554900	1435800
Special Accounts	0.00	0.00	0.00	0.03	0.11	0.23	0.9	3.0	10.2	23	53	108	334	1495	11307	147900	254900	678600
Decentralized Agencies	0.00	0.00	0.01	0.01	0.02	0.03	0.0	1.1	7.3	14	32	72	189	817	554	3700	17000	28100
Provinces & MCBA	0.00	0.00	0.00	0.00	0.00	0.00	0.0	0.0	0.0	0	0	0	0	0	0	0	0	0
Public Enterprises	0.00	0.00	0.00	0.00	0.00	0.00	0.0	0.0	0.0	0	0	0	0	0	0	0	0	0
Social Security System	0.00	0.00	0.00	0.00	0.00	0.00	0.0	0.0	0.0	0	0	0	0	0	0	0	0	0
IX- Contributions to:	0.05	0.06	0.12	0.33	0.76	2.14	10.3	34.6	124.7	323	670	1324	3182	16446	103682	1064600	1881700	2493800
Central Administration	0.00	0.00	0.00	0.00	0.00	0.00	0.0	0.0	0.0	0	0	0	0	0	0	0	0	71600
Special Accounts	0.00	0.06	0.12	0.03	0.11	0.23	0.9	3.0	10.2	23	53	108	334	1495	11307	147900	254900	678600
Decentralized Agencies	0.00	0.00	0.00	0.03	0.14	0.23	0.8	2.1	13.3	36	24	51	75	14	43	100	500	13800
Provinces & MCBA	0.01	0.00	0.00	0.10	0.23	0.96	5.5	18.5	62.9	168	372	741	1789	9221	57006	487900	1074500	0
Public Enterprises	0.04	0.00	0.00	0.18	0.29	0.72	3.1	11.0	38.3	95	220	425	984	5716	35326	428700	551800	1729800
Social Security System	0.00	0.00	0.00	0.00	0.00	0.00	0.0	0.0	0.0	0	0	0	0	0	0	0	0	0
X- Revenues (I+IV+VI+VII+VIII)	0.13	0.14	0.28	0.66	1.65	4.08	19.5	65.6	214.9	732	1566	2816	7085	33623	201271	1797000	3541300	8399500
XI- Expenditures (II+V+IX)	0.12	0.14	0.27	0.66	1.65	4.08	19.6	67.1	215.3	629	1244	2476	5881	27938	174603	1622200	3128700	8156300
XII- Financing Needs (XI-X)	-0.00	0.00	-0.01	0.00	-0.00	-0.00	0.1	1.5	0.4	-104	-322	-340	-1204	-5685	-26668	-174800	-412600	-243200
XIII-Net Financing	0.00	0.00	0.00	0.00	0.00	0.00	0.1	8.6	0.4	-7	-65	-81	-814	-3304	-18059	-15400	-12500	-33000
a. Net Use of Credit	0.00	0.00	0.00	0.00	0.00	0.00	0.1	8.6	0.4	-7	-43	-70	-696	-1452	-17113	-14300	-11000	-33000
Domestic	-0.00	0.00	0.00	-0.00	-0.00	-0.00	-0.0	-0.0	-0.0	-8	-5	-11	-4	-1440	-268	15700	14500	-38000
Credit	0.00	0.00	0.00	0.00	0.01	0.00	0.0	0.0	0.0	0	0	0	0	0	0	0	0	1700
Amortization	0.00	0.00	0.00	0.00	0.01	0.00	0.0	0.0	0.0	8	5	11	4	1440	268	15700	14500	39700
Foreign	0.00	0.00	0.00	0.00	0.00	0.01	0.1	8.6	0.4	1	-38	-59	-692	-12	-16845	1400	3500	5000
Credit	0.00	0.00	0.00	0.00	0.00	0.01	0.1	8.6	0.5	1	1	57	36	296	2362	14800	21400	117700
Amortization	0.00	0.00	0.00	0.00	0.00	0.00	0.0	0.0	0.1	0	39	116	728	308	19207	13400	17900	112700
b. Net Use of Advances	0.00	0.00	0.00	0.00	0.00	0.00	0.0	0.0	0.0	0	-23	-12	-119	-1852	-946	-1100	-1500	0
Credit	0.00	0.00	0.00	0.00	0.00	0.00	0.0	0.0	0.0	0	0	0	0	1	0	0	0	0
Debit	0.00	0.00	0.00	0.00	0.00	0.00	0.0	0.0	0.0	0	23	12	119	1853	946	1100	1500	0
XIV- Overall Balance (XIII-XII)	0.00	0.00	0.01	-0.00	0.00	0.01	0.0	0.0	-0.0	97	256	258	389	2381	8609	159400	400100	210200
Central Bank	0.00	0.00	0.00	0.00	0.00	0.00	0.0	0.0	0.0	0	0	0	0	0	0	0	0	0
Net Variation in Passive Financing	-0.00	0.00	-0.01	0.00	-0.00	-0.01	0.0	0.0	0.0	-97	-256	-258	-389	-2381	-8609	-159400	-400100	-210200

Source: Ministry of Economy.

a/ Excludes provincial governments.

June 1989

Table 5.8: ARGENTINA - SPECIAL ACCOUNTS: REVENUES, EXPENDITURES AND FINANCING, 1970-1987
(Percent of GDP)

	1970	1971	1972	1973	1974	1975	1976	1977	1978	1979	1980	1981	1982	1983	1984	1985	1986	1987 a/
Current Account																		
I- Current Revenues	1.24	1.04	1.16	1.61	2.75	1.93	1.58	2.48	3.05	3.39	3.83	2.90	2.89	3.25	2.57	3.37	3.52	3.30
Tax Revenues	0.62	0.56	0.58	1.24	2.16	1.54	1.35	1.75	2.59	2.46	2.60	1.99	1.99	2.38	1.97	2.78	2.63	2.45
Non-tax Revenues	0.62	0.48	0.58	0.37	0.60	0.39	0.23	0.73	0.46	0.93	1.23	0.90	0.90	0.87	0.59	0.59	0.89	0.85
II- Current Expenditures	0.49	0.40	0.43	0.56	0.88	0.80	0.64	0.66	0.86	1.22	1.35	1.31	1.31	1.20	1.17	1.23	1.28	2.86
Personnel Expenditures	0.21	0.16	0.14	0.25	0.35	0.34	0.29	0.25	0.25	0.28	0.32	0.31	0.23	0.27	0.28	0.25	0.27	0.29
Goods and Services	0.21	0.16	0.19	0.23	0.29	0.24	0.31	0.36	0.53	0.68	0.68	0.59	0.55	0.42	0.34	0.45	0.48	0.59
Interest on Debt	0.00	0.00	0.00	0.00	0.00	0.00	0.00	0.00	0.00	0.03	0.00	0.00	0.17	0.14	0.13	0.02	0.02	0.04
Domestic	0.00	0.00	0.00	0.00	0.00	0.00	0.00	0.00	0.00	0.03	0.00	0.00	0.00	0.05	0.00	0.00	0.01	0.01
Foreign	0.00	0.00	0.00	0.00	0.00	0.00	0.00	0.00	0.00	0.00	0.00	0.00	0.17	0.09	0.13	0.02	0.01	0.03
Other Current Expenditures	0.00	0.00	0.00	0.00	0.00	0.00	0.00	0.00	0.00	0.00	0.00	0.00	0.00	0.00	0.00	0.00	0.00	0.00
Current & Capital Transfers	0.07	0.08	0.10	0.08	0.24	0.22	0.04	0.05	0.08	0.23	0.34	0.41	0.36	0.36	0.42	0.50	0.51	2.02
Economies	0.00	0.00	0.00	0.00	0.00	0.00	0.00	0.00	0.00	0.00	0.00	0.00	0.00	0.00	0.00	0.00	0.00	-0.08
III- Savings	0.74	0.64	0.72	1.04	1.88	1.13	0.95	1.82	2.19	2.18	2.49	1.58	1.58	2.05	1.40	2.15	2.24	0.44
Capital Account																		
IV- Capital Revenues	0.03	0.00	0.05	0.00	0.11	0.01	0.01	0.00	0.01	0.01	0.02	0.03	0.03	0.03	0.02	0.02	0.02	0.04
V- Capital Expenditures	0.29	0.24	0.29	0.36	0.96	0.55	0.59	0.89	0.87	0.93	0.68	0.79	0.52	0.49	0.18	0.18	0.40	0.34
Fixed Investments	0.25	0.24	0.24	0.23	0.31	0.52	0.57	0.88	0.83	0.85	0.59	0.66	0.42	0.37	0.13	0.11	0.16	0.26
Machinery and Equipment	0.08	0.08	0.10	0.07	0.14	0.27	0.44	0.72	0.60	0.63	0.42	0.47	0.31	0.27	0.08	0.04	0.06	0.14
Construction	0.17	0.16	0.14	0.16	0.17	0.24	0.13	0.16	0.23	0.23	0.17	0.19	0.12	0.10	0.05	0.07	0.10	0.12
Changes in Inventories	0.00	0.00	0.00	0.00	0.00	0.00	0.00	0.00	0.00	0.00	0.03	0.02	0.01	0.00	0.00	0.00	0.02	0.02
Financial Investments	0.03	0.00	0.05	0.13	0.65	0.03	0.01	0.01	0.03	0.08	0.06	0.11	0.09	0.11	0.05	0.06	0.22	0.10
Economies	0.00	0.00	0.00	0.00	0.00	0.00	0.00	0.00	0.00	0.00	0.00	0.00	0.00	0.00	0.00	0.00	0.00	-0.04
VI-Financing Due to Economic Emergency	0.00	0.00	0.00	0.00	0.00	0.00	0.00	0.00	0.00	0.00	0.00	0.00	0.00	0.00	0.00	0.00	0.00	0.00
VII-Capital from Previous Period	0.00	0.00	0.00	0.00	0.00	0.00	0.00	0.00	0.00	0.32	0.34	0.57	0.24	0.07	0.04	0.03	0.11	0.20
VIII-Contributions from:	0.18	0.08	0.14	0.25	0.53	0.92	0.98	0.65	1.04	1.41	1.33	1.65	1.64	1.57	1.19	1.12	1.11	1.21
Central Administration	0.11	0.08	0.10	0.14	0.27	0.73	0.86	0.45	0.71	1.15	1.03	1.32	1.28	1.23	0.96	0.74	0.75	0.81
Special Accounts	0.02	0.00	0.00	0.08	0.23	0.16	0.12	0.15	0.20	0.16	0.19	0.20	0.23	0.22	0.21	0.37	0.34	0.38
Decentralized Agencies	0.05	0.00	0.05	0.03	0.04	0.02	0.00	0.05	0.14	0.10	0.11	0.13	0.13	0.12	0.01	0.01	0.02	0.02
Provinces & MCBA	0.00	0.00	0.00	0.00	0.00	0.00	0.00	0.00	0.00	0.00	0.00	0.00	0.00	0.00	0.00	0.00	0.00	0.00
Public Enterprises	0.00	0.00	0.00	0.00	0.00	0.00	0.00	0.00	0.00	0.00	0.00	0.00	0.00	0.00	0.00	0.00	0.00	0.00
Social Security System	0.00	0.00	0.00	0.00	0.00	0.00	0.00	0.00	0.00	0.00	0.00	0.00	0.00	0.00	0.00	0.00	0.00	0.00
IX- Contributions to:	0.61	0.48	0.58	0.94	1.57	1.49	1.36	1.65	2.38	2.26	2.36	2.42	2.16	2.41	1.96	2.69	2.53	1.41
Central Administration	0.00	0.00	0.00	0.00	0.00	0.00	0.00	0.00	0.00	0.00	0.00	0.00	0.00	0.00	0.00	0.00	0.00	0.04
Special Accounts	0.02	0.48	0.58	0.09	0.22	0.16	0.12	0.15	0.20	0.16	0.19	0.20	0.23	0.22	0.21	0.37	0.34	0.38
Decentralized Agencies	0.04	0.00	0.00	0.07	0.28	0.16	0.11	0.10	0.25	0.25	0.09	0.09	0.05	0.00	0.00	0.00	0.00	0.01
Provinces & MCBA	0.13	0.00	0.00	0.27	0.48	0.67	0.72	0.88	1.20	1.18	1.31	1.35	1.21	1.35	1.08	1.23	1.45	0.00
Public Enterprises	0.43	0.00	0.00	0.51	0.59	0.50	0.40	0.53	0.73	0.67	0.78	0.78	0.67	0.84	0.67	1.08	0.74	0.98
Social Security System	0.00	0.00	0.00	0.00	0.00	0.00	0.00	0.00	0.00	0.00	0.00	0.00	0.00	0.00	0.00	0.00	0.00	0.00
X- Revenues (I+IV+VI+VII+VIII)	**1.44**	**1.12**	**1.35**	**1.86**	**3.40**	**2.85**	**2.57**	**3.13**	**4.11**	**5.14**	**5.53**	**5.14**	**4.80**	**4.93**	**3.81**	**4.54**	**4.77**	**4.74**
XI- Expenditures (II+V+IX)	**1.39**	**1.12**	**1.30**	**1.86**	**3.40**	**2.85**	**2.58**	**3.20**	**4.11**	**4.41**	**4.39**	**4.52**	**3.98**	**4.09**	**3.31**	**4.10**	**4.21**	**4.60**
XII- Financing Needs (XI-X)	-0.05	0.00	-0.05	0.01	-0.00	-0.00	0.01	0.07	0.01	-0.73	-1.14	-0.62	-0.82	-0.83	-0.50	-0.44	-0.56	-0.14
XIII-Net Financing	0.01	0.00	0.00	0.00	0.00	0.00	0.01	0.41	0.01	-0.05	-0.23	-0.15	-0.55	-0.48	-0.34	-0.04	-0.02	-0.02
a. Net Use of Credit	0.01	0.00	0.00	0.00	0.00	0.00	0.01	0.41	0.01	-0.05	-0.15	-0.13	-0.47	-0.21	-0.32	-0.04	-0.01	-0.02
Domestic	-0.00	0.00	0.00	-0.00	-0.00	-0.00	-0.00	-0.00	-0.00	-0.05	-0.02	-0.02	-0.00	-0.21	-0.01	-0.04	-0.02	-0.02
Credit	0.00	0.00	0.00	0.00	0.01	0.00	0.00	0.00	0.00	0.00	0.00	0.00	0.00	0.00	0.00	0.00	0.00	0.00
Amortization	0.00	0.00	0.00	0.00	0.01	0.00	0.00	0.00	0.00	0.06	0.02	0.02	0.00	0.21	0.01	0.04	0.02	0.02
Foreign	0.01	0.00	0.00	0.01	0.01	0.00	0.01	0.41	0.01	0.00	-0.13	-0.11	-0.47	-0.00	-0.32	0.00	0.00	0.00
Credit	0.01	0.00	0.00	0.01	0.01	0.01	0.01	0.41	0.01	0.00	0.00	0.10	0.02	0.04	0.04	0.04	0.03	0.07
Amortization	0.00	0.00	0.00	0.00	0.00	0.00	0.00	0.00	0.00	0.00	0.14	0.21	0.49	0.05	0.36	0.03	0.02	0.06
b. Net Use of Advances	0.00	0.00	0.00	0.00	0.00	0.00	0.00	0.00	0.00	0.00	-0.08	-0.02	-0.08	-0.27	-0.02	-0.00	-0.00	0.00
Credit	0.00	0.00	0.00	0.00	0.00	0.00	0.00	0.00	0.00	0.00	0.00	0.00	0.00	0.00	0.00	0.00	0.00	0.00
Debit	0.00	0.00	0.00	0.00	0.00	0.00	0.00	0.00	0.00	0.00	0.08	0.02	0.08	0.27	0.02	0.00	0.00	0.00
XIV- Overall Balance (XIII-XII)	**0.05**	**0.00**	**0.05**	**-0.00**	**0.01**	**0.01**	**0.00**	**0.00**	**-0.00**	**0.68**	**0.90**	**0.47**	**0.26**	**0.35**	**0.16**	**0.40**	**0.54**	**0.12**
Central Bank	**0.00**	**0.00**	**0.00**	**0.00**	**0.00**	**0.00**	**0.00**	**0.00**	**0.00**	**0.00**	**0.00**	**0.00**	**0.00**	**0.00**	**0.00**	**0.00**	**0.00**	**0.00**
Net Variation in Passive Financing	**-0.05**	**0.00**	**-0.05**	**0.00**	**-0.01**	**-0.01**	**0.00**	**0.00**	**0.00**	**-0.68**	**-0.90**	**-0.47**	**-0.26**	**-0.35**	**-0.16**	**-0.40**	**-0.54**	**-0.12**

Source: Ministry of Economy.

a/ Excludes provincial governments.

June 1989

Table 5.9: ARGENTINA - SOCIAL SECURITY: REVENUES, EXPENDITURES AND FINANCING, 1970-1987
(Thousands of Australes)

	1970	1971	1972	1973	1974	1975	1976	1977	1978	1979	1980	1981	1982	1983	1984	1985	1986	1987 a/
Current Account																		
I- Current Revenues	0.40	0.57	0.77	1.64	2.46	6.15	29.91	85.0	226.2	685	1652	3338	7229	35481	141899	1765900	3143000	7176800
Tax Revenues	0.40	0.57	0.76	1.60	2.43	6.13	29.17	74.4	220.8	669	1581	2936	6358	32254	135384	1421200	2936300	6687200
Non-tax Revenues	0.00	0.00	0.01	0.04	0.03	0.02	0.74	10.7	5.4	16	71	402	871	3227	6515	344700	206700	489600
II- Current Expenditures	0.38	0.56	0.76	1.46	2.33	5.73	25.04	69.9	241.7	697	1668	3329	7174	41479	293628	2215000	4103900	8783600
Personnel Expenditures	0.00	0.00	0.00	0.00	0.00	0.00	0.00	0.0	0.0	0	0	0	0	0	0	0	0	0
Goods and Services	0.00	0.00	0.00	0.00	0.00	0.00	0.00	0.0	0.0	0	0	0	0	0	0	0	0	0
Interest on Debt	0.00	0.00	0.00	0.00	0.00	0.00	0.00	0.0	0.0	0	0	0	0	0	0	0	0	0
Domestic	0.00	0.00	0.00	0.00	0.00	0.00	0.00	0.0	0.0	0	0	0	0	0	0	0	0	0
Foreign	0.00	0.00	0.00	0.00	0.00	0.00	0.00	0.0	0.0	0	0	0	0	0	0	0	0	0
Other Current Expenditures	0.00	0.00	0.00	0.00	0.00	0.00	0.00	0.0	0.0	9	22	50	92	0	0	0	0	0
Current & Capital Transfers	0.38	0.56	0.76	1.46	2.33	5.73	25.04	69.9	241.7	688	1646	3279	7082	41479	293628	2215000	4103900	8783600
Economies	0.00	0.00	0.00	0.00	0.00	0.00	0.00	0.0	0.0	0	0	0	0	0	0	0	0	0
III- Savings	0.02	0.01	0.01	0.18	0.13	0.42	4.87	15.1	-15.5	-12	-16	9	55	-5998	-151729	-449100	-960900	-1606800
Capital Account																		
IV- Capital Revenues	0.00	0.00	0.00	0.00	0.00	0.00	0.00	0.0	14.6	9	2	1	4	0	0	0	0	0
V- Capital Expenditures	0.01	0.00	0.01	0.00	0.00	0.00	4.87	15.1	0.0	0	0	0	0	0	0	0	0	0
Fixed Investments	0.00	0.00	0.00	0.00	0.00	0.00	0.00	0.0	0.0	0	0	0	0	0	0	0	0	0
Machinery and Equipment	0.00	0.00	0.00	0.00	0.00	0.00	0.00	0.0	0.0	0	0	0	0	0	0	0	0	0
Construction	0.00	0.00	0.00	0.00	0.00	0.00	0.00	0.0	0.0	0	0	0	0	0	0	0	0	0
Changes in Inventories	0.00	0.00	0.00	0.00	0.00	0.00	0.00	0.0	0.0	0	0	0	0	0	0	0	0	0
Financial Investments	0.01	0.00	0.01	0.00	0.00	0.00	4.87	15.1	0.0	0	0	0	0	0	0	0	0	0
Economies	0.00	0.00	0.00	0.00	0.00	0.00	0.00	0.0	0.0	0	0	0	0	0	0	0	0	0
VI-Financing Due to Economic Emergency	0.00	0.00	0.00	0.00	0.00	0.00	0.00	0.0	0.0	0	0	0	0	0	0	0	0	0
VII-Capital from Previous Period	0.00	0.00	0.00	0.00	0.00	0.00	0.00	0.0	0.0	0	0	0	0	0	0	0	0	0
VIII-Contributions from:	0.00	0.00	0.00	0.04	0.01	0.09	0.24	0.3	1.4	5	10	0	0	0	154155	570500	1073700	1486000
Central Administration	0.00	0.00	0.00	0.04	0.01	0.09	0.24	0.3	1.4	5	10	0	0	0	154155	570500	1073700	1486000
Special Accounts	0.00	0.00	0.00	0.00	0.00	0.00	0.00	0.0	0.0	0	0	0	0	0	0	0	0	0
Decentralized Agencies	0.00	0.00	0.00	0.00	0.00	0.00	0.00	0.0	0.0	0	0	0	0	0	0	0	0	0
Provinces & MCBA	0.00	0.00	0.00	0.00	0.00	0.00	0.00	0.0	0.0	0	0	0	0	0	0	0	0	0
Public Enterprises	0.00	0.00	0.00	0.00	0.00	0.00	0.00	0.0	0.0	0	0	0	0	0	0	0	0	0
Social Security System	0.00	0.00	0.00	0.00	0.00	0.00	0.00	0.0	0.0	0	0	0	0	0	0	0	0	0
IX- Contributions to:	0.00	0.00	0.00	0.00	0.00	0.00	0.00	0.0	0.0	0	0	0	0	0	0	0	0	0
Central Administration	0.00	0.00	0.00	0.00	0.00	0.00	0.00	0.0	0.0	0	0	0	0	0	0	0	0	0
Special Accounts	0.00	0.00	0.00	0.00	0.00	0.00	0.00	0.0	0.0	0	0	0	0	0	0	0	0	0
Decentralized Agencies	0.00	0.00	0.00	0.00	0.00	0.00	0.00	0.0	0.0	0	0	0	0	0	0	0	0	0
Provinces & MCBA	0.00	0.00	0.00	0.00	0.00	0.00	0.00	0.0	0.0	0	0	0	0	0	0	0	0	0
Public Enterprises	0.00	0.00	0.00	0.00	0.00	0.00	0.00	0.0	0.0	0	0	0	0	0	0	0	0	0
Social Security System	0.00	0.00	0.00	0.00	0.00	0.00	0.00	0.0	0.0	0	0	0	0	0	0	0	0	0
X- Revenues (I+IV+VI+VII+VIII)	0.40	0.57	0.77	1.68	2.47	6.24	30.15	85.3	242.2	699	1664	3339	7233	35481	296054	2336400	4216700	8662800
XI- Expenditures (II+V+IX)	0.39	0.56	0.77	1.46	2.33	5.73	29.91	85.0	241.7	697	1668	3329	7174	41479	293628	2215000	4103900	8783600
XII- Financing Needs (XI-X)	-0.01	-0.01	0.00	-0.22	-0.14	-0.51	-0.24	-0.3	-0.5	-3	4	-10	-59	5998	-2426	-121400	-112800	120800
XIII-Net Financing	0.00	0.00	0.00	0.00	0.00	0.00	0.00	0.0	0.0	0	0	0	0	0	0	0	0	0
a. Net Use of Credit	0.00	0.00	0.00	0.00	0.00	0.00	0.00	0.0	0.0	0	0	0	0	0	0	0	0	0
Domestic	0.00	0.00	0.00	0.00	0.00	0.00	0.00	0.0	0.0	0	0	0	0	0	0	0	0	0
Credit	0.00	0.00	0.00	0.00	0.00	0.00	0.00	0.0	0.0	0	0	0	0	0	0	0	0	0
Amortization	0.00	0.00	0.00	0.00	0.00	0.00	0.00	0.0	0.0	0	0	0	0	0	0	0	0	0
Foreign	0.00	0.00	0.00	0.00	0.00	0.00	0.00	0.0	0.0	0	0	0	0	0	0	0	0	0
Credit	0.00	0.00	0.00	0.00	0.00	0.00	0.00	0.0	0.0	0	0	0	0	0	0	0	0	0
Amortization	0.00	0.00	0.00	0.00	0.00	0.00	0.00	0.0	0.0	0	0	0	0	0	0	0	0	0
b. Net Use of Advances	0.00	0.00	0.00	0.00	0.00	0.00	0.00	0.0	0.0	0	0	0	0	0	0	0	0	0
Credit	0.00	0.00	0.00	0.00	0.00	0.00	0.00	0.0	0.0	0	0	0	0	0	0	0	0	0
Debit	0.00	0.00	0.00	0.00	0.00	0.00	0.00	0.0	0.0	0	0	0	0	0	0	0	0	0
XIV- Overall Balance (XIII-XII)	0.01	0.01	0.00	0.22	0.14	0.51	0.24	0.3	0.5	3	-4	10	59	-5998	2426	121400	112800	-120800
Central Bank	0.00	0.00	0.00	0.00	0.00	0.00	0.00	0.0	0.0	0	0	0	0	0	36132	0	0	0
Net Variation in Passive Financing	-0.01	-0.01	0.00	-0.22	-0.14	-0.51	-0.24	-0.3	-0.5	-3	4	-10	-59	5998	-38558	-121400	-112800	120800

Source: Ministry of Economy.

a/ Excludes provincial governments.

June 1989

Table 5.10: ARGENTINA - SOCIAL SECURITY: REVENUES, EXPENDITURES AND FINANCING, 1970-1987
(Percent of GDP)

	1970	1971	1972	1973	1974	1975	1976	1977	1978	1979	1980	1981	1982	1983	1984	1985	1986	1987 a/
Current Account																		
I- Current Revenues	4.56	4.55	3.72	4.62	5.06	4.30	3.94	4.06	4.32	4.81	5.83	6.10	4.90	5.20	2.69	4.46	4.23	4.05
Tax Revenues	4.56	4.55	3.67	4.51	5.00	4.29	3.84	3.55	4.22	4.70	5.58	5.36	4.31	4.72	2.56	3.59	3.95	3.77
Non-tax Revenues	0.00	0.00	0.05	0.11	0.06	0.01	0.10	0.51	0.10	0.11	0.25	0.73	0.59	0.47	0.12	0.87	0.28	0.28
II- Current Expenditures	4.33	4.47	3.67	4.11	4.79	4.01	3.30	3.34	4.62	4.89	5.89	6.08	4.86	6.08	5.56	5.59	5.52	4.96
Personnel Expenditures	0.00	0.00	0.00	0.00	0.00	0.00	0.00	0.00	0.00	0.00	0.00	0.00	0.00	0.00	0.00	0.00	0.00	0.00
Goods and Services	0.00	0.00	0.00	0.00	0.00	0.00	0.00	0.00	0.00	0.00	0.00	0.00	0.00	0.00	0.00	0.00	0.00	0.00
Interest on Debt	0.00	0.00	0.00	0.00	0.00	0.00	0.00	0.00	0.00	0.00	0.00	0.00	0.00	0.00	0.00	0.00	0.00	0.00
Domestic	0.00	0.00	0.00	0.00	0.00	0.00	0.00	0.00	0.00	0.00	0.00	0.00	0.00	0.00	0.00	0.00	0.00	0.00
Foreign	0.00	0.00	0.00	0.00	0.00	0.00	0.00	0.00	0.00	0.00	0.00	0.00	0.00	0.00	0.00	0.00	0.00	0.00
Other Current Expenditures	0.00	0.00	0.00	0.00	0.00	0.00	0.00	0.00	0.00	0.06	0.08	0.09	0.06	0.00	0.00	0.00	0.00	0.00
Current & Capital Transfers	4.33	4.47	3.67	4.11	4.79	4.01	3.30	3.34	4.62	4.82	5.81	5.99	4.80	6.08	5.56	5.59	5.52	4.96
Economies	0.00	0.00	0.00	0.00	0.00	0.00	0.00	0.00	0.00	0.00	0.00	0.00	0.00	0.00	0.00	0.00	0.00	0.00
III- Savings	0.23	0.08	0.05	0.51	0.27	0.29	0.64	0.72	-0.30	-0.08	-0.06	0.02	0.04	-0.88	-2.87	-1.13	-1.29	-0.91
Capital Account	0.00	0.00	0.00	0.00	0.00	0.00	0.00	0.00	0.00	0.00	0.00	0.00	0.00	0.00	0.00	0.00	0.00	0.00
IV- Capital Revenues	0.00	0.00	0.00	0.00	0.00	0.00	0.00	0.00	0.28	0.07	0.01	0.00	0.00	0.00	0.00	0.00	0.00	0.00
V- Capital Expenditures	0.11	0.00	0.05	0.00	0.00	0.00	0.64	0.72	0.00	0.00	0.00	0.00	0.00	0.00	0.00	0.00	0.00	0.00
Fixed Investments	0.00	0.00	0.00	0.00	0.00	0.00	0.00	0.00	0.00	0.00	0.00	0.00	0.00	0.00	0.00	0.00	0.00	0.00
Machinery and Equipment	0.00	0.00	0.00	0.00	0.00	0.00	0.00	0.00	0.00	0.00	0.00	0.00	0.00	0.00	0.00	0.00	0.00	0.00
Construction	0.00	0.00	0.00	0.00	0.00	0.00	0.00	0.00	0.00	0.00	0.00	0.00	0.00	0.00	0.00	0.00	0.00	0.00
Changes in Inventories	0.00	0.00	0.00	0.00	0.00	0.00	0.00	0.00	0.00	0.00	0.00	0.00	0.00	0.00	0.00	0.00	0.00	0.00
Financial Investments	0.11	0.00	0.05	0.00	0.00	0.00	0.64	0.72	0.00	0.00	0.00	0.00	0.00	0.00	0.00	0.00	0.00	0.00
Economies	0.00	0.00	0.00	0.00	0.00	0.00	0.00	0.00	0.00	0.00	0.00	0.00	0.00	0.00	0.00	0.00	0.00	0.00
VI-Financing Due to Economic Emergency	0.00	0.00	0.00	0.00	0.00	0.00	0.00	0.00	0.00	0.00	0.00	0.00	0.00	0.00	0.00	0.00	0.00	0.00
VII-Capital from Previous Period	0.00	0.00	0.00	0.00	0.00	0.00	0.00	0.00	0.00	0.00	0.00	0.00	0.00	0.00	0.00	0.00	0.00	0.00
VIII-Contributions from:	0.00	0.00	0.00	0.11	0.02	0.06	0.03	0.01	0.03	0.04	0.04	0.00	0.00	0.00	2.92	1.44	1.44	0.84
Central Administration	0.00	0.00	0.00	0.11	0.02	0.06	0.03	0.01	0.03	0.04	0.04	0.00	0.00	0.00	2.92	1.44	1.44	0.84
Special Accounts	0.00	0.00	0.00	0.00	0.00	0.00	0.00	0.00	0.00	0.00	0.00	0.00	0.00	0.00	0.00	0.00	0.00	0.00
Decentralized Agencies	0.00	0.00	0.00	0.00	0.00	0.00	0.00	0.00	0.00	0.00	0.00	0.00	0.00	0.00	0.00	0.00	0.00	0.00
Provinces & MCBA	0.00	0.00	0.00	0.00	0.00	0.00	0.00	0.00	0.00	0.00	0.00	0.00	0.00	0.00	0.00	0.00	0.00	0.00
Public Enterprises	0.00	0.00	0.00	0.00	0.00	0.00	0.00	0.00	0.00	0.00	0.00	0.00	0.00	0.00	0.00	0.00	0.00	0.00
Social Security System	0.00	0.00	0.00	0.00	0.00	0.00	0.00	0.00	0.00	0.00	0.00	0.00	0.00	0.00	0.00	0.00	0.00	0.00
IX- Contributions to:	0.00	0.00	0.00	0.00	0.00	0.00	0.00	0.00	0.00	0.00	0.00	0.00	0.00	0.00	0.00	0.00	0.00	0.00
Central Administration	0.00	0.00	0.00	0.00	0.00	0.00	0.00	0.00	0.00	0.00	0.00	0.00	0.00	0.00	0.00	0.00	0.00	0.00
Special Accounts	0.00	0.00	0.00	0.00	0.00	0.00	0.00	0.00	0.00	0.00	0.00	0.00	0.00	0.00	0.00	0.00	0.00	0.00
Decentralized Agencies	0.00	0.00	0.00	0.00	0.00	0.00	0.00	0.00	0.00	0.00	0.00	0.00	0.00	0.00	0.00	0.00	0.00	0.00
Provinces & MCBA	0.00	0.00	0.00	0.00	0.00	0.00	0.00	0.00	0.00	0.00	0.00	0.00	0.00	0.00	0.00	0.00	0.00	0.00
Public Enterprises	0.00	0.00	0.00	0.00	0.00	0.00	0.00	0.00	0.00	0.00	0.00	0.00	0.00	0.00	0.00	0.00	0.00	0.00
Social Security System	0.00	0.00	0.00	0.00	0.00	0.00	0.00	0.00	0.00	0.00	0.00	0.00	0.00	0.00	0.00	0.00	0.00	0.00
X- Revenues (I+IV+VI+VII+VIII)	**4.56**	**4.55**	**3.72**	**4.73**	**5.08**	**4.36**	**3.97**	**4.08**	**4.63**	**4.91**	**5.87**	**6.10**	**4.90**	**5.20**	**5.61**	**5.90**	**5.67**	**4.89**
XI- Expenditures (II+V+IX)	**4.44**	**4.47**	**3.72**	**4.11**	**4.79**	**4.01**	**3.94**	**4.06**	**4.62**	**4.89**	**5.89**	**6.08**	**4.86**	**6.08**	**5.56**	**5.59**	**5.52**	**4.96**
XII- Financing Needs (XI-X)	-0.11	-0.08	0.00	-0.62	-0.29	-0.36	-0.03	-0.01	-0.01	-0.02	0.01	-0.02	-0.04	0.88	-0.05	-0.31	-0.15	0.07
XIII-Net Financing	0.00	0.00	0.00	0.00	0.00	0.00	0.00	0.00	0.00	0.00	0.00	0.00	0.00	0.00	0.00	0.00	0.00	0.00
a. Net Use of Credit	0.00	0.00	0.00	0.00	0.00	0.00	0.00	0.00	0.00	0.00	0.00	0.00	0.00	0.00	0.00	0.00	0.00	0.00
Domestic	0.00	0.00	0.00	0.00	0.00	0.00	0.00	0.00	0.00	0.00	0.00	0.00	0.00	0.00	0.00	0.00	0.00	0.00
Credit	0.00	0.00	0.00	0.00	0.00	0.00	0.00	0.00	0.00	0.00	0.00	0.00	0.00	0.00	0.00	0.00	0.00	0.00
Amortization	0.00	0.00	0.00	0.00	0.00	0.00	0.00	0.00	0.00	0.00	0.00	0.00	0.00	0.00	0.00	0.00	0.00	0.00
Foreign	0.00	0.00	0.00	0.00	0.00	0.00	0.00	0.00	0.00	0.00	0.00	0.00	0.00	0.00	0.00	0.00	0.00	0.00
Credit	0.00	0.00	0.00	0.00	0.00	0.00	0.00	0.00	0.00	0.00	0.00	0.00	0.00	0.00	0.00	0.00	0.00	0.00
Amortization	0.00	0.00	0.00	0.00	0.00	0.00	0.00	0.00	0.00	0.00	0.00	0.00	0.00	0.00	0.00	0.00	0.00	0.00
b. Net Use of Advances	0.00	0.00	0.00	0.00	0.00	0.00	0.00	0.00	0.00	0.00	0.00	0.00	0.00	0.00	0.00	0.00	0.00	0.00
Credit	0.00	0.00	0.00	0.00	0.00	0.00	0.00	0.00	0.00	0.00	0.00	0.00	0.00	0.00	0.00	0.00	0.00	0.00
Debit	0.00	0.00	0.00	0.00	0.00	0.00	0.00	0.00	0.00	0.00	0.00	0.00	0.00	0.00	0.00	0.00	0.00	0.00
XIV- Overall Balance (XIII-XII)	0.11	0.08	0.00	0.62	0.29	0.36	0.03	0.01	0.01	0.02	-0.01	0.02	0.04	-0.88	0.05	0.31	0.15	-0.07
Central Bank	0.00	0.00	0.00	0.00	0.00	0.00	0.00	0.00	0.00	0.00	0.00	0.00	0.00	0.00	0.68	0.00	0.00	0.00
Net Variation in Passive Financing	-0.11	-0.08	0.00	-0.62	-0.29	-0.36	-0.03	-0.01	-0.01	-0.02	0.01	-0.02	-0.04	0.88	-0.73	-0.31	-0.15	0.07

Source: Ministry of Economy.
a/ Excludes provincial governments.
June 1989

Table 5.11: ARGENTINA - PROVINCIAL GOVERNMENTS: REVENUES, EXPENDITURES AND FINANCING, 1970-1986
(Thousands of Australes)

	1970	1971	1972	1973	1974	1975	1976	1977	1978	1979	1980	1981	1982	1983	1984	1985	1986
Current Account																	
I- Current Revenues	0.59	0.71	0.96	1.86	3.05	5.51	38.3	168.8	460.5	1240	2694	4374	10765	44428	342346	2007700	3677600
Tax Revenues	0.49	0.58	0.83	1.60	2.62	4.43	34.6	147.8	391.5	1068	2352	3784	9210	35353	278199	1504700	2830300
Non-tax Revenues	0.10	0.13	0.13	0.26	0.42	1.08	3.7	21.0	69.0	172	342	590	1556	9075	64147	503000	847300
II- Current Expenditures	0.47	0.64	1.03	2.22	3.66	11.27	41.0	114.5	340.0	935	2272	4670	10390	58096	485776	3474200	6276400
Personnel Expenditures	0.33	0.46	0.74	1.62	2.60	8.23	26.7	65.6	218.7	614	1456	2634	5696	33773	305877	2210900	4067300
Goods and Services	0.07	0.09	0.13	0.24	0.37	0.95	4.8	16.2	44.8	116	300	488	1261	5710	50322	405800	768400
Interest on Debt	0.00	0.00	0.00	0.01	0.02	0.05	0.2	0.4	0.8	7	27	144	223	1232	972	19300	12900
Domestic	0.00	0.00	0.00	0.01	0.02	0.05	0.2	0.4	0.8	7	27	144	209	253	670	18000	0
Foreign	0.00	0.00	0.00	0.00	0.00	0.00	0.0	0.0	0.0	0	0	0	14	979	302	1300	12900
Other Current Expenditures	0.00	0.00	0.00	0.00	0.00	0.00	0.0	0.0	0.0	0	0	0	0	0	0	0	0
Current & Capital Transfers	0.07	0.10	0.16	0.35	0.67	2.04	9.3	32.3	75.8	197	489	1404	3210	17381	128605	838200	1427800
Economies	0.00	0.00	0.00	0.00	0.00	0.00	0.0	0.0	0.0	0	0	0	0	0	0	0	0
III- Savings	0.12	0.07	-0.06	-0.36	-0.61	-5.76	-2.8	54.3	120.5	305	422	-295	375	-13668	-143430	-1466500	-2598800
Capital Account																	
IV- Capital Revenues	0.01	0.01	0.01	0.01	0.06	0.05	0.4	0.9	3.2	7	16	26	119	234	3811	15200	15200
V- Capital Expenditures	0.20	0.24	0.44	0.64	1.18	3.89	25.4	82.1	205.2	518	1016	1897	3777	19750	142258	1028000	2007700
Fixed Investments	0.19	0.24	0.42	0.61	1.12	3.79	24.8	76.0	198.5	501	961	1518	3447	18822	128463	959600	1953700
Machinery and Equipment	0.02	0.02	0.03	0.04	0.10	0.23	1.5	6.5	13.0	21	43	61	127	725	6218	40200	47100
Construction	0.17	0.21	0.40	0.57	1.01	3.56	23.3	69.5	185.5	480	918	1457	3320	18097	122245	919400	1906600
Changes in Inventories	0.00	0.00	0.00	0.00	0.00	0.00	0.0	0.0	0.0	0	30	77	125	230	1861	14600	15200
Financial Investments	0.01	0.00	0.02	0.02	0.06	0.10	0.5	6.1	6.7	17	25	302	205	698	11934	53800	38800
Economies	0.00	0.00	0.00	0.00	0.00	0.00	0.0	0.0	0.0	0	0	0	0	0	0	0	0
VI-Financing Due to Economic Emergency	0.00	0.00	0.00	0.00	0.00	0.00	0.0	0.0	0.0	0	0	0	0	0	0	0	0
VII-Capital from Previous Period	0.00	0.00	0.00	0.00	0.00	0.00	0.0	0.0	0.0	6	30	37	37	81	298	2900	0
VIII-Contributions from:	0.07	0.13	0.32	0.84	1.48	8.81	26.1	32.6	86.9	218	488	1465	2987	34831	202171	2240500	4646500
Central Administration	0.06	0.11	0.28	0.74	1.24	7.84	20.6	13.9	23.1	48	101	710	1173	25509	144961	1748700	3566800
Special Accounts	0.01	0.02	0.03	0.10	0.23	0.96	5.5	18.4	62.9	168	372	741	1789	9221	57006	487900	1074500
Decentralized Agencies	0.00	0.00	0.00	0.00	0.01	0.00	0.0	0.3	0.9	2	15	14	25	101	204	3900	5200
Provinces & MCBA	0.00	0.00	0.00	0.00	0.00	0.00	0.0	0.0	0.0	0	0	0	0	0	0	0	0
Public Enterprises	0.00	0.00	0.00	0.00	0.00	0.00	0.0	0.0	0.0	0	0	0	0	0	0	0	0
Social Security System	0.00	0.00	0.00	0.00	0.00	0.00	0.0	0.0	0.0	0	0	0	0	0	0	0	0
IX- Contributions to:	0.00	0.00	0.00	0.00	0.00	0.00	0.0	0.0	0.0	52	0	0	0	0	0	0	0
Central Administration	0.00	0.00	0.00	0.00	0.00	0.00	0.0	0.0	0.0	46	0	0	0	0	0	0	0
Special Accounts	0.00	0.00	0.00	0.00	0.00	0.00	0.0	0.0	0.0	0	0	0	0	0	0	0	0
Decentralized Agencies	0.00	0.00	0.00	0.00	0.00	0.00	0.0	0.0	0.0	0	0	0	0	0	0	0	0
Provinces & MCBA	0.00	0.00	0.00	0.00	0.00	0.00	0.0	0.0	0.0	5	0	0	0	0	0	0	0
Public Enterprises	0.00	0.00	0.00	0.00	0.00	0.00	0.0	0.0	0.0	0	0	0	0	0	0	0	0
Social Security System	0.00	0.00	0.00	0.00	0.00	0.00	0.0	0.0	0.0	0	0	0	0	0	0	0	0
X- Revenues (I+IV+VI+VII+VIII)	0.66	0.85	1.29	2.71	4.59	14.36	64.7	202.3	550.6	1472	3229	5902	13908	79574	548626	4266300	8339300
XI- Expenditures (II+V+IX)	0.66	0.89	1.47	2.85	4.84	15.16	66.4	196.6	545.2	1504	3288	6566	14166	77846	628034	4502200	8284100
XII- Financing Needs (XI-X)	0.00	0.04	0.18	0.14	0.25	0.80	1.7	-5.8	-5.4	32	59	664	258	-1728	79408	235900	-55200
XIII-Net Financing	-0.01	0.01	0.05	0.02	-0.00	0.30	0.3	-0.2	1.8	33	38	316	16	126	1436	-38500	-76800
a. Net Use of Credit	-0.01	0.01	0.05	0.02	-0.00	0.30	0.3	-0.2	1.8	33	38	315	17	125	1440	-38300	-76800
Domestic	-0.01	0.01	0.05	0.02	-0.02	0.28	0.3	-0.2	1.8	33	38	315	33	119	449	-35100	-73000
Credit	0.00	0.02	0.06	0.04	0.04	0.37	0.7	0.3	2.5	38	49	344	41	403	981	25700	0
Amortization	0.01	0.01	0.01	0.02	0.06	0.09	0.4	0.4	0.7	5	11	30	8	284	532	60800	73000
Foreign	0.00	0.00	0.01	0.00	0.01	0.02	0.0	0.0	0.0	0	0	0	-16	6	991	-3200	-3800
Credit	0.00	0.00	0.01	0.00	0.02	0.03	0.0	0.0	0.0	0	0	0	84	34	2189	1900	110200
Amortization	0.00	0.00	0.00	0.00	0.00	0.01	0.0	0.0	0.0	0	0	0	100	28	1198	5100	114000
b. Net Use of Advances	0.00	0.00	0.00	0.00	0.00	0.00	0.0	0.0	0.0	0	0	2	-2	1	-4	-200	0
Credit	0.00	0.00	0.00	0.00	0.00	0.00	0.0	0.0	0.0	0	0	2	1	1	0	100	0
Debit	0.00	0.00	0.00	0.00	0.00	0.00	0.0	0.0	0.0	0	0	0	2	0	4	300	0
XIV- Overall Balance (XIII-XII)	-0.01	-0.03	-0.13	-0.12	-0.25	-0.50	-1.4	5.6	7.2	1	-21	-348	-243	1854	-77972	-274400	-21600
Central Bank	0.00	0.00	0.00	0.00	0.00	0.00	0.0	0.0	0.0	0	0	0	0	0	0	0	0
Net Variation in Passive Financing	0.01	0.03	0.13	0.12	0.25	0.50	1.4	-5.6	-7.2	-1	21	348	243	-1854	77972	274400	21600

Source: Ministry of Economy.
March 1988

Table 5.12: ARGENTINA - PROVINCIAL GOVERNMENTS: REVENUES, EXPENDITURES AND FINANCING, 1970-1986
(Percent of GDP)

	1970	1971	1972	1973	1974	1975	1976	1977	1978	1979	1980	1981	1982	1983	1984	1985	1986
Current Account																	
I- Current Revenues	6.70	5.67	4.66	5.24	6.26	3.85	5.04	8.07	8.80	8.70	9.51	7.99	7.29	6.51	6.48	5.07	4.95
Tax Revenues	5.55	4.64	4.03	4.50	5.39	3.09	4.56	7.06	7.48	7.49	8.30	6.91	6.24	5.18	5.27	3.80	3.81
Non-tax Revenues	1.15	1.02	0.62	0.73	0.87	0.76	0.48	1.00	1.32	1.21	1.21	1.08	1.05	1.33	1.21	1.27	1.14
II- Current Expenditures	5.32	5.14	4.97	6.25	7.52	7.88	5.41	5.47	6.50	6.56	8.02	8.53	7.04	8.51	9.20	8.77	8.45
Personnel Expenditures	3.73	3.67	3.56	4.55	5.35	5.75	3.52	3.13	4.18	4.31	5.14	4.81	3.86	4.95	5.79	5.58	5.47
Goods and Services	0.76	0.68	0.64	0.67	0.75	0.66	0.63	0.78	0.86	0.82	1.06	0.89	0.85	0.84	0.95	1.02	1.03
Interest on Debt	0.04	0.03	0.01	0.03	0.05	0.04	0.03	0.02	0.01	0.05	0.10	0.26	0.15	0.18	0.02	0.05	0.02
Domestic	0.03	0.03	0.00	0.03	0.04	0.03	0.03	0.02	0.01	0.05	0.10	0.26	0.14	0.04	0.01	0.05	0.00
Foreign	0.00	0.00	0.00	0.00	0.00	0.00	0.00	0.00	0.00	0.00	0.00	0.00	0.01	0.14	0.01	0.00	0.02
Other Current Expenditures	0.00	0.00	0.00	0.00	0.00	0.00	0.00	0.00	0.00	0.00	0.00	0.00	0.00	0.00	0.00	0.00	0.00
Current & Capital Transfers	0.80	0.76	0.75	0.99	1.37	1.43	1.22	1.54	1.45	1.38	1.72	2.56	2.17	2.55	2.44	2.12	1.92
Economies	0.00	0.00	0.00	0.00	0.00	0.00	0.00	0.00	0.00	0.00	0.00	0.00	0.00	0.00	0.00	0.00	0.00
III- Savings	1.38	0.52	-0.31	-1.01	-1.26	-4.03	-0.36	2.59	2.30	2.14	1.49	-0.54	0.25	-2.00	-2.72	-3.70	-3.50
Capital Account																	
IV- Capital Revenues	0.08	0.04	0.05	0.03	0.11	0.03	0.05	0.04	0.06	0.05	0.06	0.05	0.08	0.03	0.07	0.04	0.02
V- Capital Expenditures	2.23	1.94	2.13	1.79	2.42	2.72	3.34	3.92	3.92	3.63	3.59	3.46	2.56	2.89	2.69	2.60	2.70
Fixed Investments	2.14	1.90	2.05	1.72	2.29	2.65	3.27	3.63	3.79	3.52	3.39	2.77	2.34	2.76	2.43	2.42	2.63
Machinery and Equipment	0.25	0.20	0.14	0.12	0.21	0.16	0.20	0.31	0.25	0.15	0.15	0.11	0.09	0.11	0.12	0.10	0.06
Construction	1.89	1.70	1.91	1.60	2.08	2.49	3.07	3.32	3.54	3.37	3.24	2.66	2.25	2.65	2.31	2.32	2.57
Changes in Inventories	0.00	0.00	0.00	0.00	0.00	0.00	0.00	0.00	0.00	0.00	0.11	0.14	0.08	0.03	0.04	0.04	0.02
Financial Investments	0.09	0.04	0.08	0.07	0.12	0.07	0.07	0.29	0.13	0.12	0.09	0.55	0.14	0.10	0.23	0.14	0.05
Economies	0.00	0.00	0.00	0.00	0.00	0.00	0.00	0.00	0.00	0.00	0.00	0.00	0.00	0.00	0.00	0.00	0.00
VI- Financing Due to Economic Eme	0.00	0.00	0.00	0.00	0.00	0.00	0.00	0.00	0.00	0.00	0.00	0.00	0.00	0.00	0.00	0.00	0.00
VII-Unused Capital from Previous P	0.00	0.00	0.00	0.00	0.00	0.00	0.00	0.00	0.00	0.04	0.11	0.07	0.03	0.01	0.01	0.01	0.00
VIII-Contributions from:	0.76	1.04	1.53	2.37	3.05	6.16	3.44	1.56	1.66	1.53	1.72	2.68	2.02	5.10	3.83	5.66	6.25
Central Administration	0.63	0.91	1.36	2.09	2.56	5.48	2.71	0.66	0.44	0.34	0.36	1.30	0.79	3.74	2.74	4.42	4.80
Special Accounts	0.13	0.13	0.16	0.27	0.48	0.67	0.72	0.88	1.20	1.18	1.31	1.35	1.21	1.35	1.08	1.23	1.45
Decentralized Agencies	0.01	0.00	0.00	0.01	0.01	0.00	0.00	0.02	0.02	0.01	0.05	0.03	0.02	0.01	0.00	0.01	0.01
Provinces & MCBA	0.00	0.00	0.00	0.00	0.00	0.00	0.00	0.00	0.00	0.00	0.00	0.00	0.00	0.00	0.00	0.00	0.00
Public Enterprises	0.00	0.00	0.00	0.00	0.00	0.00	0.00	0.00	0.00	0.00	0.00	0.00	0.00	0.00	0.00	0.00	0.00
Social Security System	0.00	0.00	0.00	0.00	0.00	0.00	0.00	0.00	0.00	0.00	0.00	0.00	0.00	0.00	0.00	0.00	0.00
IX- Contributions to:	0.00	0.00	0.00	0.00	0.00	0.00	0.00	0.00	0.00	0.36	0.00	0.00	0.00	0.00	0.00	0.00	0.00
Central Administration	0.00	0.00	0.00	0.00	0.00	0.00	0.00	0.00	0.00	0.33	0.00	0.00	0.00	0.00	0.00	0.00	0.00
Special Accounts	0.00	0.00	0.00	0.00	0.00	0.00	0.00	0.00	0.00	0.00	0.00	0.00	0.00	0.00	0.00	0.00	0.00
Decentralized Agencies	0.00	0.00	0.00	0.00	0.00	0.00	0.00	0.00	0.00	0.00	0.00	0.00	0.00	0.00	0.00	0.00	0.00
Provinces & MCBA	0.00	0.00	0.00	0.00	0.00	0.00	0.00	0.00	0.00	0.04	0.00	0.00	0.00	0.00	0.00	0.00	0.00
Public Enterprises	0.00	0.00	0.00	0.00	0.00	0.00	0.00	0.00	0.00	0.00	0.00	0.00	0.00	0.00	0.00	0.00	0.00
Social Security System	0.00	0.00	0.00	0.00	0.00	0.00	0.00	0.00	0.00	0.00	0.00	0.00	0.00	0.00	0.00	0.00	0.00
X- Revenues (I+IV+VI+VII+VIII)	**7.55**	**6.75**	**6.23**	**7.64**	**9.43**	**10.04**	**8.53**	**9.67**	**10.52**	**10.33**	**11.39**	**10.78**	**9.42**	**11.66**	**10.39**	**10.78**	**11.22**
XI- Expenditures (II+V+IX)	**7.55**	**7.08**	**7.10**	**8.04**	**9.94**	**10.60**	**8.75**	**9.39**	**10.42**	**10.55**	**11.60**	**11.99**	**9.60**	**11.40**	**11.89**	**11.37**	**11.15**
XII- Financing Needs (XI-X)	**0.00**	**0.32**	**0.87**	**0.40**	**0.51**	**0.56**	**0.22**	**-0.28**	**-0.10**	**0.23**	**0.21**	**1.21**	**0.17**	**-0.25**	**1.50**	**0.60**	**-0.07**
XIII-Net Financing	-0.11	0.06	0.26	0.06	-0.01	0.21	0.04	-0.01	0.03	0.23	0.13	0.58	0.01	0.02	0.03	-0.10	-0.10
a. Net Use of Credit	-0.11	0.06	0.26	0.06	-0.01	0.21	0.04	-0.01	0.03	0.23	0.13	0.57	0.01	0.02	0.03	-0.10	-0.10
Domestic	-0.11	0.05	0.23	0.06	-0.04	0.20	0.04	-0.01	0.03	0.23	0.13	0.57	0.02	0.02	0.01	-0.09	-0.10
Credit	0.00	0.12	0.30	0.11	0.09	0.26	0.09	0.01	0.05	0.27	0.17	0.63	0.03	0.06	0.02	0.06	0.00
Amortization	0.11	0.07	0.07	0.05	0.13	0.06	0.05	0.02	0.01	0.03	0.04	0.05	0.01	0.04	0.01	0.15	0.10
Foreign	0.00	0.02	0.03	0.00	0.03	0.01	0.00	0.00	0.00	0.00	0.00	0.00	-0.01	0.00	0.02	-0.01	-0.01
Credit	0.00	0.02	0.03	0.00	0.03	0.02	0.00	0.00	0.00	0.00	0.00	0.00	0.06	0.00	0.04	0.00	0.15
Amortization	0.00	0.01	0.01	0.00	0.01	0.01	0.00	0.00	0.00	0.00	0.00	0.00	0.07	0.00	0.02	0.01	0.15
b. Net Use of Advances	0.00	0.00	0.00	0.00	0.00	0.00	0.00	0.00	0.00	0.00	0.00	0.00	-0.00	0.00	-0.00	-0.00	0.00
Credit	0.00	0.00	0.00	0.00	0.00	0.00	0.00	0.00	0.00	0.00	0.00	0.00	0.00	0.00	0.00	0.00	0.00
Debit	0.00	0.00	0.00	0.00	0.00	0.00	0.00	0.00	0.00	0.00	0.00	0.00	0.00	0.00	0.00	0.00	0.00
XIV- Overall Balance (XIII-XII)	**-0.12**	**-0.26**	**-0.61**	**-0.34**	**-0.52**	**-0.35**	**-0.18**	**0.27**	**0.14**	**0.01**	**-0.08**	**-0.64**	**-0.16**	**0.27**	**-1.48**	**-0.69**	**-0.03**
Central Bank	**0.00**	**0.00**	**0.00**	**0.00**	**0.00**	**0.00**	**0.00**	**0.00**	**0.00**	**0.00**	**0.00**	**0.00**	**0.00**	**0.00**	**0.00**	**0.00**	**0.00**
Net Variation in Passive Financi	**0.12**	**0.26**	**0.61**	**0.34**	**0.52**	**0.35**	**0.18**	**-0.27**	**-0.14**	**-0.01**	**0.08**	**0.64**	**0.16**	**-0.27**	**1.48**	**0.69**	**0.03**

Source: Ministry of Economy.

March 1988

Table 5.13: ARGENTINA - PUBLIC ENTERPRISES: REVENUES, EXPENDITURES AND FINANCING, 1970-1987
(Thousands of Australes)

	1970	1971	1972	1973	1974	1975	1976	1977	1978	1979	1980	1981	1982	1983	1984	1985	1986	1987 1/
Current Account																		
I- Current Revenues	0.75	0.96	1.65	2.84	4.16	12.03	73.7	205.6	580	1249	2447	5624	14346	75306	556351	5372800	8964500	19810300
Tax Revenues	0.00	0.00	0.00	0.00	0.00	0.00	0.0	0.0	0	0	0	0	0	0	0	0	0	0
Non-tax Revenues	0.75	0.96	1.65	2.84	4.16	12.03	73.7	205.6	580	1249	2447	5624	14346	75306	556351	5372800	8964500	19810300
II- Current Expenditures	0.64	0.92	1.53	2.71	4.07	13.28	70.8	167.4	498	1236	2797	6667	20061	92823	632441	5971500	8628600	20893200
Personnel Expenditures	0.26	0.42	0.61	1.25	2.05	6.50	23.4	56.8	158	446	948	1620	3249	20423	169502	1218400	2211300	5442600
Goods and Services	0.34	0.44	0.78	1.31	1.46	5.53	41.0	88.0	245	591	1065	2489	9014	48372	335726	3527700	4825200	10849700
Interest on Debt	0.03	0.05	0.10	0.15	0.18	0.83	4.1	18.3	59	155	448	1926	6038	16801	107089	997300	1102700	2682300
Domestic	0.01	0.02	0.06	0.10	0.06	0.28	1.6	8.0	37	96	309	1325	2737	3255	16067	219300	48900	91800
Foreign	0.02	0.03	0.04	0.05	0.12	0.55	2.5	10.2	22	58	139	601	3301	13546	91022	778000	1053800	2590500
Other Current Expenditures	0.01	0.01	0.04	0.00	0.38	0.42	2.3	4.3	37	45	336	632	1760	7227	20124	228100	489400	1918600
Current & Capital Transfers	0.00	0.00	0.00	0.00	0.00	0.00	0.0	0.0	0	0	0	0	0	0	0	0	0	0
Economies	0.00	0.00	0.00	0.00	0.00	0.00	0.0	0.0	0	0	0	0	0	0	0	0	0	0
III- Savings	0.11	0.04	0.12	0.13	0.09	-1.25	3.0	38.2	81	13	-350	-1043	-5715	-17517	-76090	-598700	335900	-1082900
Capital Account																		
IV- Capital Revenues	0.02	0.03	0.03	0.02	0.10	0.14	0.6	7.3	14	25	60	89	538	1012	7216	75000	82800	168700
V- Capital Expenditures	0.34	0.49	0.81	1.14	1.729	5.635	41.5	103.0	266	559	1010	1913	5028	26483	188570	1163100	2056000	5969100
Fixed Investments	0	0	1	1	1.62	5.34	37.6	97.6	244	527	968	1738	4857	24624	176215	1115100	1927400	5199000
Machinery and Equipment	0.19	0.25	0.49	0.69	0.78	2.34	16.7	41.8	129	294	419	815	2364	11820	103152	518100	867300	2339600
Construction	0.12	0.19	0.30	0.42	0.84	3.00	20.9	55.8	115	233	549	923	2493	12804	73063	597000	1060100	2859400
Changes in Inventories	0.00	0.00	0.00	0.00	0.00	0.00	0.0	0.0	0	0	0	0	0	0	0	0	0	0
Financial Investments	0.03	0.05	0.02	0.03	0.11	0.29	3.8	5.5	22	32	42	176	170	1859	12355	48000	128600	770100
Economies	0.00	0.00	0.00	0.00	0.00	0.00	0.0	0.0	0	0	0	0	0	0	0	0	0	0
VI-Financing Due to Economic Emergency	0.00	0.00	0.00	0.00	0.00	0.00	0.0	0.0	0	0	0	0	0	0	0	0	0	0
VII-Capital from Previous Period	0.00	0.00	0.00	0.00	0.00	0.00	0.0	0.0	0	0	0	0	0	0	0	0	0	0
VIII-Contributions from:	0.13	0.20	0.26	0.53	0.95	3.15	24.8	39.8	68	202	426	850	2817	48031	160236	1166500	1688200	6558400
Central Administration	0.08	0.15	0.19	0.35	0.67	2.43	21.8	28.8	30	102	205	424	1833	42310	124910	737800	1136400	4828600
Special Accounts	0.04	0.04	0.07	0.18	0.29	0.72	3.1	11.0	38	95	220	425	984	5716	35326	428700	551800	1729800
Decentralized Agencies	0.01	0.01	0.00	0.00	0.00	0.00	0.0	0.0	0	0	1	1	1	5	0	0	0	0
Provinces & MCBA	0.00	0.00	0.00	0.00	0.00	0.00	0.0	0.0	0	5	0	0	0	0	0	0	0	0
Public Enterprises	0.00	0.00	0.00	0.00	0.00	0.00	0.0	0.0	0	0	0	0	0	0	0	0	0	0
Social Security System	0.00	0.00	0.00	0.00	0.00	0.00	0.0	0.0	0	0	0	0	0	0	0	0	0	0
IX- Contributions to:	0.00	0.00	0.00	0.00	0.00	0.00	0.0	0.0	0	111	0	0	0	0	0	0	0	0
Central Administration	0.00	0.00	0.00	0.00	0.00	0.00	0.0	0.0	0	111	0	0	0	0	0	0	0	0
Special Accounts	0.00	0.00	0.00	0.00	0.00	0.00	0.0	0.0	0	0	0	0	0	0	0	0	0	0
Decentralized Agencies	0.00	0.00	0.00	0.00	0.00	0.00	0.0	0.0	0	0	0	0	0	0	0	0	0	0
Provinces & MCBA	0.00	0.00	0.00	0.00	0.00	0.00	0.0	0.0	0	0	0	0	0	0	0	0	0	0
Public Enterprises	0.00	0.00	0.00	0.00	0.00	0.00	0.0	0.0	0	0	0	0	0	0	0	0	0	0
Social Security System	0.00	0.00	0.00	0.00	0.00	0.00	0.0	0.0	0	0	0	0	0	0	0	0	0	0
X- Revenues (I+IV+VI+VII+VIII)	**0.90**	**1.19**	**1.94**	**3.39**	**5.21**	**15.32**	**99.1**	**252.7**	**662**	**1476**	**2933**	**6562**	**17701**	**124349**	**723803**	**6614300**	**10735500**	**26537400**
XI- Expenditures (II+V+IX)	**0.98**	**1.41**	**2.34**	**3.85**	**5.80**	**18.92**	**112.3**	**270.4**	**764**	**1907**	**3807**	**8580**	**25089**	**119306**	**821011**	**7134600**	**10684600**	**26862300**
XII- Financing Needs (XI-X)	**0.08**	**0.22**	**0.40**	**0.46**	**0.59**	**3.60**	**13.1**	**17.8**	**102**	**431**	**874**	**2018**	**7387**	**-5043**	**97208**	**520300**	**-50900**	**324900**
XIII-Net Financing	0.05	0.12	0.25	0.24	0.65	0.30	6.4	19.9	122	407	856	2732	7580	-1777	4654	139400	32800	141400
a. Net Use of Credit	0.05	0.12	0.25	0.24	0.65	0.30	6.4	19.9	122	407	881	2775	7503	-1736	3972	153100	74600	314400
Domestic	0.00	0.01	0.06	0.06	0.25	-0.01	0.8	8.5	44	200	471	1374	5437	-1577	-2427	-5100	-15200	-91500
Credit	0.02	0.03	0.12	0.22	0.38	0.21	1.6	11.0	56	347	560	2401	7135	2096	127	9100	57400	24000
Amortization	0.02	0.02	0.06	0.16	0.13	0.22	0.8	2.5	12	147	89	1027	1699	3673	2554	14200	72600	115500
Foreign	0.05	0.11	0.19	0.18	0.40	0.31	5.7	11.4	78	207	410	1401	2066	-159	6399	158200	89800	405900
Credit	0.09	0.16	0.30	0.40	0.51	0.96	8.9	21.9	106	466	564	2537	19109	43973	417861	3643500	1108400	2430700
Amortization	0.04	0.05	0.11	0.22	0.11	0.65	3.2	10.5	29	259	154	1136	17043	44132	411462	3485300	1018600	2024800
b. Net Use of Advances	0	0	0	0	0	0	0.0	0.0	0	0	-25	-43	77	-41	682	-13700	-41800	-173000
Credit	0	0	0	0	0	0	0.0	0.0	0	0	0	142	77	-41	682	0	0	0
Debit	0	0	0	0	0	0	0.0	0.0	0	0	25	185	0	0	0	13700	41800	173000
XIV- Overall Balance (XIII-XII)	**-0.03**	**-0.10**	**-0.15**	**-0.22**	**0.06**	**-3.30**	**-6.7**	**2.2**	**19**	**-24**	**-18**	**714**	**193**	**3266**	**-92554**	**-380900**	**83700**	**-183500**
Central Bank	**0.03**	**0.04**	**0.04**	**0.07**	**0.30**	**2.13**	**1.9**	**0.0**	**0**	**0**	**0**	**0**	**0**	**0**	**0**	**0**	**0**	**0**
Net Variation in Passive Financing	**0.00**	**0.06**	**0.11**	**0.15**	**-0.36**	**1.17**	**4.8**	**-2.2**	**-19**	**24**	**18**	**-714**	**-193**	**-3266**	**92554**	**380900**	**-83700**	**183500**

Source: Ministry of Economy.

1/ Excludes provincial governments.

June 1989

Table 5.14: ARGENTINA - PUBLIC ENTERPRISES: REVENUES, EXPENDITURES AND FINANCING, 1970-1987
(Percent of GDP)

	1970	1971	1972	1973	1974	1975	1976	1977	1978	1979	1980	1981	1982	1983	1984	1985	1986	1987 1/
Current Account																		
I- Current Revenues	8.55	7.67	7.97	8.00	8.55	8.41	9.72	9.82	11.07	8.76	8.64	10.27	9.72	11.03	10.53	13.57	12.06	11.18
Tax Revenues	0.00	0.00	0.00	0.00	0.00	0.00	0.00	0.00	0.00	0.00	0.00	0.00	0.00	0.00	0.00	0.00	0.00	0.00
Non-tax Revenues	8.55	7.67	7.97	8.00	8.55	8.41	9.72	9.82	11.07	8.76	8.64	10.27	9.72	11.03	10.53	13.57	12.06	11.18
II- Current Expenditures	7.29	7.35	7.39	7.64	8.37	9.29	9.33	8.00	9.52	8.68	9.87	12.18	13.59	13.60	11.98	15.08	11.61	11.79
Personnel Expenditures	2.96	3.36	2.95	3.52	4.21	4.54	3.08	2.72	3.01	3.13	3.35	2.96	2.20	2.99	3.21	3.08	2.98	3.07
Goods and Services	3.87	3.51	3.77	3.69	3.00	3.87	5.40	4.20	4.67	4.15	3.76	4.55	6.11	7.09	6.36	8.91	6.49	6.12
Interest on Debt	0.34	0.40	0.48	0.42	0.37	0.58	0.54	0.87	1.12	1.08	1.58	3.52	4.09	2.46	2.03	2.52	1.48	1.51
Domestic	0.11	0.16	0.29	0.28	0.12	0.20	0.21	0.38	0.71	0.68	1.09	2.42	1.85	0.48	0.30	0.55	0.07	0.05
Foreign	0.23	0.24	0.19	0.14	0.25	0.38	0.32	0.49	0.41	0.41	0.49	1.10	2.24	1.98	1.72	1.97	1.42	1.46
Other Current Expenditures	0.11	0.08	0.19	0.00	0.78	0.29	0.31	0.20	0.71	0.31	1.19	1.15	1.19	1.06	0.38	0.58	0.66	1.08
Current & Capital Transfers	0.00	0.00	0.00	0.00	0.00	0.00	0.00	0.00	0.00	0.00	0.00	0.00	0.00	0.00	0.00	0.00	0.00	0.00
Economies	0.00	0.00	0.00	0.00	0.00	0.00	0.00	0.00	0.00	0.00	0.00	0.00	0.00	0.00	0.00	0.00	0.00	0.00
III- Savings	1.25	0.32	0.58	0.37	0.19	-0.87	0.39	1.83	1.56	0.09	-1.24	-1.91	-3.87	-2.57	-1.44	-1.51	0.45	-0.61
Capital Account																		
IV- Capital Revenues	0.23	0.24	0.14	0.06	0.20	0.10	0.08	0.35	0.27	0.18	0.21	0.16	0.36	0.15	0.14	0.19	0.11	0.10
V- Capital Expenditures	3.87	3.91	3.91	3.21	3.55	3.94	5.47	4.92	5.08	3.92	3.57	3.49	3.41	3.88	3.57	2.94	2.77	3.37
Fixed Investments	3.53	3.51	3.82	3.13	3.32	3.74	4.96	4.66	4.66	3.70	3.42	3.17	3.29	3.61	3.34	2.82	2.59	2.93
Machinery and Equipment	2.17	2.00	2.37	1.94	1.59	1.64	2.21	2.00	2.46	2.06	1.48	1.49	1.60	1.73	1.95	1.31	1.17	1.32
Construction	1.37	1.52	1.45	1.18	1.73	2.10	2.76	2.67	2.20	1.63	1.94	1.69	1.69	1.88	1.38	1.51	1.43	1.61
Changes in Inventories	0.00	0.00	0.00	0.00	0.00	0.00	0.00	0.00	0.00	0.00	0.00	0.00	0.00	0.00	0.00	0.00	0.00	0.00
Financial Investments	0.34	0.40	0.10	0.08	0.23	0.20	0.51	0.26	0.41	0.23	0.15	0.32	0.12	0.27	0.23	0.12	0.17	0.43
Economies	0.00	0.00	0.00	0.00	0.00	0.00	0.00	0.00	0.00	0.00	0.00	0.00	0.00	0.00	0.00	0.00	0.00	0.00
VI-Financing Due to Economic Emergency	0.00	0.00	0.00	0.00	0.00	0.00	0.00	0.00	0.00	0.00	0.00	0.00	0.00	0.00	0.00	0.00	0.00	0.00
VII-Capital from Previous Period	0.00	0.00	0.00	0.00	0.00	0.00	0.00	0.00	0.00	0.00	0.00	0.00	0.00	0.00	0.00	0.00	0.00	0.00
VIII-Contributions from:	1.48	1.60	1.26	1.49	1.96	2.20	3.27	1.90	1.30	1.42	1.50	1.55	1.91	7.04	3.03	2.95	2.27	3.70
Central Administration	0.91	1.20	0.92	0.99	1.37	1.70	2.87	1.37	0.56	0.72	0.72	0.77	1.24	6.20	2.37	1.86	1.53	2.73
Special Accounts	0.46	0.32	0.34	0.51	0.59	0.50	0.40	0.53	0.73	0.67	0.78	0.78	0.67	0.84	0.67	1.08	0.74	0.98
Decentralized Agencies	0.11	0.08	0.00	0.00	0.00	0.00	0.00	0.00	0.00	0.00	0.00	0.00	0.00	0.00	0.00	0.00	0.00	0.00
Provinces & MCBA	0.00	0.00	0.00	0.00	0.00	0.00	0.00	0.00	0.00	0.04	0.00	0.00	0.00	0.00	0.00	0.00	0.00	0.00
Public Enterprises	0.00	0.00	0.00	0.00	0.00	0.00	0.00	0.00	0.00	0.00	0.00	0.00	0.00	0.00	0.00	0.00	0.00	0.00
Social Security System	0.00	0.00	0.00	0.00	0.00	0.00	0.00	0.00	0.00	0.00	0.00	0.00	0.00	0.00	0.00	0.00	0.00	0.00
IX- Contributions to:	0.00	0.00	0.00	0.00	0.00	0.00	0.00	0.00	0.00	0.78	0.00	0.00	0.00	0.00	0.00	0.00	0.00	0.00
Central Administration	0.00	0.00	0.00	0.00	0.00	0.00	0.00	0.00	0.00	0.78	0.00	0.00	0.00	0.00	0.00	0.00	0.00	0.00
Special Accounts	0.00	0.00	0.00	0.00	0.00	0.00	0.00	0.00	0.00	0.00	0.00	0.00	0.00	0.00	0.00	0.00	0.00	0.00
Decentralized Agencies	0.00	0.00	0.00	0.00	0.00	0.00	0.00	0.00	0.00	0.00	0.00	0.00	0.00	0.00	0.00	0.00	0.00	0.00
Provinces & MCBA	0.00	0.00	0.00	0.00	0.00	0.00	0.00	0.00	0.00	0.00	0.00	0.00	0.00	0.00	0.00	0.00	0.00	0.00
Public Enterprises	0.00	0.00	0.00	0.00	0.00	0.00	0.00	0.00	0.00	0.00	0.00	0.00	0.00	0.00	0.00	0.00	0.00	0.00
Social Security System	0.00	0.00	0.00	0.00	0.00	0.00	0.00	0.00	0.00	0.00	0.00	0.00	0.00	0.00	0.00	0.00	0.00	0.00
X- Revenues (I+IV+VI+VII+VIII)	10.26	9.51	9.38	9.55	10.71	10.71	13.06	12.07	12.64	10.36	10.35	11.99	11.99	18.22	13.71	16.71	14.45	14.98
XI- Expenditures (II+V+IX)	11.17	11.26	11.31	10.85	11.92	13.23	14.79	12.92	14.59	13.38	13.44	15.67	17.00	17.48	15.55	18.02	14.38	15.16
XII- Financing Needs (XI-X)	0.91	1.76	1.93	1.30	1.21	2.52	1.73	0.85	1.95	3.02	3.09	3.69	5.00	-0.74	1.84	1.31	-0.07	0.18
XIII-Net Financing	0.57	0.96	1.21	0.68	1.34	0.21	0.84	0.95	2.32	2.85	3.02	4.99	5.14	-0.26	0.09	0.35	0.04	0.08
a. Net Use of Credit	0.57	0.96	1.21	0.68	1.34	0.21	0.84	0.95	2.32	2.85	3.11	5.07	5.08	-0.25	0.08	0.39	0.10	0.18
Domestic	0.00	0.08	0.29	0.17	0.51	-0.01	0.10	0.41	0.84	1.40	1.66	2.51	3.68	-0.23	-0.05	-0.01	-0.02	-0.05
Credit	0.23	0.24	0.58	0.62	0.78	0.15	0.21	0.53	1.08	2.43	1.98	4.39	4.83	0.31	0.00	0.02	0.08	0.01
Amortization	0.23	0.16	0.29	0.45	0.27	0.15	0.11	0.12	0.24	1.03	0.31	1.88	1.15	0.54	0.05	0.04	0.10	0.07
Foreign	0.57	0.88	0.92	0.51	0.82	0.22	0.74	0.55	1.48	1.45	1.45	2.56	1.40	-0.02	0.12	0.40	0.12	0.23
Credit	1.03	1.28	1.45	1.13	1.05	0.67	1.17	1.05	2.03	3.27	1.99	4.63	12.95	6.44	7.91	9.20	1.49	1.37
Amortization	0.46	0.40	0.53	0.62	0.23	0.45	0.43	0.50	0.55	1.82	0.54	2.07	11.55	6.46	7.79	8.80	1.37	1.14
b. Net Use of Advances	0.00	0.00	0.00	0.00	0.00	0.00	0.00	0.00	0.00	0.00	-0.09	-0.08	0.05	-0.01	0.01	-0.03	-0.06	-0.10
Credit	0.00	0.00	0.00	0.00	0.00	0.00	0.00	0.00	0.00	0.00	0.00	0.26	0.05	-0.01	0.01	0.00	0.00	0.00
Debit	0.00	0.00	0.00	0.00	0.00	0.00	0.00	0.00	0.00	0.00	0.09	0.34	0.00	0.00	0.00	0.03	0.06	0.10
XIV- Overall Balance (XIII-XII)	-0.34	-0.80	-0.72	-0.62	0.12	-2.31	-0.89	0.10	0.37	-0.17	-0.06	1.30	0.13	0.48	-1.75	-0.96	0.11	-0.10
Central Bank	0.34	0.32	0.19	0.20	0.62	1.49	0.25	0.00	0.00	0.00	0.00	0.00	0.00	0.00	0.00	0.00	0.00	0.00
Net Variation in Passive Financing	0.00	0.48	0.53	0.42	-0.74	0.82	0.63	-0.10	-0.37	0.17	0.06	-1.30	-0.13	-0.48	1.75	0.96	-0.11	0.10

Source: Ministry of Economy.

1/ Excludes provincial governments.

June 1989

Table 5.15: ARGENTINA - PUBLIC EXPENDITURE BY DESTINATION, 1970-1985
(Thousands of Australes)

	1970	1971	1972	1973	1974	1975	1976	1977	1978	1979	1980	1981	1982	1983	1984	1985
TOTAL EXPENDITURES	2.95	4.12	6.65	12.31	18.89	56.59	299.7	793.0	2283.0	5862.6	12441	26895	71112	352336	2416541	18745442
A- General Administration	0.18	0.26	0.39	0.87	1.34	3.74	18.6	59.8	166.4	470.0	804	1849	4039	24335	53320	445137
B- Defense	0.17	0.23	0.38	0.65	0.81	2.92	18.0	50.9	141.9	372.8	768	1656	4807	20251	93841	635579
C- Security	0.13	0.17	0.27	0.54	0.83	2.52	13.1	35.7	93.5	236.7	615	1149	2334	12137	33594	224269
D- Health	0.14	0.19	0.31	0.61	1.11	3.51	15.9	41.0	125.2	305.4	626	1285	2745	13655	27164	214725
E- Education and Culture	0.31	0.43	0.68	1.43	2.18	6.14	20.4	55.2	189.7	499 8	1137	2255	4817	28165	90758	612945
Culture	0.01	0.02	0.02	0.03	0.05	0.12	0.4	0.8	4.4	11.4	28	64	137	698	1809	15992
Elementary Education	0.15	0.20	0.33	0.70	1.11	3.10	10.0	26.4	92.2	229.4	524	915	1944	12681	3830	24553
Secondary Education	0.08	0.11	0.17	0.37	0.55	1.71	6.0	16.5	54.1	148.8	343	695	1472	7709	50486	315846
Higher Ed. and University	0.05	0.07	0.10	0.21	0.30	1.02	3.0	8.4	28.2	76.9	178	380	803	4272	30120	218340
Unclassified	0.03	0.04	0.06	0.11	0.17	0.19	0.9	3.2	10.8	33.3	64	201	461	2805	4513	38214
F- Economic Development	1.47	2.01	3.38	5.81	8.26	25.72	150.8	389.3	1059.0	2596.7	5391	11509	33900	167851	998201	8362841
Land Improvement	0.02	0.03	0.05	0.10	0.17	0.51	2.4	9.9	20.9	39.3	117	142	336	1965	1492	3698
Agriculture and Livestock	0.04	0.04	0.06	0.10	0.13	0.57	1.6	3.9	10.1	27.6	64	121	284	1530	6651	48834
Energy	0.57	0.81	1.27	2.33	3.77	11.90	80.8	193.0	567.6	1365.5	2757	6243	21052	96875	622397	5743777
Mining	0.00	0.00	0.02	0.09	0.07	0.18	1.5	4.4	7.2	4.3	38	22	47	205	699	4397
Industry	0.17	0.22	0.45	0.65	0.44	1.32	5.9	15.6	47.2	100.0	224	456	1337	7640	29721	218914
Tourism	0.00	0.01	0.01	0.02	0.03	0.13	0.4	1.5	2.4	4.8	15	19	22	161	141	723
Railway transport	0.16	0.22	0.38	0.64	0.78	2.72	13.6	26.8	75.6	231.7	514	1093	2374	11321	101377	598996
Road Transport	0.19	0.22	0.44	0.74	1.00	2.77	16.4	50.3	118.9	268.4	529	861	2000	10758	33432	246253
Maritime Transport	0 07	0.11	0.17	0.28	0.45	1.35	8.3	22.2	42.3	119.2	209	469	1208	7947	32166	234043
Air Transport	0.04	0.05	0.09	0.13	0.26	1.00	6.2	13.3	41.5	110.7	203	489	1507	4333	43124	383704
Communications	0.13	0.21	0.31	0.55	0.67	1.92	6.8	29.1	78.4	195.6	473	1068	2217	12860	70132	519917
Trade and Storage	0.02	0.04	0.04	0.06	0.13	0.36	1.7	4.6	12.1	34.0	56	114	257	1248	8504	69091
Finance and Insurance	0.05	0.03	0.03	0.06	0.08	0.24	1.6	4.5	5.8	8 0	33	48	362	5219	37251	236422
Unclassified	0.02	0.03	0.07	0.07	0 28	0.75	3.5	10.2	29.0	87.5	160	364	897	5789	11114	54071
G- Social Welfare	0.51	0.78	1.08	2.16	3.88	10.78	49.9	132.7	394.2	1059.3	2584	5324	11792	68075	375036	2829117
Social Security	0.44	0.66	0.90	1.74	2.88	7.29	36.7	105.4	301.3	859.5	2065	4517	9883	55176	341732	2500678
Labor	0.00	0.00	0.00	0.00	0.03	0.10	0.3	0.7	2.1	5.3	10	18	38	177	1583	13303
Housing and Urban Development	0.04	0.06	0.07	0.20	0.49	1.88	7.8	10.8	35.6	89.6	333	389	1152	8253	5523	61352
Social Assistance	0.02	0.02	0.03	0.07	0.14	0.40	2.0	4.4	9.7	26.5	51	94	175	1067	8797	110555
Sports and Recreation	0.01	0.02	0.03	0.06	0.13	0.68	2.2	8.4	31.7	33.7	51	78	212	1675	6047	51109
Unclassified	0.01	0.02	0.05	0.09	0.20	0.44	0.9	2.9	13.8	44.6	74	228	332	1727	11354	92121
H- Science and Technology	0.00	0.00	0.05	0.09	0.12	0.32	1.7	5.9	18.8	48.6	107	232	562	2816	16026	144485
I- Unclassified	0.00	0.00	0.00	0.00	0.00	0.00	0.0	0.0	0.0	0.0	0	0	0	0	616267	4428100
J- Public Debt	0.04	0.06	0.11	0.17	0.36	0.94	11.3	22.4	94.3	273.5	409	1636	6116	15051	112334	848243

Source: Ministry of Economy, National Directorate of Budgetary Programming.

March 1988

Table 5.16: ARGENTINA - PUBLIC EXPENDITURE BY DESTINATION, 1970-1985
(1970 Australes)

	1970	1971	1972	1973	1974	1975	1976	1977	1978	1979	1980	1981	1982	1983	1984	1985
TOTAL EXPENDITURES	2950	2991	2957	3319	3923	3939	3989	4119	4519	4547	4892	5120	4669	4891	3186	3340
A- General Administration	181	188	174	234	279	260	248	311	329	365	316	352	265	338	70	79
B- Defense	169	164	170	174	169	203	240	265	281	289	302	315	316	281	124	113
C- Security	129	126	122	145	172	175	174	185	185	184	242	219	153	168	44	40
D- Health	141	141	137	164	230	244	211	213	248	237	246	245	180	190	36	38
E- Education and Culture	310	310	300	385	453	427	272	287	375	388	447	429	316	391	120	109
Culture	11	11	8	8	10	8	6	4	9	9	11	12	9	10	2	3
Elementary Education	153	144	145	190	230	216	134	137	182	178	206	174	128	176	5	4
Secondary Education	76	79	77	100	115	119	80	85	107	115	135	132	97	107	67	56
Higher Ed. and University	45	49	45	56	62	71	40	44	56	60	70	72	53	59	40	39
Unclassified	25	27	25	31	36	13	12	16	21	26	25	38	30	39	6	7
F- Economic Development	1470	1459	1504	1566	1715	1790	2008	2022	2096	2014	2120	2191	2226	2330	1316	1490
Land Improvement	23	21	20	26	35	35	32	52	41	30	46	27	22	27	2	1
Agriculture and Livestock	37	31	25	28	28	40	21	20	20	21	25	23	19	21	9	9
Energy	565	586	564	628	783	828	1075	1002	1123	1059	1084	1189	1382	1345	821	1023
Mining	2	3	8	24	14	13	20	23	14	3	15	4	3	3	1	1
Industry	168	156	200	174	91	92	79	81	93	78	88	87	88	106	39	39
Tourism	4	6	4	5	6	9	6	8	5	4	6	4	1	2	0	0
Railway transport	162	157	171	172	161	189	182	139	150	180	202	208	156	157	134	107
Road Transport	185	159	197	200	207	193	219	261	235	208	208	164	131	149	44	44
Maritime Transport	67	77	76	75	94	94	111	115	84	92	82	89	79	110	42	42
Air Transport	43	39	40	35	54	69	82	69	82	86	80	93	99	60	57	68
Communications	125	152	138	147	140	134	91	151	155	152	186	203	146	179	92	93
Trade and Storage	18	30	19	16	27	25	23	24	24	26	22	22	17	17	11	12
Finance and Insurance	49	20	13	16	16	17	22	23	11	6	13	9	24	72	49	42
Unclassified	22	22	29	20	59	52	47	53	57	68	63	69	59	80	15	10
G- Social Welfare	514	563	481	582	805	750	664	689	780	822	1016	1014	774	945	495	504
Social Security	441	478	400	469	599	507	489	547	596	667	812	860	649	766	451	446
Labor	0	0	0	0	7	7	4	4	4	4	4	3	2	2	2	2
Housing and Urban Development	36	41	31	54	101	131	104	56	70	70	131	74	76	115	7	11
Social Assistance	16	17	15	18	29	28	26	23	19	21	20	18	11	15	12	20
Sports and Recreation	9	12	14	17	27	48	29	44	63	26	20	15	14	23	8	9
Unclassified	12	15	21	24	42	30	12	15	27	35	29	43	22	24	15	16
															0	0
H- Science and Technology	0	0	21	23	25	23	22	31	37	38	42	44	37	39	21	26
I- Unclassified	0	0	0	0	0	0	0	0	0	0	0	0	0	0	813	789
J- Public Debt	36	40	48	46	75	66	150	116	187	212	161	311	402	209	148	151

Source: Ministry of Economy, National Directorate of Budgetary Programming.

March 1988

Table 5.17: ARGENTINA - PUBLIC EXPENDITURE BY DESTINATION, 1970-1985
(Growth Rates)

	1971	1972	1973	1974	1975	1976	1977	1978	1979	1980	1981	1982	1983	1984	1985
TOTAL EXPENDITURES	1.4	-1.1	12.2	18.2	0.4	1.3	3.3	9.7	0.6	7.6	4.7	-8.8	4.8	-34.9	4.8
A- General Administration	3.9	-7.4	34.5	19.2	-6.6	-4.8	25.2	6.0	10.7	-13.3	11.4	-24.7	27.4	-79.2	12.8
B- Defense	-3.0	3.7	2.4	-2.9	20.1	18.1	10.4	6.2	2.9	4.4	4.4	0.1	-10.9	-56.0	-8.5
C- Security	-2.3	-3.2	18.9	18.6	2.0	-0.9	6.6	-0.1	-0.8	31.8	-9.6	-29.9	9.9	-73.7	-9.8
D- Health	0.0	-2.8	19.7	40.2	6.3	-13.5	0.8	16.3	-4.4	3.9	-0.6	-26.3	5.2	-81.1	6.8
E- Education and Culture	0.0	-3.2	28.3	17.7	-5.7	-36.4	5.5	30.9	3.3	15.3	-4.0	-26.3	23.6	-69.4	-8.7
Culture	0.0	-27.3	0.0	25.0	-19.3	-30.8	-28.1	117.4	0.8	24.9	10.8	-26.2	7.7	-75.4	19.5
Elementary Education	-5.9	0.7	31.0	21.1	-6.1	-38.2	2.8	32.9	-2.5	15.8	-15.4	-26.7	37.9	-97.1	-13.4
Secondary Education	3.9	-2.5	29.9	15.0	3.7	-32.5	6.2	25.3	7.8	16.9	-2.0	-27.0	10.7	-37.8	-15.5
Higher Ed. and University	8.9	-8.2	24.4	10.7	14.2	-43.7	9.6	27.6	6.9	17.4	3.3	-27.1	12.5	-33.0	-2.1
Unclassified	8.0	-7.4	24.0	16.1	-63.3	-6.6	32.6	30.5	20.8	-3.2	53.1	-20.9	28.6	-84.7	14.4
F- Economic Development	-0.7	3.1	4.1	9.5	4.4	12.1	0.7	3.6	-3.9	5.3	3.4	1.6	4.7	-43.5	13.2
Land Improvement	-8.7	-4.8	30.0	34.6	1.2	-9.3	60.5	-20.0	-26.2	51.1	-41.2	-18.4	23.7	-92.8	-66.5
Agriculture and Livestock	-16.2	-19.4	12.0	0.0	41.2	-45.7	-5.5	-1.5	7.0	16.9	-7.9	-19.1	13.9	-58.7	-0.8
Energy	3.7	-3.8	11.3	24.7	5.8	29.8	-6.8	12.1	-5.7	2.3	9.6	16.3	-2.7	-39.0	24.7
Mining	50.0	166.7	200.0	-41.7	-9.0	54.8	15.0	-37.5	-76.2	344.9	-72.1	-26.3	-7.8	-67.6	-15.0
Industry	-7.1	28.2	-13.0	-47.7	1.2	-14.6	3.2	15.2	-17.0	13.4	-1.3	1.1	20.8	-63.1	-0.5
Tourism	50.0	-33.3	25.0	20.0	49.7	-37.9	41.3	-38.8	-22.7	60.9	-39.7	-60.1	54.7	0.0	0.0
Railway transport	-3.1	8.9	0.6	-6.4	17.4	-3.9	-23.3	7.4	20.2	12.4	3.0	-25.1	0.8	-14.9	-20.2
Road Transport	-14.1	23.9	1.5	3.5	-6.9	13.6	19.4	-10.0	-11.5	-0.1	-21.2	-19.9	13.7	-70.5	-0.5
Maritime Transport	14.9	-1.3	-1.3	25.3	0.2	17.7	3.8	-27.2	10.3	-11.3	8.9	-11.2	39.1	-61.6	-1.7
Air Transport	-9.3	2.6	-12.5	54.3	28.4	18.9	-16.1	18.7	4.6	-6.8	16.4	6.3	-39.2	-5.5	20.2
Communications	21.6	-9.2	6.5	-4.8	-4.3	-32.3	66.4	2.8	-2.3	22.6	9.3	-28.4	22.6	-48.2	0.2
Trade and Storage	66.7	-36.7	-15.8	68.8	-6.2	-11.1	6.8	-0.6	10.2	-16.6	-1.3	-22.2	2.7	-35.3	9.8
Finance and Insurance	-59.2	-35.0	23.1	0.0	5.7	27.2	8.8	-51.0	-45.7	108.7	-29.7	160.1	204.8	-32.2	-14.2
Unclassified	0.0	31.8	-31.0	195.0	-11.9	-10.2	13.7	8.3	18.1	-7.2	10.0	-15.0	36.5	-81.8	-34.3
G- Social Welfare	9.5	-14.6	21.0	38.3	-6.8	-11.5	3.8	13.2	5.3	23.7	-0.2	-23.6	22.1	-47.7	1.9
Social Security	8.4	-16.3	17.3	27.7	-15.3	-3.6	12.0	8.9	11.8	21.8	5.9	-24.5	18.0	-41.2	-1.1
Labor	0.0	0.0	0.0	0.0	-2.5	-44.4	0.4	11.4	-2.7	-3.1	-14.3	-27.2	-1.5	-15.1	13.6
Housing and Urban Development	13.9	-24.4	74.2	87.0	29.2	-20.1	-46.0	25.0	-1.3	88.5	-43.5	2.1	51.5	-93.6	50.1
Social Assistance	6.3	-11.8	20.0	61.1	-4.9	-5.6	-11.6	-16.6	7.1	-2.7	-10.5	-35.8	28.9	-21.7	69.8
Sports and Recreation	33.3	16.7	21.4	58.8	76.3	-38.3	48.5	44.1	-58.3	-23.6	-25.8	-6.3	67.1	-65.7	14.2
Unclassified	25.0	40.0	14.3	75.0	-27.4	-61.5	30.2	78.3	27.0	-16.2	49.7	-49.8	10.0	-37.6	9.6
H- Science and Technology	0.0	0.0	0.0	8.7	-9.8	-1.3	38.7	20.6	1.1	11.5	5.2	-16.5	5.9	-45.9	21.8
I- Unclassified	-	-	-	-	-	-	-	-	-	-	-	-	-	-	-
J- Public Debt	11.1	20.0	-4.2	63.0	-12.7	129.5	-22.6	60.5	13.7	-24.1	93.5	28.9	-48.0	-29.1	2.0

Source: Table 5.16.

March 1988

Table 5.18: ARGENTINA - PUBLIC EXPENDITURE BY DESTINATION, 1970-1985
(Percent of GDP)

	1970	1971	1972	1973	1974	1975	1976	1977	1978	1979	1980	1981	1982	1983	1984	1985
TOTAL EXPENDITURES	33.6	32.9	32.2	34.4	38.9	39.6	39.6	38.0	43.7	41.2	44.0	49.1	48.2	51.6	45.8	47.3
A- General Administration	2.1	2.1	1.9	2.4	2.8	2.6	2.5	2.9	3.2	3.3	2.8	3.4	2.7	3.6	1.0	1.1
B- Defense	1.9	1.8	1.8	1.8	1.7	2.0	2.4	2.4	2.7	2.6	2.7	3.0	3.3	3.0	1.8	1.6
C- Security	1.5	1.4	1.3	1.5	1.7	1.8	1.7	1.7	1.8	1.7	2.2	2.1	1.6	1.8	0.6	0.6
D- Health	1.6	1.6	1.5	1.7	2.3	2.5	2.1	2.0	2.4	2.1	2.2	2.3	1.9	2.0	0.5	0.5
E- Education and Culture	3.5	3.4	3.3	4.0	4.5	4.3	2.7	2.6	3.6	3.5	4.0	4.1	3.3	4.1	1.7	1.5
Culture	0.1	0.1	0.1	0.1	0.1	0.1	0.1	0.0	0.1	0.1	0.1	0.1	0.1	0.1	0.0	0.0
Elementary Education	1.7	1.6	1.6	2.0	2.3	2.2	1.3	1.3	1.8	1.6	1.9	1.7	1.3	1.9	0.1	0.1
Secondary Education	0.9	0.9	0.8	1.0	1.1	1.2	0.8	0.8	1.0	1.0	1.2	1.3	1.0	1.1	1.0	0.8
Higher Ed. and University	0.5	0.5	0.5	0.6	0.6	0.7	0.4	0.4	0.5	0.5	0.6	0.7	0.5	0.6	0.6	0.6
Unclassified	0.3	0.3	0.3	0.3	0.4	0.1	0.1	0.2	0.2	0.2	0.2	0.4	0.3	0.4	0.1	0.1
F- Economic Development	16.8	16.1	16.4	16.2	17.0	18.0	19.9	18.6	20.3	18.3	19.1	21.0	23.0	24.6	18.9	21.1
Land Improvement	0.3	0.2	0.2	0.3	0.3	0.4	0.3	0.5	0.4	0.3	0.4	0.3	0.2	0.3	0.0	0.0
Agriculture and Livestock	0.4	0.3	0.3	0.3	0.3	0.4	0.2	0.2	0.2	0.2	0.2	0.2	0.2	0.2	0.1	0.1
Energy	6.4	6.5	6.1	6.5	7.8	8.3	10.7	9.2	10.9	9.6	9.8	11.4	14.3	14.2	11.8	14.5
Mining	0.0	0.0	0.1	0.2	0.1	0.1	0.2	0.2	0.1	0.0	0.1	0.0	0.0	0.0	0.0	0.0
Industry	1.9	1.7	2.2	1.8	0.9	0.9	0.8	0.7	0.9	0.7	0.8	0.8	0.9	1.1	0.6	0.6
Tourism	0.0	0.1	0.0	0.1	0.1	0.1	0.1	0.1	0.0	0.0	0.1	0.0	0.0	0.0	0.0	0.0
Railway transport	1.8	1.7	1.9	1.8	1.6	1.9	1.8	1.3	1.4	1.6	1.8	2.0	1.6	1.7	1.9	1.5
Road Transport	2.1	1.8	2.1	2.1	2.1	1.9	2.2	2.4	2.3	1.9	1.9	1.6	1.4	1.6	0.6	0.6
Maritime Transport	0.8	0.8	0.8	0.8	0.9	0.9	1.1	1.1	0.8	0.8	0.7	0.9	0.8	1.2	0.6	0.6
Air Transport	0.5	0.4	0.4	0.4	0.5	0.7	0.8	0.6	0.8	0.8	0.7	0.9	1.0	0.6	0.8	1.0
Communications	1.4	1.7	1.5	1.5	1.4	1.3	0.9	1.4	1.5	1.4	1.7	2.0	1.5	1.9	1.3	1.3
Trade and Storage	0.2	0.3	0.2	0.2	0.3	0.3	0.2	0.2	0.2	0.2	0.2	0.2	0.2	0.2	0.2	0.2
Finance and Insurance	0.6	0.2	0.1	0.2	0.2	0.2	0.2	0.2	0.1	0.1	0.1	0.1	0.2	0.8	0.7	0.6
Unclassified	0.3	0.2	0.3	0.2	0.6	0.5	0.5	0.5	0.6	0.6	0.6	0.7	0.6	0.8	0.2	0.1
G- Social Welfare	5.9	6.2	5.2	6.0	8.0	7.5	6.6	6.4	7.5	7.4	9.1	9.7	8.0	10.0	7.1	7.1
Social Security	5.0	5.3	4.4	4.9	5.9	5.1	4.8	5.0	5.8	6.0	7.3	8.3	6.7	8.1	6.5	6.3
Labor	0.0	0.0	0.0	0.0	0.1	0.1	0.0	0.0	0.0	0.0	0.0	0.0	0.0	0.0	0.0	0.0
Housing and Urban Development	0.4	0.5	0.3	0.6	1.0	1.3	1.0	0.5	0.7	0.6	1.2	0.7	0.8	1.2	0.1	0.2
Social Assistance	0.2	0.2	0.2	0.2	0.3	0.3	0.3	0.2	0.2	0.2	0.2	0.2	0.1	0.2	0.2	0.3
Sports and Recreation	0.1	0.1	0.2	0.2	0.3	0.5	0.3	0.4	0.6	0.2	0.2	0.1	0.1	0.2	0.1	0.1
Unclassified	0.1	0.2	0.2	0.2	0.4	0.3	0.1	0.1	0.3	0.3	0.3	0.4	0.2	0.3	0.2	0.2
H- Science and Technology	0.0	0.0	0.2	0.2	0.2	0.2	0.2	0.3	0.4	0.3	0.4	0.4	0.4	0.4	0.3	0.4
I- Unclassified	0.0	0.0	0.0	0.0	0.0	0.0	0.0	0.0	0.0	0.0	0.0	0.0	0.0	0.0	11.7	11.2
J- Public Debt	0.4	0.4	0.5	0.5	0.7	0.7	1.5	1.1	1.8	1.9	1.4	3.0	4.1	2.2	2.1	2.1

Source: Table 5.15.

March 1988

Table 5.19: ARGENTINA - PUBLIC EXPENDITURE BY DESTINATION, 1970-1985
(Percent of Total)

	1970	1971	1972	1973	1974	1975	1976	1977	1978	1979	1980	1981	1982	1983	1984	1985
TOTAL EXPENDITURES	100.0	100.0	100.0	100.0	100.0	100.0	100.0	100.0	100.0	100.0	100.0	100.0	100.0	100.0	100.0	100.0
A- General Administration	6.1	6.3	5.9	7.1	7.1	6.6	6.2	7.5	7.3	8.0	6.5	6.9	5.7	6.9	2.2	2.4
B- Defense	5.7	5.5	5.7	5.2	4.3	5.2	6.0	6.4	6.2	6.4	6.2	6.2	6.8	5.7	3.9	3.4
C- Security	4.4	4.2	4.1	4.4	4.4	4.5	4.4	4.5	4.1	4.0	4.9	4.3	3.3	3.4	1.4	1.2
D- Health	4.8	4.7	4.6	4.9	5.9	6.2	5.3	5.2	5.5	5.2	5.0	4.8	3.9	3.9	1.1	1.1
E- Education and Culture	10.5	10.4	10.1	11.6	11.5	10.8	6.8	7.0	8.3	8.5	9.1	8.4	6.8	8.0	3.8	3.3
Culture	0.4	0.4	0.3	0.2	0.3	0.2	0.1	0.1	0.2	0.2	0.2	0.2	0.2	0.2	0.1	0.1
Elementary Education	5.2	4.8	4.9	5.7	5.9	5.5	3.3	3.3	4.0	3.9	4.2	3.4	2.7	3.6	0.2	0.1
Secondary Education	2.6	2.6	2.6	3.0	2.9	3.0	2.0	2.1	2.4	2.5	2.8	2.6	2.1	2.2	2.1	1.7
Higher Ed. and University	1.5	1.6	1.5	1.7	1.6	1.8	1.0	1.1	1.2	1.3	1.4	1.4	1.1	1.2	1.2	1.2
Unclassified	0.8	0.9	0.8	0.9	0.9	0.3	0.3	0.4	0.5	0.6	0.5	0.7	0.6	0.8	0.2	0.2
F- Economic Development	49.8	48.8	50.9	47.2	43.7	45.4	50.3	49.1	46.4	44.3	43.3	42.8	47.7	47.6	41.3	44.6
Land Improvement	0.8	0.7	0.7	0.8	0.9	0.9	0.8	1.3	0.9	0.7	0.9	0.5	0.5	0.6	0.1	0.0
Agriculture and Livestock	1.3	1.0	0.8	0.8	0.7	1.0	0.5	0.5	0.4	0.5	0.5	0.4	0.4	0.4	0.3	0.3
Energy	19.2	19.6	19.1	18.9	20.0	21.0	26.9	24.3	24.9	23.3	22.2	23.2	29.6	27.5	25.8	30.6
Mining	0.1	0.1	0.3	0.7	0.4	0.3	0.5	0.6	0.3	0.1	0.3	0.1	0.1	0.1	0.0	0.0
Industry	5.7	5.2	6.8	5.2	2.3	2.3	2.0	2.0	2.1	1.7	1.8	1.7	1.9	2.2	1.2	1.2
Tourism	0.1	0.2	0.1	0.2	0.2	0.2	0.1	0.2	0.1	0.1	0.1	0.1	0.0	0.0	0.0	0.0
Railway transport	5.5	5.2	5.8	5.2	4.1	4.8	4.6	3.4	3.3	4.0	4.1	4.1	3.3	3.2	4.2	3.2
Road Transport	6.3	5.3	6.7	6.0	5.3	4.9	5.5	6.3	5.2	4.6	4.3	3.2	2.8	3.1	1.4	1.3
Maritime Transport	2.3	2.6	2.6	2.3	2.4	2.4	2.8	2.8	1.9	2.0	1.7	1.7	1.7	2.3	1.3	1.2
Air Transport	1.5	1.3	1.4	1.1	1.4	1.8	2.1	1.7	1.8	1.9	1.6	1.8	2.1	1.2	1.8	2.0
Communications	4.2	5.1	4.7	4.4	3.6	3.4	2.3	3.7	3.4	3.3	3.8	4.0	3.1	3.6	2.9	2.8
Trade and Storage	0.6	1.0	0.6	0.5	0.7	0.6	0.6	0.6	0.5	0.6	0.4	0.4	0.4	0.4	0.4	0.4
Finance and Insurance	1.7	0.7	0.4	0.5	0.4	0.4	0.5	0.6	0.3	0.1	0.3	0.2	0.5	1.5	1.5	1.3
Unclassified	0.7	0.7	1.0	0.6	1.5	1.3	1.2	1.3	1.3	1.5	1.3	1.4	1.3	1.6	0.5	0.3
G- Social Welfare	17.4	18.8	16.3	17.5	20.5	19.0	16.6	16.7	17.3	18.1	20.8	19.8	16.6	19.3	15.5	15.1
Social Security	14.9	16.0	13.5	14.1	15.3	12.9	12.3	13.3	13.2	14.7	16.6	16.8	13.9	15.7	14.1	13.3
Labor	0.0	0.0	0.0	0.0	0.2	0.2	0.1	0.1	0.1	0.1	0.1	0.1	0.1	0.1	0.1	0.1
Housing and Urban Development	1.2	1.4	1.0	1.6	2.6	3.3	2.6	1.4	1.6	1.5	2.7	1.4	1.6	2.3	0.2	0.3
Social Assistance	0.5	0.6	0.5	0.5	0.7	0.7	0.7	0.6	0.4	0.5	0.4	0.3	0.2	0.3	0.4	0.6
Sports and Recreation	0.3	0.4	0.5	0.5	0.7	1.2	0.7	1.1	1.4	0.6	0.4	0.3	0.3	0.5	0.3	0.3
Unclassified	0.4	0.5	0.7	0.7	1.1	0.8	0.3	0.4	0.6	0.8	0.6	0.8	0.5	0.5	0.5	0.5
H- Science and Technology	0.0	0.0	0.7	0.7	0.6	0.6	0.6	0.8	0.8	0.8	0.9	0.9	0.8	0.8	0.7	0.8
I- Unclassified	0.0	0.0	0.0	0.0	0.0	0.0	0.0	0.0	0.0	0.0	0.0	0.0	0.0	0.0	25.5	23.6
J- Public Debt	1.2	1.3	1.6	1.4	1.9	1.7	3.8	2.8	4.1	4.7	3.3	6.1	8.6	4.3	4.6	4.5

Source: Table 5.15.

March 1988

Table 5.20: ARGENTINA - TAX REVENUES, 1970-1987
(Thousands of Australes)

	1970	1971	1972	1973	1974	1975	1976	1977	1978	1979	1980	1981	1982	1983	1984	1985	1986	1987 a/
I- NATIONAL TAX REVENUE BY SOURCE (Gross)	1.41	1.78	2.64	4.87	8.26	16.76	99.7	324.2	886.3	2514	5617	9621	24606	114556	814985	7852644	14679688	29890648
NATIONAL TAX REVENUE BY SOURCE (Net of Deductions)	1.41	1.78	2.64	4.87	8.02	16.41	96.4	316.2	866.1	2467	5526	9382	23607	112496	602541	7752767	14076273	28303973
A. Income Taxes	0.20	0.23	0.33	0.64	0.85	1.14	9.0	37.5	90.7	170	446	924	2069	7606	31313	401383	949701	2777146
Personal and Corporate Income	0.19	0.22	0.31	0.53	0.71	1.07	8.6	35.4	86.0	158	418	866	1973	6885	27085	378795	890021	2612226
Tax on Foreign Investment Income	0.00	0.00	0.00	0.00	0.00	0.04	0.1	0.7	1.2	2	5	24	28	155	297	2402	13643	35824
Capital Gains Tax	0.01	0.01	0.02	0.03	0.02	0.00	0.1	1.0	1.8	5	13	15	25	105	666	6506	12457	55520
Other Income Taxes	0.00	0.00	0.00	0.08	0.12	0.03	0.1	0.5	1.6	4	9	19	42	459	3265	13680	33580	73575
B. Property Taxes	0.05	0.07	0.09	0.18	0.26	0.06	1.8	11.5	33.3	110	237	445	1628	7753	40558	341592	688482	1717201
Capital Tax	0.00	0.00	0.00	0.00	0.05	0.02	1.2	10.1	29.2	100	218	424	1517	6119	28635	247931	457227	1041193
Net Worth Tax	0.00	0.00	0.00	0.00	0.02	0.00	0.0	0.3	0.5	2	5	10	20	100	964	8792	108970	258784
Tax for Areas Affected by Flood	0.00	0.00	0.00	0.00	0.00	0.00	0.0	0.0	0.0	0	0	0	0	1194	8913	62137	76854	260199
Tax on Financial Assets	0.00	0.00	0.00	0.00	0.00	0.00	0.5	0.4	1.1	3	4	11	91	340	2046	22731	45431	142696
Other Property Taxes	0.05	0.07	0.09	0.18	0.19	0.04	0.2	0.7	2.4	5	9	0	0	0	0	0	0	14329
	0.05	0.07	0.09	0.18	0.19	0.04	0.6	1.1	3.5	8	13	11	91	1534	10959	84869	122285	417223
C. Sales and Excise Taxes	0.50	0.65	0.93	1.59	3.18	6.69	38.7	153.7	446.5	1153	2530	5555	14101	59298	439356	3561850	7020632	13317276
Value Added Tax	0.16	0.22	0.30	0.43	1.02	2.70	20.7	68.5	175.8	510	1223	2815	6627	24992	152413	1272844	2459996	5648349
Unified Excise Tax	0.14	0.18	0.26	0.41	0.55	1.00	5.5	23.3	77.9	227	493	990	2647	10383	62902	591615	1248336	2906421
Tax on Bank Debits	0.00	0.00	0.00	0.00	0.00	0.00	1.0	4.8	12.7	1	0	0	1	635	15275	170027	408933	740737
Oils and Fuel Tax	0.09	0.13	0.19	0.43	1.02	1.80	5.5	26.8	99.3	188	354	1087	3283	16207	167871	1098944	2101622	3097328
Stamp Duty	0.04	0.05	0.06	0.13	0.19	0.35	1.8	7.0	15.9	51	117	187	441	1560	8129	76398	198321	345369
Foreign Exchange Transaction Tax	0.01	0.01	0.02	0.04	0.06	0.18	0.8	2.4	6.5	16	38	95	219	970	5243	54796	89947	206001
Electricity Consumption Tax	0.01	0.01	0.01	0.02	0.05	0.06	0.4	3.5	12.6	32	66	125	261	1146	4111	77119	175431	0
Petroleum Production Tax	0.01	0.01	0.01	0.03	0.12	0.20	0.6	2.7	10.3	19	47	98	264	1725	10107	81512	64350	0
Other Taxes	0.05	0.05	0.08	0.11	0.18	0.41	2.4	14.7	35.6	109	193	157	358	1681	13306	138593	273696	373071
D. Foreign Trade Taxes	0.17	0.23	0.48	0.74	0.97	2.37	18 6	40.4	77.2	238	561	921	2337	15966	102887	1256279	1868531	3422302
Import Tax	0.11	0.14	0.22	0.23	0.46	1.07	6.8	24.3	55.5	184	475	798	1450	5720	37361	380677	975259	2694989
Export Tax	0.06	0.08	0.25	0.49	0.46	1.19	11.2	13.4	15.5	29	45	93	819	10131	64688	836054	835124	611268
Other Trade Tax	0.00	0.00	0.01	0.02	0.05	0.12	0.6	2.7	6.1	24	40	29	67	116	838	39547	58147	116045
E. Social Security and Contributory Taxes	0.42	0.59	0.81	1.71	2.67	6.31	31.4	81.0	238.4	843	1843	1775	4450	22355	189658	1998517	3735027	8274926
Social Security System	0.41	0.58	0.80	1.70	2.65	6.27	31.2	80.5	236.8	720	1533	1447	3328	16406	151992	1661765	3029335	6705410
Contributions for Public Housing	0.00	0.00	0.00	0.00	0.00	0.00	0.0	0.0	0.0	119	301	320	1121	5935	37592	336546	704846	1568580
Unemployment	0.00	0.00	0.00	0.00	0.00	0.00	0.0	0.0	0.1	0	1	1	0	10	74	206	846	936
Other Contributory Taxes	0.00	0.00	0.01	0.01	0.02	0.03	0.1	0.4	1.5	4	8	6	2	5	0	0	0	0
F. Other Taxes	0.07	0.01	0.00	0.01	0.33	0.19	0.1	0.1	0.2	1	1	2	23	1605	11213	14007	3137	357129
G. Forced Saving	0.00	0.00	0.00	0.00	0.00	0.00	0.0	0.0	0.0	0	0	0	0	0	0	279015	414179	24668
H. Total Deductions (-)	0.00	0.00	0.00	0.00	0.24	0.35	3.3	8.1	20.3	47	91	239	1001	2090	12444	99877	603415	1586675
II-PROVINCIAL TAX REVENUE BY SOURCE	0.24	0.29	0.41	0.68	1.08	1.40	12.3	60.3	183.0	535	1235	2028	4904	18269	169209	1345563	2845627	..
A. Property Taxes	0.05	0.05	0.09	0.14	0.20	0.31	1.5	10.8	37.1	106	248	457	1123	4032	45291	338156	639084	..
B. Gross Income Tax	0.10	0.12	0.16	0.26	0.44	0.19	6.7	32.6	95.8	264	594	1005	2563	10009	87871	688747	1551380	..
C. Automotive License Tax	0.02	0.02	0.03	0.05	0.08	0.18	0.5	2.9	10.2	46	102	181	412	1726	18570	162605	254728	..
D. Stamp Duty	0.05	0.06	0.08	0.14	0.23	0.45	2.0	9.5	28.2	83	199	285	574	2164	14712	127088	341954	..
E. Other Taxes	0.03	0.03	0.05	0.08	0.13	0.27	1.6	4.6	11.8	36	93	99	233	338	2765	28967	58481	..
TOTAL TAX REVENUE	1.65	2.07	3.05	5.55	9.34	18.16	111.9	384.6	1069.4	3050	6852	11648	29513	132855	984194	9198207	17525315	29890648

Source: Ministry of Economy, National Directorate of Budgetary Programming.
a/ Excludes tax revenues of provincial governments.
May 1989

Table 5.21: ARGENTINA - TAX REVENUES BY SOURCE, 1970-1987
(Percent of GDP)

	1970	1971	1972	1973	1974	1975	1976	1977	1978	1979	1980	1981	1982	1983	1984	1985	1986	1987 a/
I- NATIONAL TAX REVENUE BY SOURCE (Gross)	16.01	14.20	12.77	13.71	16.98	11.72	13.14	15.49	16.93	17.64	19.82	17.57	16.67	15.78	15.43	19.83	19.75	16.87
NATIONAL TAX REVENUE BY SOURCE (Net of Deduction	16.01	14.20	12.77	13.71	16.49	11.48	12.70	15.10	16.55	17.31	19.50	17.13	15.99	15.48	[illegible]	19.58	18.94	15.97
A. Income Taxes	2.26	1.87	1.61	1.79	1.75	0.80	1.19	1.79	1.73	1.19	1.57	1.69	1.40	1.11	0.59	1.01	1.28	1.57
Personal and Corporate Income	2.12	1.77	1.50	1.48	1.46	0.75	1.14	1.69	1.64	1.11	1.47	1.58	1.34	1.01	0.51	0.96	1.20	1.47
Tax on Foreign Investment Income	0.00	0.00	0.00	0.00	0.01	0.03	0.01	0.03	0.02	0.02	0.02	0.04	0.02	0.02	0.01	0.01	0.02	0.02
Capital Gains Tax	0.14	0.10	0.11	0.09	0.04	0.00	0.02	0.05	0.03	0.04	0.05	0.03	0.02	0.02	0.01	0.02	0.02	0.03
Other Income Taxes	0.00	0.00	0.00	0.22	0.24	0.02	0.02	0.02	0.03	0.03	0.03	0.03	0.03	0.07	0.06	0.03	0.05	0.04
B. Property Taxes	0.62	0.53	0.43	0.50	0.53	0.04	0.24	0.55	0.64	0.77	0.83	0.81	1.10	1.14	0.77	0.86	0.93	0.97
Capital Tax	0.00	0.00	0.00	0.00	0.10	0.01	0.16	0.48	0.56	0.70	0.77	0.77	1.03	0.90	0.54	0.63	0.62	0.59
Net Worth Tax	0.00	0.00	0.00	0.00	0.04	0.00	0.00	0.01	0.01	0.01	0.02	0.02	0.01	0.01	0.02	0.02	0.15	0.15
Tax for Areas Affected by Flood	0.00	0.00	0.00	0.00	0.00	0.00	0.00	0.00	0.00	0.00	0.00	0.00	0.00	0.17	0.17	0.16	0.10	0.15
Tax on Financial Assets	0.00	0.00	0.00	0.00	0.00	0.00	0.06	0.02	0.02	0.02	0.02	0.02	0.06	0.05	0.04	0.06	0.06	0.08
Other Property Taxes	0.62	0.53	0.43	0.49	0.40	0.03	0.02	0.00	0.00	0.04	0.03	0.00	0.00	0.00	0.00	0.00	0.00	0.01
	0.62	0.53	0.43	0.50	0.40	0.03	0.09	0.02	0.02	0.06	0.05	0.02	0.06	0.22	0.21	0.21	0.16	0.24
C. Sales and Excise Taxes	5.74	5.18	4.48	4.48	6.53	4.68	5.10	7.34	8.53	8.09	8.93	10.14	9.55	8.69	8.32	9.00	9.45	7.52
Value Added Tax	1.85	1.77	1.45	1.20	2.10	1.89	2.73	3.27	3.36	3.58	4.32	5.14	4.49	3.66	2.89	3.21	3.31	3.19
Unified Excise Tax	1.58	1.43	1.23	1.15	1.13	0.70	0.73	1.12	1.49	1.59	1.74	1.81	1.79	1.52	1.19	1.49	1.68	1.64
Tax on Bank Debits	0.00	0.00	0.00	0.00	0.00	0.00	0.14	0.23	0.24	0.01	0.00	0.00	0.00	0.09	0.29	0.43	0.55	0.42
Oils and Fuel Tax	1.08	1.03	0.91	1.22	2.09	1.26	0.73	1.28	1.90	1.32	1.25	1.99	2.22	2.37	3.18	2.78	2.83	1.75
Stamp Duty	0.45	0.37	0.30	0.36	0.38	0.24	0.24	0.33	0.30	0.36	0.41	0.34	0.30	0.23	0.15	0.19	0.27	0.19
Foreign Exchange Transaction Tax	0.08	0.09	0.08	0.10	0.12	0.12	0.10	0.12	0.13	0.11	0.13	0.17	0.15	0.14	0.10	0.14	0.12	0.12
Electricity Consumption Tax	0.12	0.09	0.07	0.07	0.11	0.04	0.05	0.16	0.24	0.22	0.23	0.23	0.18	0.17	0.08	0.19	0.24	0.00
Petroleum Production Tax	0.07	0.05	0.06	0.08	0.24	0.14	0.08	0.13	0.20	0.13	0.16	0.18	0.18	0.25	0.19	0.21	0.09	0.00
Other Taxes	0.52	0.37	0.38	0.30	0.36	0.28	0.31	0.70	0.68	0.77	0.68	0.29	0.24	0.25	0.25	0.35	0.37	0.21
	0.71	0.50	0.52	0.45	0.71	0.46	0.44	1.00	1.12	1.12	1.08	0.69	0.60	0.67	0.52	0.75	0.69	0.21
D. Foreign Trade Taxes	1.91	1.80	2.33	2.09	1.99	1.66	2.46	1.93	1.47	1.67	1.98	1.68	1.58	2.34	1.95	3.17	2.51	1.93
Import Tax	1.23	1.14	1.07	0.66	0.94	0.75	0.90	1.16	1.06	1.29	1.68	1.46	0.98	0.84	0.71	0.96	1.31	1.52
Export Tax	0.66	0.64	1.20	1.39	0.95	0.83	1.48	0.64	0.30	0.21	0.16	0.17	0.55	1.48	1.22	2.11	1.12	0.34
Other Trade Tax	0.02	0.02	0.06	0.05	0.09	0.08	0.08	0.13	0.12	0.17	0.14	0.05	0.05	0.02	0.02	0.10	0.08	0.07
E. Social Security and Contributory Taxes	4.74	4.70	3.91	4.82	5.50	4.41	4.13	3.87	4.56	5.91	6.50	3.24	3.01	3.27	3.59	5.05	5.03	4.67
Social Security System	4.70	4.67	3.88	4.79	5.46	4.39	4.12	3.85	4.52	5.05	5.41	2.64	2.25	2.40	2.88	4.20	4.08	3.78
Contributions for Public Housing	0.00	0.00	0.00	0.00	0.00	0.00	0.00	0.00	0.00	0.84	1.06	0.58	0.76	0.87	0.71	0.85	0.95	0.89
Unemployment	0.00	0.00	0.00	0.00	0.00	0.00	0.00	0.00	0.00	0.00	0.00	0.00	0.00	0.00	0.00	0.00	0.00	0.00
Other Contributory Taxes	0.04	0.03	0.03	0.03	0.04	0.02	0.01	0.02	0.03	0.03	0.03	0.01	0.00	0.00	0.00	0.00	0.00	0.00
F. Other Taxes	0.74	0.12	0.01	0.03	0.68	0.13	0.01	0.01	0.00	0.00	0.00	0.00	0.02	0.24	0.21	0.04	0.00	0.20
G. Forced Saving	0.00	0.00	0.00	0.00	0.00	0.00	0.00	0.00	0.00	0.00	0.00	0.00	0.00	0.00	0.00	0.70	0.56	0.01
H. Total Deductions (-)	0.00	0.00	0.00	0.00	0.49	0.24	0.43	0.39	0.39	0.33	0.32	0.44	0.68	0.31	0.24	0.25	0.81	0.90
II- PROVINCIAL TAX REVENUE BY SOURCE	2.79	2.30	1.97	1.92	2.23	0.98	1.62	2.88	3.50	3.76	4.36	3.70	3.32	2.68	3.20	3.40	3.83	..
A. Property Taxes	0.53	0.43	0.42	0.41	0.41	0.22	0.20	0.52	0.71	0.74	0.87	0.84	0.76	0.59	0.86	0.85	0.86	..
B. Gross Income Tax	1.14	0.95	0.78	0.74	0.91	0.13	0.88	1.56	1.83	1.85	2.10	1.84	1.74	1.47	1.66	1.74	2.09	..
C. Automotive License Tax	0.22	0.19	0.15	0.15	0.16	0.13	0.07	0.14	0.19	0.32	0.36	0.33	0.28	0.25	0.35	0.41	0.34	..
D. Stamp Duty	0.52	0.45	0.39	0.41	0.48	0.31	0.26	0.45	0.54	0.58	0.70	0.52	0.39	0.32	0.28	0.32	0.46	..
E. Other Taxes	0.37	0.28	0.22	0.21	0.26	0.19	0.21	0.22	0.23	0.25	0.33	0.18	0.16	0.05	0.05	0.07	0.08	..
TOTAL TAX REVENUE	18.80	16.50	14.74	15.63	19.21	12.70	14.75	18.37	20.43	21.40	24.18	21.27	19.99	19.46	18.64	23.23	23.58	16.87

Source: Ministry of Economy, National Directorate of Budgetary Programming.
a/ Excludes tax revenues of provincial governments.
May 1989

Table 6.1: ARGENTINA - PRIVATE SECTOR HOLDINGS OF FINANCIAL ASSETS, QUARTERLY 1970-1989 a/
(Thousands of Australes; stocks at the end of each quarter)

(continues...)

	Money (M1)			Quasi-Money			M2 b/	Accep-tances c/	M3 d/
	Total	Currency	Demand Deposits	Total	Savings Accounts	Time Deposits			
1970	1.5	0.8	0.7	1.0	0.9	0.1	2.5	0.0	2.5
1971 I	1.5	0.7	0.8	1.1	1.0	0.1	2.6	0.0	2.6
II	1.6	0.7	0.9	1.2	1.1	0.1	2.8	0.0	2.8
III	1.7	0.8	0.9	1.3	1.1	0.2	3.0	0.0	3.0
IV	2.0	1.0	1.0	1.4	1.2	0.2	3.4	0.0	3.4
1972 I	2.1	0.9	1.2	1.5	1.3	0.2	3.6	0.0	3.6
II	2.2	1.0	1.2	1.7	1.4	0.3	3.9	0.0	3.9
III	2.3	1.0	1.3	1.8	1.5	0.3	4.1	0.0	4.1
IV	2.9	1.3	1.6	2.1	1.7	0.4	5.0	0.5	5.5
1973 I	3.2	1.3	1.9	2.4	1.9	0.5	5.6	0.6	6.2
II	3.8	1.6	2.2	2.7	2.2	0.5	6.5	0.8	7.3
III	4.5	1.9	2.6	3.3	2.7	0.6	7.8	0.9	8.7
IV	5.6	2.6	3.0	3.9	3.3	0.6	9.5	1.0	10.5
1974 I	5.8	2.4	3.4	4.6	3.9	0.7	10.4	1.1	11.5
II	6.7	2.8	3.9	5.3	4.4	0.9	12.0	1.2	13.2
III	7.4	3.2	4.2	5.7	4.7	1.0	13.1	1.4	14.5
IV	8.9	4.3	4.6	5.9	4.8	1.1	14.8	1.5	16.3
1975 I	9.4	4.1	5.3	6.3	5.0	1.3	15.7	1.7	17.4
II	11.3	4.9	6.4	6.3	5.0	1.3	17.6	1.9	19.5
III	16.2	7.0	9.2	6.5	5.2	1.3	22.7	2.6	25.3
IV	26.0	12.4	13.6	7.8	6.5	1.3	33.8	3.3	37.1
1976 I	36.0	15.0	21.0	9.7	8.2	1.5	45.7	4.7	50.4
II	54.1	19.6	34.5	18.8	12.1	6.7	72.9	9.9	82.8
III	68.6	25.4	43.2	31.9	17.9	14.0	100.5	15.0	115.5
IV	92.6	41.0	51.6	53.1	24.6	28.5	145.7	21.2	166.9
1977 I	103.8	41.8	62.0	76.6	26.3	50.3	180.4	28.4	208.8
II	142.4	59.9	82.5	138.2	37.6	100.6	280.6	21.0	301.6
III	156.4	69.7	86.7	210.6	34.2	176.4	367.0	9.6	376.6
IV	208.4	107.3	101.1	291.6	37.4	254.2	500.0	8.0	508.0
1978 I	239.4	116.2	123.2	400.0	52.8	347.2	639.4	6.4	645.8
II	330.8	162.2	168.6	546.8	68.6	478.2	877.6	3.4	881.0
III	362.5	188.7	173.8	743.0	87.9	655.1	1105.5	2.6	1108.1
IV	563.4	333.2	230.2	868.1	104.3	763.8	1431.5	3.3	1434.8
1979 I	633.1	323.3	309.8	1203.5	131.3	1072.2	1836.6	4.3	1840.9
II	825.0	421.6	403.4	1649.4	170.2	1479.2	2474.4	4.9	2479.3
III	918.6	476.3	442.3	2252.7	184.2	2068.5	3171.3	6.1	3177.4
IV	1382.9	787.0	595.9	2914.8	254.1	2660.7	4297.7	7.1	4304.8

Table 6.1: ARGENTINA - PRIVATE SECTOR HOLDINGS OF FINANCIAL ASSETS, QUARTERLY 1970-1989 a/
(Thousands of Australes; stocks at the end of each quarter)

	Money (M1)			Quasi-Money			M2 b/	Accep-tances c/	M3 d/
	Total	Currency	Demand Deposits	Total	Savings Accounts	Time Deposits			
1980 I	1550.4	803.5	746.9	3557.7	328.1	3229.6	5110.8	9.9	5120.7
II	1854.5	996.8	857.7	3862.0	441.6	3420.4	5716.5	10.0	5726.5
III	1998.6	1099.6	899.0	4772.7	520.6	4252.1	6771.3	7.0	6778.3
IV	2735.0	1641.8	1093.2	5282.2	627.6	4654.6	8017.2	5.2	8022.4
1981 I	2166.3	1332.4	833.9	5986.5	637.5	5349.0	8152.8	0.4	8153.2
II	2625.7	1610.0	1015.7	7011.0	713.8	6297.2	9636.7	5.3	9642.0
III	2941.7	1737.1	1204.6	9582.5	938.9	8643.6	12524.2	14.7	12538.9
IV	4609.5	3020.7	1588.8	11368.2	1195.8	10172.4	15977.7	11.4	15989.1
1982 I	4338.8	2550.8	1788.0	14840.9	1500.5	13340.4	19179.7	1.3	19181.0
II	6446.7	3914.7	2532.0	17106.6	2238.2	14868.4	23553.3	4.3	23557.6
III	8780.6	4760.2	4020.4	18282.3	2368.5	15913.8	27062.9	0.0	27062.9
IV	14864.3	8736.0	6128.3	23458.4	2977.4	20481.0	38322.7	0.0	38322.7
1983 I	16557	9811	6747	36239	4146	32093	52800	0	52800
II	25060	13951	11109	49863	8344	41519	74923	0	74923
III	33902	19910	13992	72269	14525	57744	106171	0	106171
IV	69953	46342	23611	123645	30842	92803	193598	20960	214558
1984 I	103823	65652	38171	205658	52820	152838	309481	4750	314231
II	170790	109235	61555	292345	76656	215689	463135	22358	485493
III	232682	147965	84717	473961	130122	343839	706643	69039	775682
IV	434804	313400	121404	758691	240231	518460	1193495	95947	1289442
1985 I	625684	433901	191783	1337949	502381	835568	1963633	172921	2136554
II	1470616	980176	490440	3074244	1087798	1986446	4544860	13	4544873
III	2076272	1272467	803805	3941708	1282061	2659647	6017980	0	6017980
IV	3014321	2022272	992049	4610147	1476689	3133458	7624468	0	7624468
1986 I	3029811	1861442	1168369	5737789	1606535	4131254	8767600	0	8767600
II	3747440	2387367	1360073	6731689	1904256	4827433	10479634	0	10479634
III	3865811	2533439	1332372	8221209	2164159	6057050	12087020	0	12087020
IV	5587480	3989617	1597863	10641529	2562314	8079215	16229029	258333	16487362
1987 I	7174000	3782000	3392000	14829000	2884000	11945000	22003000	124000	22127000
II	8375000	4386000	3989000	17754000	3212000	14542000	26129000	150000	26279000
III	9161000	4834000	4327000	22924000	4298000	18626000	32085000	214000	32299000
IV	13980000	7683000	6297000	30453000	5423000	25030000	44433000	210000	44643000
1988 I	15527000	8518000	7009000	46024000	6313000	39711000	61551000	109000	61660000
II	22970000	11627000	11343000	71453000	8658000	62795000	94423000	33000	94456000
III	37885000	19461000	18424000	129939000	14446000	115493000	167824000	59000	167883000
IV	57648000	33913000	23735000	182705000	20122000	162583000	240353000	218000	240571000
1989 I	75124000	43375000	31749000	266210000	25878000	240332000	341334000	84000	341418000
II	262359000	126956000	135403000	845227000	57268000	787959000	1107586000	69464000	1177050000

Source: Central Bank of the Republic of Argentina (BCRA).

a/ Does not include deposits issued by non-bank financial institutions.

b/ M1 plus quasi-money.

c/ Includes acceptances issued by finance companies.

d/ M2 plus acceptances.

August 1989

Table 6.2: ARGENTINA - LIQUIDITY COEFFICIENTS, QUARTERLY, 1970-1988

	M1/ GDP	M2/ GDP	M3/ GDP
1970	15.4%	25.7%	25.7%
1971	14.0%	23.8%	23.8%
1972	11.6%	20.1%	22.1%
1973 I	10.6%	18.6%	20.6%
II	10.4%	17.8%	20.0%
III	12.4%	21.6%	24.1%
IV	14.3%	24.3%	26.8%
1974 I	14.2%	25.5%	28.1%
II	14.4%	25.8%	28.3%
III	14.9%	26.4%	29.2%
IV	15.5%	25.7%	28.3%
1975 I	13.8%	23.0%	25.5%
II	12.1%	18.9%	20.9%
III	9.6%	13.5%	15.0%
IV	10.7%	14.0%	15.3%
1976 I	9.6%	12.2%	13.5%
II	7.6%	10.2%	11.6%
III	8.0%	11.7%	13.5%
IV	8.5%	13.4%	15.4%
1977 I	7.4%	12.9%	15.0%
II	8.0%	15.8%	16.9%
III	6.8%	16.1%	16.5%
IV	7.2%	17.2%	17.4%
1978 I	7.1%	19.0%	19.2%
II	7.3%	19.4%	19.5%
III	6.4%	19.4%	19.5%
IV	7.7%	19.5%	19.5%
1979 I	6.7%	19.6%	19.6%
II	6.6%	19.9%	19.9%
III	5.7%	19.7%	19.8%
IV	7.2%	22.5%	22.6%
1980 I	7.2%	23.7%	23.8%
II	7.2%	22.3%	22.4%
III	6.6%	22.3%	22.4%
IV	7.6%	22.3%	22.4%
1981 I	5.9%	22.2%	22.2%
II	5.6%	20.5%	20.5%
III	5.0%	21.2%	21.2%
IV	6.1%	21.0%	21.0%
1982 I	5.1%	22.5%	22.5%
II	6.5%	23.8%	23.8%
III	5.5%	17.1%	17.1%
IV	6.0%	15.5%	15.5%
1983 I	5.7%	18.2%	18.2%
II	5.8%	17.4%	17.4%
III	4.6%	14.4%	14.4%
IV	5.5%	15.2%	16.8%
1984 I	5.2%	15.6%	15.8%
II	5.1%	13.7%	14.4%
III	3.9%	11.9%	13.1%
IV	4.4%	12.2%	13.1%
1985 I	3.6%	11.4%	12.4%
II	3.9%	12.2%	12.2%
III	4.1%	12.0%	12.0%
IV	5.6%	14.2%	14.2%
1986 I	5.4%	15.6%	15.6%
II	5.7%	16.0%	16.0%
III	4.8%	15.0%	15.0%
IV	5.9%	17.1%	17.4%
1987 I	6.5%	19.8%	19.9%
II	6.2%	19.3%	19.4%
III	5.0%	17.5%	17.6%
IV	5.0%	15.9%	16.0%
1988 I	4.6%	18.1%	18.1%
II	4.2%	17.1%	17.1%
III	3.7%	16.6%	16.6%
IV	4.2%	17.4%	17.4%
1989 I	4.5%	20.2%	20.2%
II	3.9%	16.5%	17.6%

Source: Central Bank of Argentina (BCRA) and Table 6.1.

a/ Stocks in current australes at the end of each quarter divided by four times the level of GDP for the quarter. Monetary aggregate definitions are as in Table 6.1.

August 1989

Table 6.3 : ARGENTINA - SOURCES AND USES OF BANK CREDIT, 1970-1988
(Percent of GDP, 1970 prices)

	1970	1971	1972	1973	1974	1975	1976	1977	1978	1979	1980	1981	1982	1983	1984	1985	1986	1987	1988
Uses of Credit	27.89	26.20	23.73	57.37	61.12	36.07	35.60	31.35	33.68	38.96	43.74	57.99	75.34	58.54	52.67	61.05	60.25	63.47	80.16
Cash in Vault	1.74	1.11	1.11	1.85	2.08	1.82	1.54	1.72	1.51	1.43	1.42	1.36	1.23	1.59	1.63	2.20	1.60	1.20	1.27
Net Foreign Assets	-0.65	-0.40	-0.28	-0.19	0.03	-0.71	-0.79	-1.92	-2.40	-3.83	-3.76	-9.10	-13.00	-12.36	-10.93	-10.32	-8.80	-11.15	-9.73
Credit to the Government of which:	3.36	2.28	2.47	4.37	5.06	3.36	2.51	3.38	5.10	6.04	5.87	9.44	9.77	6.13	6.18	14.89	17.14	16.82	18.13
Forced Investments																8.41	9.87	8.49	10.33
Credit to the Private Sector	19.43	19.11	16.60	19.89	20.48	10.90	10.13	14.44	18.61	24.38	29.12	32.56	29.74	23.33	19.68	20.43	21.00	21.84	20.11
Unremunerated Reserves with Central Bank	1.04	1.25	1.25	3.83	3.84	3.44	4.72	7.30	3.99	1.89	1.21	1.65	12.53	9.84	5.82	4.83	2.28	0.72	1.04
Other	2.97	2.86	2.56	27.61	29.62	17.27	17.50	6.42	6.86	9.07	9.89	22.07	35.07	30.01	30.29	29.03	27.02	34.03	49.35
Sources of Credit	27.89	26.20	23.73	57.37	61.12	36.07	35.60	31.35	33.68	38.96	43.74	57.99	75.34	58.54	52.67	61.05	60.25	63.47	80.16
Deposits	21.14	19.31	17.49	25.26	27.83	15.67	15.32	19.95	22.14	26.32	28.51	25.73	19.42	17.63	16.32	20.92	22.96	20.77	22.91
Public Sector	4.07	2.93	2.37	3.88	5.20	3.43	3.73	4.54	4.36	4.10	4.93	4.26	3.66	2.94	2.18	5.51	5.04	2.90	3.01
Private Sector	17.07	16.38	15.12	21.38	22.63	12.25	11.59	15.41	17.78	22.22	23.59	21.48	15.75	14.69	14.14	15.41	17.92	17.87	19.90
Rediscounts a/	1.38	1.74	1.32	26.62	26.02	15.13	13.45	1.39	1.95	1.63	3.74	4.42	15.77	5.97	3.55	10.86	11.23	13.57	19.96
Net Worth	2.75	2.33	1.87	2.69	2.79	1.38	2.12	3.55	4.16	4.93	5.52	7.07	8.86	7.98	7.77	7.84	7.89	7.73	6.66
Other	2.62	2.81	3.05	2.80	4.48	3.89	4.71	6.45	5.43	6.08	5.97	20.77	31.29	26.95	25.01	21.43	18.17	21.40	30.63

Source: Central Bank of the Republic of Argentina (BCRA).
a/ Before 1985 interest on rediscounts is included under "Other".

June 1989

Table 6.4: ARGENTINA - BANKING SYSTEM REAL CREDIT EXPANSION, 1970-1988

	Total Domestic Credit Expansion (% of GDP)	Percentage of Domestic Credit Expansion to:	
		Public Sector	Private Sector
1971	-0.59	163.02	-64.53
1972	-2.11	-11.47	100.35
1973	31.02	6.41	12.54
1974	5.94	15.28	27.09
1975	-23.96	7.18	40.51
1976	-1.40	61.47	54.79
1977	-4.08	-25.22	-120.69
1978	5.51	29.14	66.85
1979	10.92	11.62	64.03
1980	5.99	-1.24	85.18
1981	15.97	19.73	8.49
1982	7.15	-2.35	-63.28
1983	-13.00	25.91	42.83
1984	-1.89	-10.43	163.45
1985	5.62	149.87	-2.70
1986	4.12	73.34	39.25
1987	8.86	0.30	14.34
1988	12.02	5.36	-21.62

Source: Central Bank of the Republic of Argentina (BCRA).
a/ Private, public and other credit of the banking system.

July 1989

Table 6.5: ARGENTINA - FINANCIAL INSTITUTIONS BY ACTIVITY AND OWNERSHIP, 1979-1987
(Number of institutions, year-end)

	1979		1980		1981		1982		1983		1984		1985		1986		1987	
	HQ	Branches	HQ	Branches	HQ	Branches	HQ	Branches	HQ	Branches	HQ	Branches	HQ	Branches	HQ	Branches	HQ	Branches
Commercial Banks	211	3720	207	3714	199	3767	197	3957	203	4236	203	4420	191	4364	185	4354	172	4241
Federal Government owned	1	573	1	570	1	572	2	553	2	553	2	554	2	546	2	547	2	546
State owned	24	1056	24	1083	24	1104	24	1122	24	1133	24	1152	25	1183	25	1198	24	1216
Municipality owned	5	59	5	62	5	65	5	67	5	69	5	69	5	67	5	69	5	68
Private Domestic	161	1814	151	1784	137	1705	133	1869	140	2140	140	2305	128	2225	122	2200	109	1989
Private Foreign	20	218	26	215	32	341	33	346	32	341	32	340	31	341	31	340	32	422
Development Banks	2	33	2	33	2	33	2	32	2	33	2	33	2	33	2	33	2	33
Federal Government owned	1	33	1	33	1	33	1	32	1	33	1	33	1	33	1	33	1	33
State owned	1	0	1	0	1	0	1	0	1	0	1	0	1	0	1	..	1	..
Investment Banks	4	0	3	0	3	0	3	0	3	0	3	0	3	0	2	..	2	..
State owned	1	0	1	0	1	0	1	0	1	0	1	0	1	0	1	..	1	..
Private Domestic	2	0	1	0	1	0	1	0	1	0	1	0	1	0	..	..	..	..
Private Foreign	1	0	1	0	1	0	1	0	1	0	1	0	1	0	1	..	1	..
Mortgage Banks																		
Federal Government owned	1	52	1	52	1	53	1	53	1	53	1	53	1	53	1	53	1	53
Savings Banks																		
Federal Government owned	1	40	1	40	1	49	1	50	1	51	1	53	1	53	1	53	1	53
Credit Cooperatives a/	104	13	92	24	92	26	76	23	71	24	48	16	33	10	30	9	25	7
Finance Companies a/	142	205	135	216	126	218	111	214	102	208	87	183	71	155	64	137	56	114
Savings and Loans Assoc. a/	31	43	28	40	25	33	22	35	19	36	15	32	13	28	11	18	8	17
Consumer Credit Assoc. a/	..	..	..	..	..	..	..	..	..	..	..	..	..	..	..	..	..	..
Total	496	4106	469	4119	449	4199	413	4364	402	4641	360	4790	315	4696	296	4657	267	4518

Source: Central Bank of the Republic of Argentina, Departmento de Expansion y Servicios de Entidades.
a/ Private sector.

March 1988

Table 6.6: ARGENTINA - INTEREST RATES, 1977-1989
(Quarterly Average of Monthly Rates, Percent)

Year	Quarter	Deposit Rate a/		Lending Rate b/		Spread
		Nominal	Real c/	Nominal	Real c/	
1977	I	..	..	4.5	-1.0	4.5
	II	..	..	5.5	-0.7	5.5
	III	7.3	-3.4	8.2	-2.6	0.9
	IV	10.1	2.5	13.2	5.4	3.1
1978	I	8.6	0.7	11.3	3.2	2.7
	II	7.3	-3.4	8.2	-2.6	0.9
	III	6.6	-1.7	7.7	-0.6	1.1
	IV	6.7	-1.5	7.6	-0.6	0.9
1979	I	6.5	-0.9	7.2	-0.2	0.7
	II	6.5	2.3	7.2	-1.7	0.7
	III	7.2	0.5	7.7	-1.1	0.5
	IV	6.4	3.0	7.1	3.7	0.7
1980	I	5.3	1.2	6.0	1.9	0.7
	II	4.8	-0.7	5.5	-0.0	0.7
	III	5.1	1.4	6.2	2.5	1.1
	IV	4.8	2.7	5.6	3.5	0.8
1981	I	6.8	2.6	8.8	4.6	2.0
	II	8.5	-4.5	9.8	-3.2	1.3
	III	9.9	0.2	11.7	2.0	1.8
	IV	7.1	-2.1	8.5	-0.7	1.4
1982	I	7.1	-1.0	8.4	0.3	1.3
	II	7.2	-2.9	8.5	-1.6	1.3
	III	5.7	-15.3	6.6	-14.4	0.9
	IV	8.0	-3.3	8.5	-2.8	0.5
1983	I	10.2	-2.7	11.0	-1.9	0.8
	II	9.6	-0.6	10.6	0.4	1.0
	III	12.0	-5.9	13.0	-4.9	1.0
	IV	14.5	-2.7	15.5	-1.7	1.0
1984	I	10.5	-4.7	11.5	-3.7	1.0
	II	13.0	-5.4	14.0	-4.4	1.0
	III	15.5	-5.2	17.0	-3.7	1.5
	IV	17.0	-0.7	19.0	1.3	2.0
1985	I	18.5	-3.7	20.5	-1.7	2.0
	II	23.3	-11.7	25.3	-9.7	2.0
	III	3.5	3.1	5.0	4.6	1.5
	IV	3.1	2.3	4.5	3.7	1.4
1986	I	3.1	2.4	4.5	3.8	1.4
	II	3.2	-0.3	4.6	1.1	1.4
	III	4.4	-2.7	5.9	-1.2	1.5
	IV	5.3	0.9	6.8	2.4	1.5
1987	I	4.8	-1.9	6.0	-0.7	1.2
	II	5.1	0.6	6.1	1.6	-1.0
	III	11.0	-2.6	10.4	-3.1	0.6
	IV	13.1	0.8	12.2	-0.1	0.9
1988	I	14.6	0.8	14.1	0.2	0.6
	II	18.1	-3.3	17.6	-3.7	0.5
	III	14.6	-6.4	14.1	-6.9	0.5
	IV	10.7	5.9	10.6	5.8	0.1
1989	I	18.6	7.1	17.6	6.1	1.0
	II	116.2	18.0	102.5	4.2	13.8

Source: CEPAL.

a/ Until second quarter 1982, weighted average of rate offered by banks on 30 day deposits; from third quarter 1982 through third quarter 1987, regulated rates; from fourth quarter 1987, average of deposit rates between enterprises with Government dollar bond (BONEX) guarantee.

b/ Until second quarter 1982, average of rates charged by banks to primary clients; from third quarter 1982 through third quarter 1987, regulated rates; from fourth quarter 1987, average of lending rates with Government dollar bond (BONEX) guarantee.

c/ Deflated by WPI.

August 1989

Table 7.1: ARGENTINA - BEEF-WHEAT PRICE RATIO IN US AND ARGENTINA, 1961-1989 a/

Year	United States	Argentina
1961	7.11	3.70
1962	7.13	3.85
1963	7.00	3.70
1964	6.50	4.76
1965	9.67	6.25
1966	9.13	5.88
1967	10.75	4.76
1968	12.63	4.55
1969	12.88	4.35
1970	13.71	6.25
1971	14.43	7.69
1972	12.83	9.09
1973	6.50	6.25
1974	5.46	4.83
1975	5.96	3.03
1976	7.50	3.85
1977	9.67	5.26
1978	10.64	4.35
1979	12.36	6.67
1980	10.22	5.56
1981	9.43	5.00
1982	10.57	5.00
1983	10.72	5.56
1984	..	6.92
1985	..	5.12
1986	..	6.98
1987	..	7.83
1988	..	
January	..	6.70
February	..	6.30
March	..	5.90
April	..	5.60
May	..	4.00
June	..	3.40
July	..	3.70
August	..	4.00
September	..	5.30
October	..	7.10
November	..	5.90
December	..	5.30
1989		
January		5.90
February		5.90
March		5.60

Source: Ministry of Agriculture, S.N.E.S.R.

a/ One kg. wheat equivalent to 1 kg. of beef liveweight.

August 1989

Table 7.2: ARGENTINA - PRINCIPAL CROPS, AREA PLANTED, 1970-1989 a/
(Thousand Hectares)

	1969/ 70	1970/ 71	1971/ 72	1972/ 73	1973/ 74	1974/ 75	1975/ 76	1976/ 77	1977/ 78	1978/ 79	1979/ 80	1980/ 81	1981/ 82	1982/ 83	1983/ 84	1984/ 85	1985/ 86	1986/ 87	1987/ 88	1988/ 89	Growth Rates 1969/70 - 1988/89 b/
PRINCIPAL CEREAL CROPS																					
Wheat	6239	4468	4986	5627	4252	5183	5753	7192	4600	5230	5000	6196	6566	7410	7210	6000	5700	5000	4935	4530	0.36
Corn	4666	4993	4439	4251	4134	3871	3696	2980	3100	3300	3310	4000	3695	3440	3484	3620	3820	3650	2825	2485	-2.03
Grain sorghum	2568	3122	2759	2974	3114	2602	2358	2780	2650	2530	1884	2400	2712	2657	2550	2040	1400	1127	1075	820	-5.09
Oats	1129	1026	1098	1222	1154	1200	1341	1471	1480	1545	1680	1718	1615	1856	1800	1775	1739	2530	1960	1830	3.68
PRINCIPAL INDUSTRIAL CROPS																					
Cotton	464	388	435	536	557	513	433	543	621	702	585	343	404	373	486	462	353	292	495	465	-1.07
Sugarcane	203	227	256	299	350	348	351	360	356	351	337	353	351	354	354	358	355	356	355	..	2.11
Tobacco	76	71	74	78	89	93	82	81	70	78	64	53	50	67	66	53	55	60	55	..	-2.44
PRINCIPAL OIL CROPS																					
Sunflower	1472	1614	1533	1652	1342	1196	1411	1460	2200	1766	2000	1925	1733	1930	2131	2380	3140	1890	2117	2265	2.90
Linseed	952	973	538	509	415	520	471	722	950	893	1070	780	851	910	810	620	750	758	670	570	0.57
Peanuts	215	314	321	389	350	383	335	368	452	400	287	201	180	125	146	146	175	212	194	152	-4.73
Soybeans	31	38	80	169	377	370	443	710	1200	1640	2100	1925	2040	2226	2920	3300	3340	3700	4413	4600	25.02
VEGETABLE CROPS																					
Potatoes	199	86	55	124	29	3	0	5	9	3	20	2	104	109	115	107	109	105	113	..	6.06
Sweet potatoes	48	46	44	44	45	43	40	37	35	35	34	34	32	29	31	32	32	35	29	21	-3.08
Dry beans	46	64	63	83	9	5	15	185	54	236	243	2	236	229	199	193	230	..	171	..	11.94
Tomatoes	26	28	29	33	37	35	33	32	33	3	3	23	30	32	34	39	39	38	27	28	0.43

Source: Ministry of Agriculture, SNESR.

a/ Crop year: July through June.

b/ Where series are incomplete, growth rates are for the longest continuous series.

August 1989

Table 7.3: ARGENTINA - PRINCIPAL CROPS, AREA HARVESTED, 1970-1989 a/
(Thousand Hectares)

	1969/70	1970/71	1971/72	1972/73	1973/74	1974/75	1975/76	1976/77	1977/78	1978/79	1979/80	1980/81	1981/82	1982/83	1983/84	1984/85	1985/86	1986/87	1987/88	1988/89	Growth Rates 1969/70 - 1988/89 b/
PRINCIPAL CEREAL CROPS																					
Wheat	5191	3701	4295	4965	3958	4233	5271	6428	3910	4685	4787	5023	5923	7320	7073	5900	5382	4893	4875	4468	1.24%
Corn	4017	4066	3147	3565	3486	3070	2766	2532	2660	2800	2490	3394	3170	2720	3024	3497	3340	3231	2437	1627	-1.87%
Grain Sorghum	1872	2235	1419	2131	2324	1937	1834	2377	2254	2044	1279	2100	2510	2520	2370	1965	1280	977	956	636	-3.41%
Oats	327	300	357	399	395	282	338	383	430	500	410	350	298	408	410	434	333	312	476	446	0.98%
PRINCIPAL INDUSTRIAL CROPS																					
Cotton	452	367	398	457	474	505	414	518	607	669	568	282	399	343	469	447	339	273	492	460	-0.80%
Sugarcane	192	211	243	272	298	293	339	350	343	306	314	319	308	309	313	318	286	296	297	..	1.51%
Tobacco	69	65	68	74	83	88	79	75	62	75	57	47	55	60	62	49	49	51	53	..	-2.48%
PRINCIPAL OIL CROPS																					
Sunflower	1347	1313	1267	1338	1190	1005	1258	1233	2000	1557	1855	1280	1673	1902	1989	2360	3046	1735	2032	2129	3.64%
Linseed	791	834	451	441	390	501	446	674	884	817	978	726	818	864	804	603	688	744	655	557	1.26%
Peanuts	211	310	294	379	345	357	309	367	428	393	279	197	179	125	146	146	173	212	190	152	-4.49%
Soybeans	26	36	68	157	344	356	434	660	1150	1600	2030	1880	1986	2116	2910	3269	3314	3510	4373	3906	25.38%
VEGETABLE CROPS																					
Potatoes	190	179	147	117	127	111	108	111	115	110	112	117	102	108	113	106	107	103	113	..	-2.23%
Sweet Potatoes	44	42	35	42	33	41	38	36	34	34	34	24	32	271	31	32	32	34	29	21	-1.13%
Dry Beans	41	61	62	79	108	137	147	171	136	231	205	211	230	200	172	191	227	..	171	..	9.25%
Tomatoes	20	24	26	30	34	33	27	29	30	26	29	19	27	31	30	37	38	37	27	28	1.22%

Source: Ministry of Agriculture, SNESR.
a/ Crop year: July through June.
b/ Where series are incomplete growth rates are for longest continuous series.
August 1989

Table 7.4: **ARGENTINA - YIELD OF PRINCIPAL CROPS, 1970-1989 a/**
(Metric Tons Per Harvested Hectare)

	1969/70	1970/71	1971/72	1972/73	1973/74	1974/75	1975/76	1976/77	1977/78	1978/79	1979/80	1980/81	1981/82	1982/83	1983/84	1984/85	1985/86	1986/87	1987/88	1988/89	Growth Rates 1969/70 - 1988/89 b/
PRINCIPAL CEREAL CROPS																					
Wheat	1.35	1.33	1.27	1.59	1.66	1.41	1.63	1.71	1.36	1.73	1.69	1.55	1.40	2.05	1.84	2.30	1.62	2.06	2.20	1.70	2.06%
Corn	2.33	2.44	1.86	2.72	2.84	2.51	2.12	3.28	3.65	3.11	2.57	3.80	3.03	2.98	3.14	3.56	3.74	3.19	3.77	2.80	2.22%
Grain sorghum	2.04	2.09	1.66	2.33	2.54	2.49	2.76	2.78	3.19	3.03	2.31	3.60	3.19	3.27	2.91	3.15	3.12	3.07	3.35	2.52	2.32%
Oats	1.30	1.20	1.33	1.42	1.42	1.16	1.28	1.38	1.32	1.35	1.27	1.24	1.14	1.56	1.45	1.65	1.20	1.59	1.51	1.30	0.62%
PRINCIPAL INDUSTRIAL CROPS																					
Cotton	1.00	0.78	0.73	0.93	0.88	1.07	1.07	0.99	1.18	0.86	0.85	1.00	1.21	1.09	1.30	1.20	1.11	1.18	1.73	..	2.60%
Sugarcane	50.55	48.53	53.00	62.00	52.08	53.21	42.19	45.71	39.65	46.20	54.78	46.47	48.77	48.77	48.09	48.62	49.05	48.82	49.76	..	-0.36%
Tobacco	0.95	0.95	1.10	0.96	1.18	1.11	1.20	1.20	1.02	0.93	1.09	1.12	1.26	1.24	1.25	1.23	1.35	1.39	1.36	..	1.75%
PRINCIPAL OIL CROPS																					
Sunflower	0.85	0.63	0.64	0.66	0.82	0.73	0.86	0.73	0.80	0.92	0.89	1.00	1.18	1.21	1.10	1.45	1.35	1.27	1.43	1.30	4.28%
Linseed	0.81	0.82	0.70	0.75	0.76	0.76	0.85	0.92	0.92	0.73	0.76	0.81	0.73	0.85	0.82	0.83	0.67	0.84	0.82	0.74	-0.02%
Peanuts	0.78	0.88	0.60	0.81	0.59	0.74	0.77	1.14	0.61	1.20	0.74	0.86	1.15	1.28	1.61	1.64	1.50	1.33	1.63	1.20	4.59%
Soybeans	1.03	1.62	1.14	1.73	1.44	1.36	1.60	2.12	2.17	2.31	1.72	2.01	2.19	1.69	2.40	2.00	2.14	1.99	2.26	1.65	2.59%
VEGETABLE CROPS																					
Potatoes	12.32	11.00	9.10	13.09	17.16	12.18	14.18	15.89	13.80	15.47	13.94	19.26	17.77	18.60	18.60	21.11	18.90	18.44	25.34	..	3.82%
Sweet potatoes	9.91	10.80	9.44	11.33	8.99	10.21	9.21	9.17	9.41	9.53	9.99	10.11	11.54	11.20	10.50	11.77	12.74	12.35	16.10	14.50	1.92%
Dry beans	0.96	0.97	0.93	0.93	1.07	0.79	1.16	0.97	0.98	1.02	0.71	1.06	1.11	1.08	0.99	1.03	1.05	..	1.01	..	0.58%
Tomatoes	18.00	17.30	18.70	16.60	18.90	18.30	18.27	18.23	19.13	19.49	18.89	19.72	21.31	19.71	19.63	20.60	21.20	21.6	24.93	24.80	1.55%

Source: Ministry of Agriculture.
a/ Crop Year, July through June.
b/ Where data are not available for the entire series, growth rates are for the longest continuous series.
August 1989

Table 7.5: ARGENTINA - PRODUCTION, IMPORTS AND SALES OF PRINCIPAL FARM INPUTS, 1970-1988

	THOUSAND TONS					THOUSAND UNITS
	Seed Production					
	Maize	Soy beans	Fertilizer Consumption a/	Pesticide Imports b/	Fencing wire Sales	Tractor Sales
1970	..	..	..	7.4	8.3	11.3
1971	107.6	..	75.4	7.5	11.8	13.7
1972	73.0	0.2	76.1	7.9	13.1	14.2
1973	87.6	1.4	71.4	9.0	9.0	18.8
1974	78.6	9.0	62.6	13.5	5.8	20.7
1975	68.6	8.4	54.1	9.6	5.7	15.2
1976	66.0	11.4	34.7	9.0	8.2	21.0
1977	85.7	4.8	55.9	9.8	6.7	22.0
1978	87.1	18.5	63.3	8.2	3.9	6.5
1979	98.3	82.3	99.0	15.5	6.9	6.9
1980	102.3	82.4	94.4	10.7	5.1	5.0
1981	124.8	60.1	66.2	9.2	4.2	3.1
1982	87.4	94.4	76.2	13.3	6.5	4.4
1983	95.9	97.4	96.4	16.0	8.6	8.0
1984	80.6	123.3	127.8	17.8	6.7	12.4
1985	122.2	96.6	143.8	11.8	3.1	6.7
1986	94.7	109.6	101.2	15.8	..	7.5
1987	69.3	..	133.1	16.2	..	3.9
1988	56.6	131.3	140.9	15.7	..	5.3

Source: Ministry of Agriculture, SNESR.

a/ In plant nutrients of N, P and K.

b/ Insecticides, fungicides and herbicides.

August 1989

Table 8.1: ARGENTINA - VALUE ADDED BY SUBSECTORS, 1970-1987
(1970 Australes)

	1970	1971	1972	1973	1974	1975	1976	1977	1978	1979	1980	1981	1982	1983	1984	1985	1986	1987 I	1987 II	1987 III
MANUFACTURING INDUSTRY	2496.0	2641.5	2751.3	2865.9	3011.9	2945.2	2866.8	3071.6	2759.7	3043.0	2941.0	2501.1	2388.8	2641.2	2751.9	2485.2	2789.2	2488.1	2891.4	2873.7
1. Food products	372.9	357.6	378.9	383.5	393.0	405.0	423.2	428.9	422.4	433.1	433.4	427.1	396.9	408.5	420.2	427.8	448.4	425.2	447.8	441.3
2. Beverages	124.3	127.8	130.6	131.1	162.1	154.1	135.8	120.5	109.5	128.7	141.8	128.8	106.9	120.2	142.9	135.2	170.0	213.0	159.8	133.9
3. Tobacco	95.5	97.5	103.2	108.6	123.2	123.8	120.4	120.2	120.1	124.3	123.6	115.3	105.7	112.6	127.0	127.4	130.6	124.3	121.7	123.0
4. Textiles	219.0	228.6	225.7	236.9	259.5	247.3	233.6	248.8	213.9	241.9	201.8	165.9	165.9	195.5	203.2	154.3	194.4	152.5	204.3	204.1
5. Clothing	60.3	64.3	69.8	67.0	80.1	75.3	69.9	65.9	51.3	60.2	53.7	33.8	29.7	32.0	33.0	23.2	26.5	17.7	24.2	22.1
6. Leather products	23.4	23.0	27.6	24.7	22.8	24.5	26.5	28.9	29.5	25.9	21.8	20.9	23.6	20.6	17.3	19.3	20.8	18.9	19.6	17.3
7. Footwear	20.7	19.7	15.1	15.8	17.7	15.5	12.9	12.0	10.3	10.3	10.0	8.9	8.0	7.5	8.5	7.0	6.7	4.9	7.0	5.2
8. Wood products	39.9	37.7	38.0	38.5	43.0	40.2	34.8	34.3	32.2	33.6	32.0	26.7	21.7	22.3	22.1	17.8	20.4	14.5	20.6	21.3
9. Furniture	16.5	14.9	15.1	15.0	18.3	16.0	11.2	13.8	15.3	19.2	19.3	17.9	16.0	13.8	11.6	11.1	13.9	15.5	22.1	15.7
10. Paper	57.4	60.5	64.3	70.1	75.9	72.2	62.6	66.0	68.3	76.0	62.5	46.7	58.6	65.8	65.7	60.8	66.8	66.8	62.0	67.7
11. Printing	71.1	70.6	72.4	74.3	77.4	82.1	72.6	69.2	71.3	66.2	70.0	61.7	55.8	56.7	58.1	58.9	58.7	54.0	54.6	54.7
12. Basic chemicals	66.1	76.7	85.0	92.8	94.9	88.8	84.7	91.6	82.2	99.7	86.1	77.6	93.9	103.8	108.1	92.0	117.3	105.9	115.4	127.6
13. Other chemicals	137.9	155.9	155.9	180.3	162.9	195.7	192.6	179.7	156.8	171.9	183.2	187.3	165.4	177.9	206.5	207.1	225.1	172.2	257.1	246.1
14. Petroleum refineries	199.2	209.5	215.8	219.4	207.7	189.4	202.9	213.1	215.4	225.7	239.0	231.6	229.0	239.4	232.5	239.2	238.3	217.8	219.4	218.8
15. Other petroleum products	12.7	11.3	11.5	11.8	10.8	9.6	9.5	11.7	9.6	11.4	11.4	12.1	10.5	11.3	10.9	11.8	11.4	12.3	12.6	11.9
16. Rubber products	55.7	62.7	70.0	74.1	72.2	76.0	83.4	83.6	70.3	88.0	82.9	55.4	56.4	78.2	81.4	58.6	72.5	54.8	73.6	83.1
17. Plastic products	24.5	26.9	34.5	37.3	34.1	31.3	27.4	29.6	26.7	33.3	33.4	25.2	26.5	30.4	35.1	27.1	29.4	24.2	24.8	25.2
18. Clay products	13.7	14.3	15.1	15.7	16.4	15.0	15.0	12.5	11.1	13.0	12.9	6.5	7.9	10.8	9.9	6.2	9.2	5.9	10.0	9.4
19. Glassware	27.4	26.7	27.3	24.7	26.5	32.8	28.4	26.2	28.9	32.2	29.3	22.7	20.8	24.2	23.9	15.9	21.6	20.8	22.0	25.9
20. Nonmetalic minerals	89.4	99.1	102.1	95.6	103.6	101.4	100.6	103.4	102.6	107.5	105.2	92.4	81.7	88.2	80.4	66.8	78.0	78.1	80.8	93.2
21. Iron and steel	108.3	119.4	128.5	134.9	137.4	134.8	125.1	143.8	137.9	156.6	138.9	125.3	141.5	149.0	140.2	136.6	151.2	151.8	180.8	186.6
22. Nonferrous metals	26.8	31.9	33.6	34.2	36.3	31.9	25.6	29.2	25.7	34.8	37.2	28.0	30.8	34.5	37.1	28.9	35.0	28.6	38.5	39.4
23. Metal products	179.7	202.3	211.7	205.2	225.4	232.5	226.8	256.8	221.5	245.5	235.3	192.2	179.6	217.9	234.5	197.7	226.7	199.0	264.3	235.9
24. Nonelectric machinery	125.1	136.6	140.1	172.2	184.4	166.0	205.0	254.9	187.3	205.5	180.5	128.4	119.2	130.6	123.3	88.1	99.2	66.5	100.9	107.5
25. Electric machinery	95.2	104.2	107.3	109.8	115.2	115.1	93.0	107.5	94.0	100.3	96.8	75.5	70.3	75.2	83.1	71.9	94.8	68.8	101.6	109.7
26. Transport equipment	222.1	249.7	255.4	274.0	284.4	245.8	218.8	294.9	220.2	270.3	280.2	173.7	155.3	200.0	218.4	179.2	207.6	159.5	224.8	225.7
27. Scientific equipment	11.2	12.1	16.8	18.4	26.7	23.1	24.5	24.6	25.4	27.9	18.8	13.5	11.2	14.3	17.0	15.3	14.7	14.6	21.1	21.4

Source: Central Bank of the Republic of Argentina (BCRA).

July 1989

Table 8.2: ARGENTINA - VALUE ADDED BY SUBSECTOR, 1970-1987
(Percent)

	1970	1971	1972	1973	1974	1975	1976	1977	1978	1979	1980	1981	1982	1983	1984	1985	1986	1987 I	1987 II	1987 III
MANUFACTURING INDUSTRY	100.0	100.0	100.0	100.0	100.0	100.0	100.0	100.0	100.0	100.0	100.0	100.0	100.0	100.0	100.0	100.0	100.0	100.0	100.0	100.0
1. Food products	14.9	13.5	13.8	13.4	13.0	13.8	14.8	14.0	15.3	14.2	14.7	17.1	16.6	15.5	15.3	17.2	16.1	17.1	15.5	15.4
2. Beverages	5.0	4.8	4.7	4.6	5.4	5.2	4.7	3.9	4.0	4.2	4.8	5.1	4.5	4.6	5.2	5.4	6.1	8.6	5.5	4.7
3. Tobacco	3.8	3.7	3.8	3.8	4.1	4.2	4.2	3.9	4.4	4.1	4.2	4.6	4.4	4.3	4.6	5.1	4.7	5.0	4.2	4.3
4. Textiles	8.8	8.7	8.2	8.3	8.6	8.4	8.1	8.1	7.8	7.9	6.9	6.6	6.9	7.4	7.4	6.2	7.0	6.1	7.1	7.1
5. Clothing	2.4	2.4	2.5	2.3	2.7	2.6	2.4	2.1	1.9	2.0	1.8	1.4	1.2	1.2	1.2	0.9	1.0	0.7	0.8	0.8
6. Leather products	0.9	0.9	1.0	0.9	0.8	0.8	0.9	0.9	1.1	0.9	0.7	0.8	1.0	0.8	0.6	0.8	0.7	0.8	0.7	0.6
7. Footwear	0.8	0.7	0.5	0.6	0.6	0.5	0.4	0.4	0.4	0.3	0.3	0.4	0.3	0.3	0.3	0.3	0.2	0.2	0.2	0.2
8. Wood products	1.6	1.4	1.4	1.3	1.4	1.4	1.2	1.1	1.2	1.1	1.1	1.1	0.9	0.8	0.8	0.7	0.7	0.6	0.7	0.7
9. Furniture	0.7	0.6	0.5	0.5	0.6	0.5	0.4	0.4	0.6	0.6	0.7	0.7	0.7	0.5	0.4	0.4	0.5	0.6	0.8	0.5
10. Paper	2.3	2.3	2.3	2.4	2.5	2.5	2.2	2.1	2.5	2.5	2.1	1.9	2.5	2.5	2.4	2.4	2.4	2.7	2.1	2.4
11. Printing	2.8	2.7	2.6	2.6	2.6	2.8	2.5	2.3	2.6	2.2	2.4	2.5	2.3	2.1	2.1	2.4	2.1	2.2	1.9	1.9
12. Basic chemicals	2.6	2.9	3.1	3.2	3.2	3.0	3.0	3.0	3.0	3.3	2.9	3.1	3.9	3.9	3.9	3.7	4.2	4.3	4.0	4.4
13. Other chemicals	5.5	5.9	5.7	6.3	5.4	6.6	6.7	5.9	5.7	5.6	6.2	7.5	6.9	6.7	7.5	8.3	8.1	6.9	8.9	8.6
14. Petroleum refineries	8.0	7.9	7.8	7.7	6.9	6.4	7.1	6.9	7.8	7.4	8.1	9.3	9.6	9.1	8.4	9.6	8.5	8.8	7.6	7.6
15. Other petroleum products	0.5	0.4	0.4	0.4	0.4	0.3	0.3	0.4	0.3	0.4	0.4	0.5	0.4	0.4	0.4	0.5	0.4	0.5	0.4	0.4
16. Rubber products	2.2	2.4	2.5	2.6	2.4	2.6	2.9	2.7	2.5	2.9	2.8	2.2	2.4	3.0	3.0	2.4	2.6	2.2	2.5	2.9
17. Plastic products	1.0	1.0	1.3	1.3	1.1	1.1	1.0	1.0	1.0	1.1	1.1	1.0	1.1	1.2	1.3	1.1	1.1	1.0	0.9	0.9
18. Clay products	0.5	0.5	0.5	0.5	0.5	0.5	0.5	0.4	0.4	0.4	0.4	0.3	0.3	0.4	0.4	0.2	0.3	0.2	0.3	0.3
19. Glassware	1.1	1.0	1.0	0.9	0.9	1.1	1.0	0.9	1.0	1.1	1.0	0.9	0.9	0.9	0.9	0.6	0.8	0.8	0.8	0.9
20. Nonmetalic minerals	3.6	3.8	3.7	3.3	3.4	3.4	3.5	3.4	3.7	3.5	3.6	3.7	3.4	3.3	2.9	2.7	2.8	3.1	2.8	3.2
21. Iron and steel	4.3	4.5	4.7	4.7	4.6	4.6	4.4	4.7	5.0	5.1	4.7	5.0	5.9	5.6	5.1	5.5	5.4	6.1	6.3	6.5
22. Nonferrous metals	1.1	1.2	1.2	1.2	1.2	1.1	0.9	1.0	0.9	1.1	1.3	1.1	1.3	1.3	1.3	1.2	1.3	1.1	1.3	1.4
23. Metal products	7.2	7.7	7.7	7.2	7.5	7.9	7.9	8.4	8.0	8.1	8.0	7.7	7.5	8.3	8.5	8.0	8.1	8.0	9.1	8.2
24. Nonelectric machinery	5.0	5.2	5.1	6.0	6.1	5.6	7.2	8.3	6.8	6.8	6.1	5.1	5.0	4.9	4.5	3.5	3.6	2.7	3.5	3.7
25. Electric machinery	3.8	3.9	3.9	3.8	3.8	3.9	3.2	3.5	3.4	3.3	3.3	3.0	2.9	2.8	3.0	2.9	3.4	2.8	3.5	3.8
26. Transport equipment	8.9	9.5	9.3	9.6	9.4	8.3	7.6	9.6	8.0	8.9	9.5	6.9	6.5	7.6	7.9	7.2	7.4	6.4	7.8	7.9
27. Scientific equipment	0.4	0.5	0.6	0.6	0.9	0.8	0.9	0.8	0.9	0.9	0.6	0.5	0.5	0.5	0.6	0.6	0.5	0.6	0.7	0.7

Source: Central Bank of the Republic of Argentina (BCRA).

July 1989

Table 8.3: ARGENTINA - INDEX OF VALUE ADDED BY MANUFACTURING SUBSECTOR, 1970-1987
(1970=100)

	1970	1971	1972	1973	1974	1975	1976	1977	1978	1979	1980	1981	1982	1983	1984	1985	1986	1987 I	1987 II	1987 III
MANUFACTURING INDUSTRY	100.0	105.8	110.2	114.8	120.7	118.0	114.9	123.1	110.6	121.9	117.8	100.2	95.7	105.8	110.3	99.6	111.7	99.7	115.8	115.1
1. Food products	100.0	95.9	101.6	102.8	105.4	108.6	113.5	115.0	113.3	116.1	116.2	114.5	106.4	109.5	112.7	114.7	120.2	114.0	120.1	118.3
2. Beverages	100.0	102.8	105.1	105.5	130.4	124.0	109.3	96.9	88.1	103.5	114.1	103.6	86.0	96.7	115.0	108.8	136.8	171.4	128.6	107.7
3. Tobacco	100.0	102.1	108.1	113.7	129.0	129.6	126.1	125.9	125.8	130.2	129.4	120.7	110.7	117.9	133.0	133.4	136.8	130.2	127.4	128.8
4. Textiles	100.0	104.4	103.1	108.2	118.5	112.9	106.7	113.6	97.7	110.5	92.1	75.8	75.8	89.3	92.8	70.5	88.8	69.6	93.3	93.2
5. Clothing	100.0	106.6	115.8	111.1	132.8	124.9	115.9	109.3	85.1	99.8	89.1	56.1	49.3	53.1	54.7	38.5	43.9	29.4	40.1	36.7
6. Leather products	100.0	98.3	117.9	105.6	97.4	104.7	113.2	123.5	126.1	110.7	93.2	89.3	100.9	88.0	73.9	82.5	88.9	80.8	83.8	73.9
7. Footwear	100.0	95.2	72.9	76.3	85.5	74.9	62.3	58.0	49.8	49.8	48.3	43.0	38.6	36.2	41.1	33.8	32.4	23.7	33.8	25.1
8. Wood products	100.0	94.5	95.2	96.5	107.8	100.8	87.2	86.0	80.7	84.2	80.2	66.9	54.4	55.9	55.4	44.6	51.1	36.3	51.6	53.4
9. Furniture	100.0	90.3	91.5	90.9	110.9	97.0	67.9	83.6	92.7	116.4	117.0	108.5	97.0	83.6	70.3	67.3	84.2	93.9	133.9	95.2
10. Paper	100.0	105.4	112.0	122.1	132.2	125.8	109.1	115.0	119.0	132.4	108.9	81.4	102.1	114.6	114.5	105.9	116.4	116.4	108.0	117.9
11. Printing	100.0	99.3	101.8	104.5	108.9	115.5	102.1	97.3	100.3	93.1	98.5	86.8	78.5	79.7	81.7	82.8	82.6	75.9	76.8	76.9
12. Basic chemicals	100.0	116.0	128.6	140.4	143.6	134.3	128.1	138.6	124.4	150.8	130.3	117.4	142.1	157.0	163.5	139.2	177.5	160.2	174.6	193.0
13. Other chemicals	100.0	113.1	113.1	130.7	118.1	141.9	139.7	130.3	113.7	124.7	132.8	135.8	119.9	129.0	149.7	150.2	163.2	124.9	186.4	178.5
14. Petroleum refineries	100.0	105.2	108.3	110.1	104.3	95.1	101.9	107.0	108.1	113.3	120.0	116.3	115.0	120.2	116.7	120.1	119.6	109.3	110.1	109.8
15. Other petroleum products	100.0	89.0	90.6	92.9	85.0	75.6	74.8	92.1	75.6	89.8	89.8	95.3	82.7	89.0	85.8	92.9	89.8	96.9	99.2	93.7
16. Rubber products	100.0	112.6	125.7	133.0	129.6	136.4	149.7	150.1	126.2	158.0	148.8	99.5	101.3	140.4	146.1	105.2	130.2	98.4	132.1	149.2
17. Plastic products	100.0	109.8	140.8	152.2	139.2	127.8	111.8	120.8	109.0	135.9	136.3	102.9	108.2	124.1	143.3	110.6	120.0	98.8	101.2	102.9
18. Clay products	100.0	104.4	110.2	114.6	119.7	109.5	109.5	91.2	81.0	94.9	94.2	47.4	57.7	78.8	72.3	45.3	67.2	43.1	73.0	68.6
19. Glassware	100.0	97.4	99.6	90.1	96.7	119.7	103.6	95.6	105.5	117.5	106.9	82.8	75.9	88.3	87.2	58.0	78.8	75.9	80.3	94.5
20. Nonmetalic minerals	100.0	110.9	114.2	106.9	115.9	113.4	112.5	115.7	114.8	120.2	117.7	103.4	91.4	98.7	89.9	74.7	87.2	87.4	90.4	104.3
21. Iron and steel	100.0	110.2	118.7	124.6	126.9	124.5	115.5	132.8	127.3	144.6	128.3	115.7	130.7	137.6	129.5	126.1	139.6	140.2	166.9	172.3
22. Nonferrous metals	100.0	119.0	125.4	127.6	135.4	119.0	95.5	109.0	95.9	129.9	138.8	104.5	114.9	128.7	138.4	107.8	130.6	106.7	143.7	147.0
23. Metal products	100.0	112.6	117.8	114.2	125.4	129.4	126.2	142.9	123.3	136.6	130.9	107.0	99.9	121.3	130.5	110.0	126.2	110.7	147.1	131.3
24. Nonelectric machinery	100.0	109.2	112.0	137.6	147.4	132.7	163.9	203.8	149.7	164.3	144.3	102.6	95.3	104.4	98.6	70.4	79.3	53.2	80.7	85.9
25. Electric machinery	100.0	109.5	112.7	115.3	121.0	120.9	97.7	112.9	98.7	105.4	101.7	79.3	73.8	79.0	87.3	75.5	99.6	72.3	106.7	115.2
26. Transport equipment	100.0	112.4	115.0	123.4	128.1	110.7	98.5	132.8	99.1	121.7	126.2	78.2	69.9	90.0	98.3	80.7	93.5	71.8	101.2	101.6
27. Scientific equipment	100.0	108.0	150.0	164.3	238.4	206.3	218.8	219.6	226.8	249.1	167.9	120.5	100.0	127.7	151.8	136.6	131.3	130.4	188.4	191.1

Source: Central Bank of the Republic of Argentina (BCRA).

July 1989

Table 8.4: ARGENTINA - EMPLOYMENT INDEX BY MANUFACTURING SUBSECTOR, 1970-1989
(1970 = 100)

	1970	1971	1972	1973	1974	1975	1976	1977	1978	1979	1980	1981	1982	1983	1984	1985	1986	1987	1988	1989 a/
MANUFACTURING INDUSTRY	100.0	103.0	105.3	109.2	115.0	119.4	115.1	107.9	97.8	95.8	88.1	77.1	73.0	75.4	77.6	74.7	71.7	71.3	72.1	69.3
1. Food products	100.0	106.2	108.1	113.4	122.2	120.3	123.2	117.7	105.0	104.7	99.4	93.6	98.5	101.9	101.7	102.4	96.0	94.9	92.4	77.0
2. Beverages	100.0	100.3	106.1	108.3	117.8	126.5	121.0	119.1	111.1	114.6	116.9	108.3	95.1	93.4	99.9	102.4	106.9	108.3	125.6	137.3
3. Tobacco	100.0	97.5	101.9	100.0	102.8	112.6	117.6	97.8	92.7	94.4	90.8	81.3	80.3	79.7	89.5	77.3	77.2	75.2	87.0	63.4
4. Textiles	100.0	100.5	98.2	97.7	105.6	109.2	105.3	96.8	84.8	75.8	60.4	45.5	48.4	52.6	56.8	51.7	50.3	50.2	52.7	46.7
5. Clothing	100.0	104.8	107.9	110.4	118.0	119.1	111.9	99.2	87.9	82.7	65.0	57.4	52.2	54.4	57.4	52.1	47.3	41.4	42.7	40.8
6. Leather products	100.0	109.6	123.3	139.8	154.3	149.2	152.2	159.8	161.9	146.3	118.1	99.4	98.2	99.9	105.7	98.1	94.6	94.6	89.1	79.1
7. Footwear	100.0	97.8	88.4	91.7	94.1	90.8	83.1	69.3	58.7	60.3	53.7	51.5	43.3	42.8	41.8	39.7	37.5	33.9	33.0	39.2
8. Wood products	100.0	103.6	105.4	106.8	113.8	139.3	136.3	130.2	114.3	103.6	93.0	83.7	79.1	79.6	79.8	77.5	74.3	72.8	70.7	71.5
9. Furniture	100.0	91.7	93.3	94.8	94.3	89.1	71.9	61.8	58.1	57.9	64.0	65.0	60.0	62.6	56.1	53.0	55.6	52.4	48.0	38.4
10. Paper	100.0	103.9	108.3	116.0	117.3	122.0	121.1	115.9	119.0	117.1	100.8	93.8	92.6	95.0	97.9	96.7	95.9	99.5	100.5	104.7
11. Printing	100.0	98.8	94.6	94.4	97.3	97.7	85.4	74.4	70.6	69.3	75.0	72.8	66.8	66.9	67.4	67.2	65.2	61.4	57.0	57.2
12. Basic chemicals	100.0	99.9	106.7	109.5	113.4	121.3	124.6	115.2	106.6	102.2	95.9	82.9	79.9	82.8	90.0	82.3	82.9	84.0	85.2	85.9
13. Other chemicals	100.0	107.0	107.5	106.1	106.2	114.3	111.6	106.1	95.5	91.4	88.6	79.9	73.6	71.5	68.5	66.7	66.2	62.8	58.9	51.9
14. Petroleum refineries	100.0	101.6	113.3	117.6	126.2	137.4	161.1	145.3	136.1	120.6	108.7	102.8	100.5	98.1	98.7	96.2	100.0	100.8	104.3	115.8
15. Other petroleum products	100.0	91.7	91.5	87.5	90.2	96.1	98.6	96.7	87.5	84.3	81.1	78.1	74.4	75.4	74.5	71.4	67.6	60.9	61.8	69.3
16. Rubber products	100.0	106.7	116.1	121.5	131.6	141.5	141.8	140.7	125.7	134.3	125.2	106.2	94.4	104.0	114.0	108.8	102.6	100.0	102.4	91.8
17. Plastic products	100.0	99.5	110.7	190.3	200.7	207.2	179.5	154.9	136.5	144.2	149.2	134.0	132.1	152.8	156.0	147.4	142.5	141.4	136.1	120.8
18. Clay products	100.0	106.5	110.7	112.3	114.0	113.7	115.5	100.1	87.0	96.3	95.4	64.0	60.5	73.8	76.5	69.3	78.5	75.6	78.8	75.3
19. Glassware	100.0	95.9	105.1	103.3	104.9	113.6	103.8	108.6	109.0	99.6	89.1	71.2	63.5	64.7	72.2	56.4	48.9	47.7	53.9	50.3
20. Nonmetalic minerals	100.0	101.7	100.2	96.5	103.6	107.0	102.9	95.2	91.9	91.2	85.6	80.1	73.1	75.8	75.7	70.2	65.4	71.5	92.9	89.4
21. Iron and steel	100.0	107.8	111.0	121.4	128.5	134.2	129.1	126.0	117.5	117.3	112.6	97.9	91.6	96.4	103.6	105.2	98.0	96.5	94.0	86.7
22. Nonferrous metals	100.0	123.3	134.4	137.6	137.3	147.0	140.6	137.1	127.2	128.4	126.2	119.2	111.4	113.3	117.7	114.5	112.6	108.1	106.1	103.7
23. Metal products	100.0	104.2	107.7	112.9	118.2	119.0	111.1	105.7	97.8	99.9	91.6	77.5	72.5	75.4	74.7	70.2	67.7	68.9	73.3	64.6
24. Nonelectric machinery	100.0	99.4	102.6	107.2	114.9	116.5	109.4	106.3	95.6	89.2	71.5	48.9	41.5	46.2	51.8	48.8	45.3	44.1	39.2	35.0
25. Electric machinery	100.0	102.3	103.2	101.5	101.4	99.4	91.8	88.5	81.3	78.6	71.7	56.9	49.8	49.1	52.0	48.5	45.9	43.9	44.8	37.5
26. Transport equipment	100.0	103.5	110.5	117.3	121.7	137.6	131.8	118.8	101.7	104.3	101.8	87.0	72.1	73.0	76.5	71.8	69.5	71.5	68.7	61.5
27. Scientific equipment	100.0	101.0	105.8	123.8	144.1	163.4	158.0	145.5	143.1	131.5	95.9	79.0	67.9	62.0	66.4	67.4	69.5	75.6	77.1	66.9

Source: INDEC.

a/ First quarter average.

August 1989

Table 8.5: ARGENTINA - MANUFACTURING INDUSTRY: EXPORTS BY SUBSECTOR, 1970-1987 a/
(Millions of 1970 US dollars)

	1970	1971	1972	1973	1974	1975	1976	1977	1978	1979	1980	1981	1982	1983	1984	1985	1986	1987 b/
MANUFACTURING INDUSTRY	478.7	479.8	522.8	831.3	1,084.3	665.7	879.8	1,220.1	1,284.2	1,292.9	1,247.4	1,116.3	1,518.0	1,803.2	1,858.1	1,845.1	1,451.0	1,185.6
1. Food products	255.6	240.8	216.9	255.5	422.0	202.5	316.8	476.0	427.7	487.1	441.8	378.7	336.3	804.4	863.2	628.5	499.9	409.7
2. Beverages	3.7	11.9	10.6	13.7	12.0	9.0	15.1	26.5	36.4	35.9	27.7	20.0	18.9	12.9	11.3	11.0	9.0	6.2
3. Tobacco	0.2	0.2	0.4	4.5	10.2	0.5	0.4	0.7	0.4	0.5	0.1	0.1	0.1	24.9	21.5	27.4	20.0	17.4
4. Textiles	4.6	2.5	3.6	18.7	12.4	1.9	23.0	35.4	35.6	16.4	21.1	13.3	16.6	110.5	137.0	139.8	100.1	92.3
5. Clothing	15.7	5.2	7.8	26.0	29.9	13.1	30.0	63.4	105.3	120.7	90.6	44.2	24.1	1.5	1.1	17.4	27.7	34.5
6. Leather products	1.2	1.3	2.2	3.4	5.5	3.1	7.5	13.0	14.0	15.7	23.0	11.8	11.6	149.8	148.1	13.8	20.5	24.7
7. Footwear	0.3	1.1	2.9	17.6	21.9	3.3	4.3	14.5	12.4	3.5	1.0	1.2	3.4	2.8	1.1	1.1	4.4	15.2
8. Wood products	0.0	0.1	0.1	0.1	0.2	0.1	0.1	0.2	0.3	0.2	0.1	0.5	0.6	0.4	1.8	2.1	3.2	3.8
9. Furniture	0.1	0.1	0.2	0.5	0.9	1.1	1.5	1.8	2.2	2.2	2.7	0.8	0.2	0.0	0.0	0.0	0.0	0.0
10. Paper	1.8	2.7	4.2	13.0	25.6	2.7	4.1	11.8	11.5	10.2	7.5	4.6	6.1	10.5	12.0	3.9	3.5	10.9
11. Printing	15.0	13.9	16.1	21.6	17.9	15.0	15.1	63.6	18.3	18.5	22.5	18.3	17.9	9.0	7.2	9.2	10.7	8.7
12. Basic chemicals	28.4	29.0	38.8	33.0	41.6	42.9	53.6	62.5	70.8	79.2	120.0	100.0	123.5	132.8	102.2	122.1	104.3	95.2
13. Other chemicals	24.3	26.7	33.6	37.7	36.6	24.9	25.5	28.5	38.5	37.6	55.8	50.2	56.4	15.7	15.9	14.1	23.8	10.7
14. Petroleum refineries	0.7	2.6	1.5	0.5	1.6	0.2	3.7	4.4	8.8	6.8	36.5	71.7	265.9	0.0	0.0	0.0	0.0	0.0
15. Other petroleum products	8.0	5.4	4.1	4.3	4.2	10.9	11.5	13.2	19.3	17.8	21.8	17.8	59.8	174.1	154.9	308.2	82.8	37.1
16. Rubber products	4.1	0.6	3.1	7.2	4.6	2.1	2.3	4.7	8.7	3.9	3.6	10.0	11.7	13.2	12.4	21.1	15.7	13.9
17. Plastic products	2.4	2.9	4.5	6.7	4.4	2.1	3.5	6.9	6.0	6.5	5.5	3.8	17.7	26.6	26.9	23.6	21.7	28.4
18. Clay products	0.1	0.1	0.2	0.3	0.6	0.3	0.5	0.5	0.5	0.6	0.5	0.9	1.2	1.6	1.9	0.9	2.0	4.1
19. Glassware	1.6	1.9	2.6	5.5	7.0	2.0	5.1	9.9	14.6	12.9	10.0	7.1	8.1	3.0	2.3	3.8	6.4	8.0
20. Nonmetalic minerals	0.7	0.6	0.9	2.1	1.9	1.5	1.9	3.6	2.8	2.2	0.9	0.5	1.4	3.8	3.7	1.2	1.7	1.6
21. Iron and steel	28.0	32.9	34.4	93.3	86.0	12.7	47.6	40.6	99.5	80.8	54.3	100.7	181.2	111.0	90.7	173.5	172.7	138.7
22. Nonferrous metals	2.0	1.8	2.8	4.9	3.9	0.4	2.5	3.3	9.7	26.5	57.0	56.1	57.3	43.3	40.8	63.4	57.0	53.8
23. Metal products	9.3	11.5	15.0	20.0	26.1	23.6	22.7	23.5	34.2	24.5	22.7	11.4	26.7	12.2	8.9	9.5	9.0	6.8
24. Nonelectric machinery	48.1	55.8	67.9	119.6	150.2	148.3	119.2	133.7	145.1	140.6	89.6	108.6	129.1	76.0	76.5	102.5	116.5	84.8
25. Electric machinery	8.2	8.5	13.5	29.5	34.5	24.9	23.0	27.7	36.5	47.5	31.7	20.2	23.6	14.2	13.7	27.6	24.6	19.1
26. Transport equipment	10.1	15.2	29.3	82.9	113.1	108.3	129.0	138.3	112.7	79.1	80.2	56.1	111.6	46.0	78.8	114.5	106.8	54.2
27. Scientific equipment	4.5	4.8	5.5	9.1	9.3	8.0	10.5	12.1	12.3	15.7	19.2	7.7	7.0	3.0	4.2	4.7	7.2	5.6

Source: INDEC.

a/ 1970-84 figures in current US dollars are adjusted by US implicit deflator for manufacturing value added; 1985-87 figures adjusted by US industrial WPI deflator.

b/ Estimate based on data through November 1987.

July 1989

Table 8.5: ARGENTINA - MANUFACTURING INDUSTRY: EXPORTS BY SUBSECTOR, 1970-1987 a/
(Millions of US Dollars)

	1970	1971	1972	1973	1974	1975	1976	1977	1978	1979	1980	1981	1982	1983	1984 a/	1985	1986	1987 a/
MANUFACTURING INDUSTRY	478.7	488.0	564.0	1,024.1	1,500.4	1,026.7	1,388.7	1,996.9	2,369.1	2,671.6	2,902.3	3,087.6	2,943.4	3,614.9	3,819.0	3,806.4	2,886.4	2,421.5
1. Food products	255.6	241.3	243.6	369.5	595.9	334.0	495.1	748.3	802.0	1,018.0	927.4	812.0	652.0	1611.2	1813.3	1295.9	993.1	835.5
2. Beverages	3.7	12.2	11.1	14.8	14.9	13.0	23.2	47.1	64.5	67.0	57.2	43.5	36.7	25.8	23.2	22.7	17.9	12.7
3. Tobacco	0.2	0.2	0.4	5.3	13.9	0.7	0.6	1.1	0.7	0.9	0.3	0.2	0.2	50.0	44.2	56.6	39.7	35.5
4. Textiles	4.6	2.5	3.8	21.6	16.1	2.4	31.7	50.8	53.0	25.8	36.1	24.7	32.1	221.4	281.3	288.2	198.8	188.3
5. Clothing	15.7	5.3	8.1	27.9	34.9	15.8	37.8	84.1	144.6	174.4	140.7	73.6	46.8	3.0	2.3	36.5	56.1	71.7
6. Leather products	1.2	1.4	2.9	5.1	7.9	4.4	13.1	24.2	31.0	51.9	66.4	35.3	22.4	300.1	304.0	28.5	40.7	50.4
7. Footwear	0.3	1.1	3.2	20.3	27.1	4.3	6.0	21.7	20.1	6.7	2.0	2.9	6.6	5.5	2.3	2.4	8.9	31.6
8. Wood products	0.0	0.1	0.1	0.1	0.3	0.2	0.2	0.4	0.7	0.5	0.2	1.2	1.2	0.7	3.6	4.2	6.1	7.5
9. Furniture	0.1	0.1	0.2	0.6	1.1	1.5	2.0	2.6	3.4	3.6	4.9	1.6	0.4	3.1	4.2	1.4	2.8	2.5
10. Paper	1.8	2.8	4.4	14.2	34.3	4.2	6.7	20.6	21.3	21.2	17.4	11.5	11.9	20.9	24.6	8.1	7.0	22.1
11. Printing	15.0	14.1	16.9	24.4	25.1	23.6	25.1	109.5	33.1	37.5	51.8	46.2	34.8	18.0	14.9	19.2	21.5	17.8
12. Basic chimicals	28.4	29.2	36.2	46.9	80.0	70.0	87.9	111.9	140.2	181.2	242.1	235.9	239.5	266.1	209.9	251.8	207.2	194.2
13. Other chimicals	24.3	27.2	34.3	40.6	52.6	44.2	46.6	53.7	74.9	81.7	142.2	141.2	109.4	31.5	32.6	29.0	47.3	21.8
14. Petroleum refineries	0.7	2.7	1.6	0.7	3.6	0.4	10.0	13.5	28.0	30.1	243.8	587.0	515.6	0.0	0.0	..	..	..
15. Other petroleum products	8.0	5.8	4.6	5.4	8.3	25.3	28.8	37.8	58.7	68.7	118.0	117.0	116.0	348.8	318.0	635.5	164.5	75.7
16. Rubber products	4.1	0.6	3.1	7.4	5.6	2.9	3.4	7.3	14.3	7.4	7.8	22.9	22.6	26.5	25.4	43.4	31.2	28.3
17. Plastic products	2.4	2.9	4.5	6.9	5.5	2.9	5.2	10.6	9.7	11.6	11.0	11.3	34.3	53.2	55.3	48.7	43.1	57.9
18. Clay products	0.1	0.1	0.2	0.3	0.7	0.4	0.7	0.8	0.9	1.1	1.1	2.1	2.3	3.2	3.9	1.8	3.9	8.3
19. Glasswear	1.6	2.0	2.9	6.3	9.5	3.1	8.4	17.5	28.8	28.3	25.0	19.5	15.8	6.0	4.6	7.6	12.4	15.9
20. Nonmetalic minerals	0.7	0.6	1.0	2.4	2.5	2.2	2.9	6.2	5.3	4.7	2.1	1.3	2.8	7.6	7.6	2.5	3.4	3.3
21. Iron and steel	28.0	34.8	38.4	110.4	133.4	22.2	89.2	81.2	219.2	198.2	143.9	288.7	351.3	222.4	186.3	357.8	343.2	283.0
22. Nonferrous metals	2.0	1.7	2.6	5.3	5.9	0.6	3.6	5.2	16.2	55.5	139.1	127.9	111.1	86.7	83.7	130.7	113.2	109.7
23. Metal products	9.3	11.7	15.9	22.8	38.5	37.5	38.1	42.1	66.5	54.4	55.8	29.3	51.7	24.5	18.2	19.6	17.8	13.9
24. Nonelectric Machinery	48.1	57.9	71.9	130.7	188.0	214.8	182.9	218.1	255.5	270.0	192.8	256.5	250.3	152.2	157.0	211.3	231.3	172.8
25. Electric machinery	8.2	8.7	14.3	32.9	47.1	37.8	36.9	47.4	67.7	98.5	73.0	49.9	45.7	28.4	28.2	57.1	48.9	39.1
26. Transport equipment	10.1	16.0	31.9	91.3	135.8	146.7	186.5	213.4	187.1	142.3	158.8	126.3	216.3	92.2	161.8	236.2	212.1	110.6
27. Scientific equipment	4.5	5.0	5.9	10.0	11.9	11.6	16.1	19.8	21.7	30.4	41.4	18.1	13.6	5.9	8.6	9.7	14.3	11.4

Source: INDEC.

a/ 1984 figure based on data through October 1984; 1987 figure based on data through November 1987.

July 1989

Table 8.7: ARGENTINA - MANUFACTURING INDUSTRY: IMPORTS BY SUBSECTOR OF ORIGIN, 1970-1987
(Millions of 1970 US dollars)

	1970	1971	1972	1973	1974	1975	1976	1977	1978	1979	1980	1981	1982	1983	1984	1985	1986	1987 a/
MANUFACTURING INDUSTRY	1,357.0	1,420.2	1,480.3	1,465.9	1,799.5	1,853.5	1,314.4	1,771.1	1,598.2	2,448.9	3,796.1	3,211.8	1,700.7	1,603.9	1,591.8	1,288.4	1,671.9	1,636.7
1. Food products	14.4	12.4	8.1	6.7	13.9	10.9	6.9	22.3	28.3	72.8	87.6	35.0	9.9	20.0	31.7	21.5	31.0	27.2
2. Beverages	6.5	7.4	6.0	7.3	5.0	7.7	6.0	4.2	7.4	22.4	23.1	15.7	4.6	2.8	1.7	3.6	2.5	2.7
3. Tobacco	0.6	0.4	0.1	0.1	0.2	0.1	0.1	0.6	1.2	1.8	4.6	4.3	0.8	0.8	0.5	0.5	0.1	0.2
4. Textiles	25.7	20.3	16.9	14.2	30.6	23.2	19.7	14.4	17.7	73.3	134.3	154.2	64.9	61.9	68.3	33.3	55.2	46.0
5. Clothing	2.4	1.1	0.1	0.3	0.3	0.8	0.9	1.1	3.3	24.3	100.7	93.2	16.0	5.2	7.4	3.6	4.1	2.7
6. Leather products	0.2	0.1	0.0	0.0	0.0	0.0	0.0	0.0	0.2	1.0	2.9	2.8	0.5	0.5	0.9	0.4	0.8	0.3
7. Footwear	0.1	0.0	0.0	0.0	0.0	0.1	0.1	0.2	0.5	4.7	17.7	13.2	1.7	0.8	0.8	0.4	0.6	0.6
8. Wood products	1.1	1.9	1.6	1.4	1.6	4.0	1.6	1.2	0.8	3.9	12.4	46.5	23.1	24.9	24.0	12.6	21.9	17.0
9. Furniture	0.2	0.3	0.2	0.1	0.2	0.1	0.2	0.4	0.7	2.2	7.8	8.7	1.7	0.6	1.1	0.6	0.9	0.7
10. Paper	66.2	57.4	53.2	56.8	81.9	70.3	47.1	59.0	60.4	61.5	100.8	86.0	38.1	36.4	21.3	18.6	33.0	32.7
11. Printing	11.6	10.7	8.2	7.4	9.4	10.1	5.8	9.1	14.8	23.4	41.4	31.1	9.4	4.5	2.6	3.1	2.9	2.0
12. Basic chemicals	124.4	144.6	186.9	126.7	227.3	268.5	211.4	216.2	164.2	251.4	317.2	269.8	247.9	252.4	254.3	198.6	292.9	231.6
13. Other chemicals	79.6	91.8	100.4	101.1	135.3	108.8	89.0	101.4	92.5	110.5	162.6	135.5	108.2	172.7	180.6	142.9	184.8	156.6
14. Petroleum refineries	0.1	11.9	1.2	24.4	20.4	26.2	23.5	24.9	9.7	98.1	16.5	13.4	13.7	63.7	65.0	60.8	58.0	79.5
15. Other petroleum products	32.4	19.1	10.0	7.2	8.9	9.7	10.8	8.9	8.6	13.0	12.1	9.9	9.2	0.0	0.0	0.0	0.0	0.0
16. Rubber products	5.8	5.8	5.9	5.7	7.8	12.1	7.4	12.4	13.1	20.2	33.5	43.5	36.3	47.8	39.9	20.8	30.9	24.5
17. Plastic products	23.3	26.7	33.2	34.0	85.0	79.2	48.5	46.8	57.3	122.3	122.2	101.3	83.9	82.5	77.9	57.2	95.8	86.9
18. Clay products	5.9	8.6	8.5	5.7	5.5	9.8	11.5	8.6	10.3	22.8	32.2	28.7	9.5	9.9	8.3	6.9	8.3	7.7
19. Glassware	11.3	10.7	9.7	10.8	11.9	11.7	8.2	7.6	8.5	15.0	22.0	16.8	5.8	6.3	4.8	3.2	4.5	4.1
20. Nonmetalic minerals	10.9	4.5	3.2	3.3	4.5	5.0	3.7	4.5	5.6	16.7	25.9	4.5	2.7	3.2	3.0	3.2	5.8	5.5
21. Iron and steel	243.5	224.1	222.0	317.0	327.7	427.5	190.1	178.6	99.1	149.7	224.4	165.4	127.8	119.9	122.0	75.0	86.8	100.0
22. Nonferrous metals	86.1	98.7	110.0	113.0	151.3	119.0	61.3	51.9	43.9	68.9	66.2	50.3	61.0	45.9	51.9	32.7	56.1	56.2
23. Metal products	38.4	31.7	33.2	26.5	31.0	25.2	30.9	35.2	46.1	42.9	92.8	39.8	15.0	14.0	13.1	10.1	10.7	10.0
24. Nonelectric machinery	345.5	366.5	392.6	310.9	329.3	355.6	321.7	498.8	479.8	565.0	911.2	726.2	360.9	292.1	258.7	267.2	280.5	337.6
25. Electric machinery	97.1	125.9	122.6	140.2	120.0	105.6	90.5	148.7	180.2	212.3	543.2	484.1	226.2	146.3	149.6	131.7	194.9	201.1
26. Transport equipment	78.8	91.2	106.4	95.6	113.8	98.8	92.0	267.4	179.0	372.1	518.0	439.0	142.8	120.1	141.5	118.6	118.5	115.2
27. Scientific equipment	45.0	46.3	40.1	49.3	76.8	63.7	25.7	46.5	65.0	76.6	162.9	192.8	79.1	68.6	61.2	61.1	90.5	87.9

Source: INDEC.
a/ Based on data through October 1987.
July 1989

Table 8.8: ARGENTINA - MANUFACTURING INDUSTRY: IMPORTS BY SUBSECTORS, 1970-1987
(Millions of US dollars)

	1970	1971	1972	1973	1974	1975	1976	1977	1978	1979	1980	1981	1982	1983	1984	1985	1986	1987 a/
MANUFACTURING INDUSTRY	1,357.0	1,459.8	1,541.6	1,682.8	2,621.9	2,927.5	2,166.2	3,053.2	2,935.5	5,166.8	8,351.4	7,797.2	4,329.2	4369.1	4431.3	3654.5	4448.9	4591.1
1. Food products	14.4	12.4	9.1	9.7	19.7	18.0	10.8	35.1	53.1	152.2	183.9	75.1	23.1	47.3	76.7	52.3	72.4	65.3
2. Beverages	6.5	7.6	6.3	7.9	6.2	11.0	9.2	7.5	13.2	41.9	47.7	34.2	10.4	6.5	4.0	8.4	5.6	6.4
3. Tobacco	0.6	0.5	0.1	0.1	0.2	0.2	0.2	0.9	2.2	3.4	9.4	9.7	2.0	1.9	1.2	1.4	0.3	0.6
4. Textiles	25.7	20.6	17.9	16.4	39.8	29.8	27.2	20.7	26.4	115.3	229.9	287.2	123.9	119.4	134.6	66.0	105.3	90.0
5. Clothing	2.4	1.2	0.1	0.4	0.4	0.9	1.1	1.5	4.5	35.1	156.4	155.4	27.9	9.2	13.2	6.5	7.2	4.8
6. Leather products	0.2	0.1	0.0	0.0	0.0	0.0	0.0	0.1	0.4	3.4	8.2	8.2	1.5	1.4	2.5	1.1	2.3	1.0
7. Footwear	0.1	0.1	0.0	0.0	0.0	0.1	0.1	0.2	0.8	9.1	36.5	31.9	3.7	1.8	1.7	1.0	1.2	1.3
8. Wood products	1.1	2.1	2.0	2.1	2.6	6.2	3.0	2.6	1.9	10.4	31.8	120.4	58.1	63.4	62.4	32.8	55.1	44.0
9. Furniture	0.2	0.3	0.2	0.1	0.2	0.1	0.3	0.6	1.1	3.7	14.3	17.0	3.1	1.2	2.1	1.2	1.6	1.4
10. Paper	66.2	59.0	55.8	62.2	109.7	109.6	77.3	103.2	112.1	127.3	233.2	216.0	99.0	95.4	57.0	50.2	85.5	87.0
11. Printing	11.6	10.9	8.6	8.4	13.1	15.9	9.7	15.7	26.7	47.3	95.3	78.6	25.0	12.0	7.2	8.5	7.7	5.5
12. Basic chemicals	124.4	145.7	174.6	179.9	437.5	437.7	346.5	387.1	325.3	575.4	640.1	636.4	610.6	628.3	646.6	507.1	720.8	584.9
13. Other chemicals	79.6	93.6	102.3	108.8	194.3	193.0	162.9	191.3	180.0	240.3	414.1	380.8	309.6	499.2	533.2	423.7	528.1	459.2
14. Petroleum refineries	0.1	12.6	1.3	31.0	45.1	66.8	64.2	75.9	30.7	431.8	110.4	110.0	100.0	470.1	490.1	460.4	423.4	595.8
15. Other petroleum products	32.4	20.6	11.3	9.1	17.5	22.4	27.1	25.4	26.1	50.0	65.8	65.0	60.0	..	..	..	..	..
16. Rubber products	5.8	5.8	5.9	5.9	9.5	16.5	11.0	19.3	21.5	38.1	72.9	100.2	85.0	113.2	96.6	50.5	72.3	58.9
17. Plastic products	23.3	26.8	33.4	35.2	106.6	109.6	71.1	72.2	92.3	218.9	244.5	302.1	186.4	185.4	178.7	131.9	212.7	198.2
18. Clay products	5.9	9.0	9.1	6.4	6.8	13.4	17.1	14.1	18.5	45.3	67.9	65.3	22.3	23.6	20.1	16.8	19.6	18.7
19. Glassware	11.3	11.6	10.8	12.4	16.1	18.0	13.5	13.5	16.7	33.0	55.0	46.0	16.5	18.0	14.1	9.5	12.7	12.0
20. Nonmetalic minerals	10.9	4.7	3.5	3.8	5.8	7.3	5.7	7.7	10.5	34.9	59.8	11.3	7.0	8.5	7.9	8.7	14.9	14.5
21. Iron and steel	243.5	237.2	247.6	375.2	508.4	746.1	356.4	357.5	218.5	367.4	594.9	474.2	375.7	356.1	370.4	228.7	255.0	301.5
22. Nonferrous metals	86.1	91.6	102.9	122.1	226.4	163.4	89.0	81.1	73.0	144.3	161.5	114.7	123.9	94.1	108.8	68.8	113.7	117.0
23. Metal products	38.4	32.3	35.2	30.1	45.6	40.1	51.8	63.0	89.8	95.3	227.6	102.4	38.8	36.6	35.0	27.2	27.7	26.4
24. Nonelectric machinery	345.5	380.0	415.6	339.7	412.1	515.2	493.5	813.7	844.6	1,084.9	1,961.4	1,715.0	903.1	738.7	668.4	693.5	701.4	866.4
25. Electric machinery	97.1	129.4	129.7	156.3	163.7	160.6	145.4	254.5	334.3	440.3	1,252.6	1,195.3	575.2	375.9	392.7	347.2	495.2	524.4
26. Transport equipment	78.8	96.3	115.7	105.4	136.7	133.8	133.0	412.8	297.2	669.8	1,026.1	988.9	341.2	289.9	349.2	293.8	282.8	282.2
27. Scientific equipment	45.0	48.1	42.8	54.4	97.9	91.9	39.4	75.9	114.5	148.2	350.3	455.9	196.2	172.0	156.9	157.3	224.4	223.7

Source: INDEC.

a/ Based on data through October 1987.

July 1989

Table 9.1: ARGENTINA - PRINCIPAL PRICE INDICATORS, ANNUAL AVERAGES 1960-1988
(1970 = 100)

Year	WHOLESALE PRICE INDEX						CONSUMER PRICE INDEX		CONSTRUCTION COST
				Non-agriculture					
	General	Total Domestic	Agriculture Domestic	Total	Domestic	Imported	General	Food	
1960	17.6	17.6	17.3	17.7	17.7	18.3	14.7	15.3	13.5
1961	19.1	19.1	18.3	19.4	19.5	17.7	16.4	16.9	16.4
1962	24.9	24.9	25.1	24.7	24.8	23.5	20.9	21.5	21.4
1963	32.0	32.1	34.0	31.2	31.4	28.6	26.2	26.9	26.6
1964	40.4	40.7	43.5	39.1	39.5	32.6	32.0	33.9	32.1
1965	50.0	50.4	47.7	51.0	51.6	42.4	41.3	43.4	45.7
1966	60.0	60.5	57.8	60.9	61.6	50.7	54.7	54.1	58.9
1967	75.4	75.6	72.8	76.5	76.8	71.7	70.7	69.8	75.8
1968	82.6	82.9	80.0	83.7	84.2	77.5	81.8	81.0	81.6
1969	87.7	87.7	86.4	88.2	88.3	85.8	88.0	86.0	89.4
1970	100.0	100.0	100.0	100.0	100.0	100.0	100.0	100.0	100.0
1971	139.5	140.3	148.3	135.9	136.7	123.1	134.7	141.7	130.9
1972	246.9	246.9	288.8	229.8	228.6	247.0	213.8	231.4	201.9
1973	370.5	417.3	346.0	351.1	389.7	383.0	342.2	359.1	347.7
1974	444.7	439.8	452.6	441.4	434.1	550.3	444.4	413.2	482.6
1975	1,300.7	1,269.7	1,108.7	1,379.3	1,339.9	1,967.2	1,202.7	1,134.7	1,696.2
1976	7,791.7	7,430.5	6,980.0	8,123.9	7,626.9	15,548.6	6,543.1	6,552.5	7,812.6
1977	19,436.0	18,702.7	18,399.9	19,859.6	18,834.4	35,175.4	18,060.9	18,621.5	15,423.6
1978	47,810.1	47,155.3	44,456.0	49,181.6	48,331.8	61,877.1	49,759.1	49,009.5	36,547.6
1979	119,189.1	119,179.2	111,486.3	122,338.4	122,532.7	119,434.6	129,130.2	131,671.5	95,451.0
1980	209,090.3	209,132.3	181,737.3	220,285.5	221,082.1	208,382.0	259,248.4	256,872.3	199,492.5
1981	436,936.4	432,184.5	348,249.2	469,673.1	463,928.9	562,660.1	530,101.3	511,571.1	393,559.0
1982	1,561,073.1	1,514,496.9	1,386,209.4	1,632,630.3	1,570,436.7	2,561,755.9	1,403,584.9	1,424,129.8	1,072,448.4
1983	7,195,435.4	7,010,734.3	6,563,856.1	7,453,802.9	7,205,531.3	11,162,788.4	6,229,234.2	6,252,514.5	5,743,988.5
1984	48,469,294.1	47,352,561.7	42,850,310.1	51,196,506.2	49,618,283.1	74,669,045.4	45,269,187.6	46,196,884.8	42,946,287.1
1985	370,557,710.8	356,735,436.5	252,885,271.6	406,620,854.9	390,059,230.1	647,197,073.7	349,559,533.7	334,528,550.9	280,806,886.2
1986	607,209,698.6	585,925,669.8	534,824,512.5	642,976,984.5	615,821,335.8	1,036,891,840.2	664,490,246.8	662,747,665.5	476,202,594.8
1987	1,353,310,616.0	1,301,184,096.8	1,156,077,994.4	1,439,977,923.9	1,374,148,886.8	2,392,105,796.3	1,537,171,784.9	1,542,630,231.0	1,083,226,347.3
1988	6,936,110,282.9	6,632,630,858.9	5,530,173,714.3	7,456,755,816.0	7,079,990,896.9	12,887,815,985.5	6,808,979,626.1	6,756,514,351.9	5,119,556,333.3

Source: INDEC.

August 1989

Table 9.2: ARGENTINA - PRINCIPAL PRICE INDICATORS, MONTHLY, 1978-1989 (1985=100)

(continues ...)

Year	WHOLESALE PRICE INDEX						CONSUMER PRICE INDEX		CONSTRUCTION COST
				Non-agriculture					
	General	Total Domestic	Agriculture Domestic	Total	Domestic	Imported	General	Food	
1978	0.013	0.013	0.018	0.012	0.012	0.010	0.014	0.015	0.013
January	0.008	0.009	0.011	0.008	0.008	0.008	0.009	0.009	0.009
February	0.009	0.009	0.011	0.009	0.009	0.008	0.009	0.010	0.009
March	0.010	0.010	0.013	0.009	0.009	0.008	0.010	0.011	0.010
April	0.011	0.011	0.014	0.010	0.010	0.009	0.012	0.012	0.011
May	0.012	0.012	0.015	0.011	0.011	0.009	0.013	0.013	0.012
June	0.012	0.012	0.016	0.012	0.012	0.010	0.013	0.014	0.013
July	0.013	0.013	0.016	0.012	0.012	0.010	0.014	0.014	0.013
August	0.014	0.014	0.019	0.013	0.016	0.008	0.015	0.016	0.014
September	0.015	0.015	0.021	0.014	0.014	0.010	0.016	0.017	0.014
October	0.016	0.017	0.023	0.015	0.015	0.011	0.018	0.019	0.015
November	0.017	0.018	0.025	0.016	0.017	0.011	0.019	0.020	0.016
December	0.019	0.019	0.026	0.017	0.018	0.012	0.021	0.022	0.019
									0.000
1979	0.032	0.033	0.044	0.030	0.031	0.018	0.037	0.039	0.034
January	0.021	0.021	0.028	0.019	0.020	0.013	0.024	0.026	0.021
February	0.022	0.023	0.030	0.021	0.022	0.013	0.026	0.028	0.022
March	0.024	0.025	0.032	0.023	0.024	0.014	0.028	0.030	0.024
April	0.025	0.026	0.034	0.024	0.025	0.015	0.030	0.031	0.026
May	0.028	0.029	0.037	0.026	0.027	0.016	0.032	0.033	0.030
June	0.031	0.032	0.042	0.029	0.030	0.018	0.035	0.037	0.033
July	0.033	0.034	0.046	0.031	0.032	0.019	0.037	0.039	0.036
August	0.038	0.039	0.055	0.035	0.036	0.021	0.042	0.046	0.039
September	0.040	0.042	0.058	0.036	0.038	0.021	0.044	0.049	0.041
October	0.040	0.042	0.054	0.038	0.040	0.023	0.046	0.049	0.044
November	0.042	0.043	0.056	0.039	0.041	0.024	0.049	0.051	0.045
December	0.043	0.044	0.056	0.040	0.042	0.024	0.051	0.053	0.047
									0.000
1980	0.056	0.059	0.072	0.054	0.057	0.032	0.074	0.077	0.071
January	0.045	0.046	0.059	0.042	0.044	0.026	0.055	0.057	0.050
February	0.046	0.048	0.060	0.044	0.046	0.027	0.058	0.061	0.052
March	0.048	0.050	0.061	0.046	0.048	0.029	0.061	0.064	0.054
April	0.050	0.052	0.064	0.048	0.050	0.029	0.065	0.068	0.056
May	0.053	0.055	0.069	0.050	0.052	0.030	0.068	0.072	0.060
June	0.057	0.059	0.076	0.053	0.056	0.031	0.072	0.077	0.065
July	0.058	0.061	0.077	0.055	0.058	0.033	0.076	0.079	0.077
August	0.060	0.062	0.079	0.057	0.060	0.034	0.078	0.081	0.081
September	0.062	0.064	0.081	0.059	0.061	0.035	0.082	0.084	0.085
October	0.065	0.068	0.079	0.063	0.066	0.037	0.088	0.089	0.087
November	0.067	0.069	0.081	0.065	0.068	0.038	0.092	0.094	0.091
December	0.067	0.070	0.077	0.067	0.070	0.039	0.096	0.096	0.095

Table 9.2: ARGENTINA - PRINCIPAL PRICE INDICATORS, MONTHLY, 1978-1989 (1985=100)

(continues ...)

Year	WHOLESALE PRICE INDEX						CONSUMER PRICE INDEX		CONSTRUCTION COST
				Non-agriculture					
	General	Total Domestic	Agriculture Domestic	Total	Domestic	Imported	General	Food	
1981	0.118	0.121	0.138	0.116	0.119	0.087	0.152	0.153	0.140
January	0.069	0.071	0.076	0.069	0.072	0.040	0.100	0.101	0.098
February	0.072	0.075	0.080	0.073	0.076	0.044	0.105	0.105	0.103
March	0.076	0.079	0.083	0.076	0.080	0.047	0.111	0.111	0.108
April	0.085	0.088	0.093	0.086	0.089	0.058	0.120	0.118	0.115
May	0.092	0.095	0.098	0.093	0.097	0.062	0.129	0.129	0.121
June	0.109	0.112	0.126	0.108	0.111	0.082	0.141	0.143	0.131
July	0.123	0.126	0.143	0.122	0.125	0.098	0.155	0.158	0.143
August	0.135	0.138	0.159	0.133	0.136	0.101	0.167	0.170	0.150
September	0.145	0.148	0.173	0.141	0.146	0.104	0.179	0.181	0.160
October	0.153	0.158	0.184	0.150	0.155	0.109	0.190	0.187	0.171
November	0.170	0.175	0.217	0.163	0.168	0.119	0.203	0.204	0.185
December	0.188	0.194	0.240	0.181	0.187	0.131	0.221	0.227	0.196
1982	0.421	0.425	0.548	0.402	0.403	0.396	0.402	0.426	0.382
January	0.215	0.220	0.266	0.208	0.213	0.166	0.248	0.258	0.237
February	0.227	0.232	0.275	0.222	0.227	0.173	0.261	0.252	0.248
March	0.237	0.242	0.279	0.234	0.239	0.187	0.273	0.280	0.262
April	0.252	0.257	0.305	0.246	0.251	0.200	0.284	0.289	0.271
May	0.275	0.278	0.331	0.269	0.271	0.254	0.293	0.297	0.285
June	0.317	0.320	0.413	0.302	0.304	0.290	0.316	0.327	0.301
July	0.405	0.409	0.538	0.384	0.381	0.416	0.368	0.397	0.367
August	0.471	0.472	0.641	0.442	0.439	0.469	0.422	0.450	0.426
September	0.561	0.566	0.771	0.525	0.526	0.519	0.494	0.542	0.530
October	0.617	0.622	0.829	0.582	0.583	0.574	0.556	0.608	0.617
November	0.703	0.704	0.939	0.664	0.659	0.721	0.619	0.669	0.706
December	0.775	0.776	0.993	0.742	0.739	0.781	0.685	0.724	0.796
1983	1.941	1.965	2.595	1.833	1.847	1.725	1.782	1.869	2.047
January	0.888	0.893	1.142	0.850	0.850	0.862	0.795	0.842	0.813
February	1.006	1.013	1.329	0.954	0.954	0.954	0.898	0.964	0.904
March	1.114	1.119	1.404	1.074	1.072	1.093	0.999	1.059	1.035
April	1.191	1.196	1.488	1.151	1.059	1.170	1.102	1.136	1.198
May	1.312	1.318	1.624	1.273	1.272	1.289	1.202	1.216	1.337
June	1.503	1.515	1.961	1.431	1.435	1.405	1.392	1.457	1.523
July	1.675	1.696	2.192	1.594	1.608	1.485	1.565	1.603	1.755
August	1.976	2.007	2.799	1.826	1.844	1.671	1.835	1.921	2.008
September	2.459	2.514	3.636	2.233	2.273	1.882	2.227	2.425	2.494
October	2.875	2.911	4.203	2.622	2.641	2.475	2.605	2.792	2.963
November	3.338	3.379	4.475	3.149	3.172	2.963	3.107	3.286	3.771
December	3.961	4.015	4.889	3.844	3.890	3.452	3.656	3.728	4.758

Table 9.2: ARGENTINA - PRINCIPAL PRICE INDICATORS, MONTHLY, 1978-1989
(1985=100)

Year	WHOLESALE PRICE INDEX						CONSUMER PRICE INDEX		CONSTRUCTION COST
	General	Total Domestic	Agriculture Domestic	Non-agriculture Total	Non-agriculture Domestic	Imported	General	Food	
1984	13.078	13.273	16.942	12.592	12.718	11.539	12.950	13.810	15.302
January	4.414	4.448	5.702	4.219	4.229	4.158	4.114	4.241	5.140
February	5.116	5.166	6.934	4.807	4.823	4.699	4.813	5.211	5.837
March	6.056	6.128	8.299	5.666	5.700	5.399	5.788	6.484	6.914
April	7.247	7.358	9.821	6.809	6.886	6.165	6.859	7.533	8.233
May	8.256	8.756	11.538	8.122	8.237	7.154	8.030	8.619	9.506
June	10.040	10.175	13.309	9.509	9.599	8.773	9.468	9.971	11.713
July	11.600	11.765	14.192	11.287	11.438	10.018	11.200	11.490	13.851
August	14.141	14.363	18.037	13.671	13.868	12.020	13.758	14.394	16.647
September	17.639	17.964	24.112	16.859	17.135	14.538	17.548	19.372	20.369
October	20.348	20.648	27.124	19.532	19.775	17.488	20.939	22.686	24.189
November	23.334	23.499	29.112	22.638	22.742	21.766	24.075	25.400	28.096
December	28.747	29.005	35.119	27.980	28.181	26.290	28.812	30.315	33.132
1985	100.000	100.000	100.000	100.000	100.000	100.000	100.000	100.000	100.000
January	34.828	35.145	41.258	34.054	34.321	31.811	36.052	38.766	38.815
February	41.042	41.240	45.137	40.548	40.715	39.145	43.504	46.376	46.066
March	52.412	52.731	52.582	52.392	52.751	49.376	55.032	57.980	58.221
April	68.923	69.329	64.735	69.428	69.948	65.049	71.247	74.115	74.132
May	90.433	90.719	75.103	92.280	92.824	87.706	89.142	88.826	96.602
June	128.665	128.341	102.928	131.765	131.766	131.751	116.361	113.951	133.469
July	127.452	126.828	112.166	129.293	128.805	133.401	123.569	118.586	125.590
August	129.412	129.050	129.076	129.452	129.046	132.864	127.355	125.340	125.756
September	130.177	129.993	136.465	129.420	129.121	131.930	129.896	128.924	125.067
October	131.141	131.084	140.450	130.020	129.822	131.686	132.424	130.379	124.827
November	132.115	132.092	146.543	130.377	130.144	132.335	135.561	135.206	125.576
December	133.400	133.447	153.556	130.972	130.737	132.947	139.859	141.549	125.880
1986	163.864	164.247	211.489	158.127	157.879	160.213	190.094	198.114	169.197
January	133.367	133.443	152.474	131.065	130.878	132.640	144.096	147.055	128.737
February	134.412	134.283	155.703	131.847	131.396	135.634	146.531	150.197	129.178
March	136.309	136.015	157.709	133.731	133.091	139.111	153.339	159.609	130.847
April	140.370	140.276	163.036	137.640	137.208	141.269	160.598	167.493	139.284
May	144.218	144.300	174.072	140.622	140.288	143.432	167.065	173.727	143.604
June	150.797	150.940	187.181	146.415	146.055	149.441	174.659	181.486	152.584
July	158.491	159.145	193.815	154.236	154.472	152.259	186.470	192.466	172.490
August	173.348	174.113	234.720	165.956	165.944	166.056	202.851	214.032	182.840
September	185.101	185.746	266.239	175.328	174.896	178.959	217.520	230.910	194.155
October	194.839	195.895	290.403	183.328	183.156	184.768	230.683	242.070	210.939
November	204.477	205.476	294.726	193.606	193.446	194.951	242.899	255.022	217.737
December	210.634	211.327	267.788	203.750	203.717	204.033	254.412	263.300	227.971

Table 9.2: ARGENTINA - PRINCIPAL PRICE INDICATORS, MONTHLY, 1978-1989
(1985=100)

Year	WHOLESALE PRICE INDEX						CONSUMER PRICE INDEX		CONSTRUCTION COST
				Non-agriculture					
	General	Total Domestic	Agriculture Domestic	Total	Domestic	Imported	General	Food	
1987	365.209	364.753	457.201	354.133	352.292	369.610	439.745	461.136	385.755
January	221.829	222.704	278.189	215.040	215.226	213.480	273.644	284.207	243.499
February	237.128	238.200	299.740	229.587	229.906	226.901	291.439	302.995	255.675
March	255.752	256.520	332.523	246.504	246.276	248.426	315.336	331.829	265.535
April	260.674	261.654	340.685	251.036	251.002	251.324	325.945	341.516	271.677
May	273.392	274.614	372.692	261.498	261.394	262.367	339.551	356.567	292.484
June	291.712	293.193	391.282	279.718	279.972	277.580	366.722	389.952	329.872
July	319.163	320.648	426.551	306.228	306.374	305.000	403.832	429.359	341.533
August	365.760	367.791	505.024	348.985	349.294	346.389	459.225	500.628	382.744
September	426.531	427.228	571.957	409.013	407.721	419.882	512.891	565.320	449.741
October	556.444	550.988	637.399	546.688	539.341	608.473	613.144	637.400	571.771
November	580.455	575.138	651.427	571.902	564.855	631.161	676.107	683.606	599.110
December	593.670	588.357	678.946	583.394	576.147	644.337	699.108	710.250	625.416
1988	1871.803	1859.285	2187.052	1833.835	1815.107	1991.328	1947.874	2019.712	1823.159
January	665.286	657.455	777.767	651.854	641.239	741.113	762.675	788.113	710.385
February	754.333	745.804	855.812	742.105	730.977	835.682	842.239	870.247	776.688
March	876.925	867.815	952.256	867.846	856.434	963.822	966.378	995.621	908.790
April	1024.541	1012.595	1057.480	1020.568	1006.545	1138.487	1132.905	1154.818	1028.783
May	1262.993	1250.349	1239.619	1265.803	1251.794	1383.605	1311.011	1293.491	1275.890
June	1566.699	1552.619	1698.548	1550.811	1532.949	1701.016	1546.530	1530.831	1500.899
July	1958.699	1948.918	2243.465	1924.396	1909.217	2052.039	1943.075	1982.857	1822.503
August	2584.139	2558.918	2875.524	2549.029	2516.240	2824.750	2479.855	2566.511	2361.112
September	2749.888	2742.796	3467.943	2663.395	2645.056	2817.608	2769.807	2889.418	2600.758
October	2875.584	2864.105	3709.640	2775.115	2750.140	2985.134	3018.908	3197.561	2746.098
November	2986.698	2969.767	3596.527	2913.236	2885.288	3148.254	3191.414	3371.091	2917.000
December	3155.852	3140.282	3770.039	3081.866	3055.399	3304.427	3409.692	3595.988	3229.002
1989									
January	3374.891	3356.988	4134.926	3283.324	3252.132	3545.516	3713.909	3947.682	3333.944
February	3658.685	3635.514	4686.020	3534.916	3493.922	3879.642	4070.117	4336.181	3655.340
March	4350.312	4292.830	5894.917	4154.221	4076.806	4898.467	4762.228	5134.184	4211.907
April	6873.079	6493.805	8413.676	6687.235	6235.035	10489.857	6351.518	6804.585	6490.452
May	14052.762	12985.774	17089.484	13686.216	12432.655	24227.621	11335.411	12155.040	13651.502
June	32639.896	31012.566	41096.122	31620.127	29653.453	48158.202	24314.457	27871.506	31960.342

Source: INDEC.

August 1989

Table 9.3: ARGENTINA - PRICE INDICATORS, MONTHLY, 1978-1989 (continues...)

Year	WPI 1985=100 Monthly	CPI 1985=100 Monthly	WPI 1985=100 Quarterly	CPI 1985=100 Quarterly	Combined Price Index
1978	0.013	0.014			
January	0.008	0.009			
February	0.009	0.009			
March	0.010	0.010	0.009	0.010	0.009
April	0.011	0.012			
May	0.012	0.013			
June	0.012	0.013	0.011	0.012	0.012
July	0.013	0.014			
August	0.014	0.015			
September	0.015	0.016	0.014	0.015	0.015
October	0.016	0.018			
November	0.017	0.019			
December	0.019	0.021	0.017	0.020	0.018
1979	0.032	0.037			
January	0.021	0.024			
February	0.022	0.026			
March	0.024	0.028	0.022	0.026	0.024
April	0.025	0.030			
May	0.028	0.032			
June	0.031	0.035	0.028	0.032	0.030
July	0.033	0.037			
August	0.038	0.042			
September	0.040	0.044	0.037	0.041	0.039
October	0.040	0.046			
November	0.042	0.049			
December	0.043	0.051	0.042	0.049	0.045
1980	0.056	0.074			
January	0.045	0.055			
February	0.046	0.058			
March	0.048	0.061	0.046	0.058	0.052
April	0.050	0.065			
May	0.053	0.068			
June	0.057	0.072	0.053	0.068	0.061
July	0.058	0.076			
August	0.060	0.078			
September	0.062	0.082	0.060	0.079	0.069
October	0.065	0.088			
November	0.067	0.092			
December	0.067	0.096	0.066	0.092	0.079
1981	0.118	0.152			
January	0.069	0.100			
February	0.072	0.105			
March	0.076	0.111	0.072	0.105	0.089
April	0.085	0.120			
May	0.092	0.129			
June	0.109	0.141	0.096	0.130	0.113
July	0.123	0.155			
August	0.135	0.167			
September	0.145	0.179	0.134	0.167	0.151
October	0.153	0.190			
November	0.170	0.203			
December	0.188	0.221	0.171	0.205	0.188

Table 9.3: ARGENTINA - PRICE INDICATORS, MONTHLY 1978-1989 (continues...)

Year	WPI 1985=100 Monthly	CPI 1985=100 Monthly	WPI 1985=100 Quarterly	CPI 1985=100 Quarterly	Combined Price Index
1982	0.421	0.402			
January	0.215	0.248			
February	0.227	0.261			
March	0.237	0.273	0.226	0.260	0.243
April	0.252	0.284			
May	0.275	0.293			
June	0.317	0.316	0.281	0.298	0.290
July	0.405	0.368			
August	0.471	0.422			
September	0.561	0.494	0.479	0.428	0.453
October	0.617	0.556			
November	0.703	0.619			
December	0.775	0.685	0.698	0.620	0.659
1983	1.941	1.782			
January	0.888	0.795			
February	1.006	0.898			
March	1.114	0.999	1.003	0.897	0.950
April	1.191	1.102			
May	1.312	1.202			
June	1.503	1.392	1.335	1.232	1.284
July	1.675	1.565			
August	1.976	1.835			
September	2.459	2.227	2.037	1.876	1.956
October	2.875	2.605			
November	3.338	3.107			
December	3.961	3.656	3.391	3.123	3.257
1984	13.078	12.950			
January	4.414	4.114			
February	5.116	4.813			
March	6.056	5.788	5.195	4.905	5.050
April	7.247	6.859			
May	8.256	8.030			
June	10.040	9.468	8.515	8.119	8.317
July	11.600	11.200			
August	14.141	13.758			
September	17.639	17.548	14.460	14.169	14.314
October	20.348	20.939			
November	23.334	24.075			
December	28.747	28.812	24.143	24.609	24.376
1985	100.000	100.000			
January	34.828	36.052			
February	41.042	43.504			
March	52.412	55.032	42.761	44.863	43.812
April	68.923	71.247			
May	90.433	89.142			
June	128.665	116.361	96.007	92.250	94.128
July	127.452	123.569			
August	129.412	127.355			
September	130.177	129.896	129.014	126.940	127.977
October	131.141	132.424			
November	132.115	135.561			
December	133.400	139.859	132.219	135.948	134.083

Table 9.3: ARGENTINA - PRICE INDICATORS, MONTHLY, 1978-1989

Year	WPI 1985=100 Monthly	CPI 1985=100 Monthly	WPI 1985=100 Quarterly	CPI 1985=100 Quarterly	Combined Price Index
1986	163.864	190.094			
January	133.367	144.096			
February	134.412	146.531			
March	136.309	153.359	134.696	147.988	141.342
April	140.370	160.598			
May	144.218	167.065			
June	150.797	174.659	145.129	167.441	156.285
July	158.491	186.470			
August	173.348	202.851			
September	185.101	217.520	172.314	202.280	187.297
October	194.839	230.683			
November	204.477	242.899			
December	210.634	254.412	203.317	242.665	222.991
1987					
January	221.829	273.644			
February	237.128	291.439			
March	255.752	315.336	238.236	293.473	265.855
April	260.674	325.945			
May	273.392	339.551			
June	291.712	366.722	275.259	344.073	309.666
July	319.163	403.832			
August	365.760	459.225			
September	426.531	512.891	370.485	458.649	414.567
October	556.444	613.144			
November	580.455	676.107			
December	593.670	699.108	576.856	662.786	619.821
1988					
January	665.286	762.675			
February	754.333	842.239			
March	876.925	966.378	765.514	857.097	811.306
April	1024.541	1132.905			
May	1262.993	1311.011			
June	1566.699	1546.530	1284.744	1330.149	1307.446
July	1958.699	1943.075			
August	2584.139	2479.855			
September	2749.888	2769.807	2430.909	2397.579	2414.244
October	2875.584	3018.908			
November	2986.698	3191.414			
December	3155.852	3409.692	3006.044	3206.671	3106.358
1989					
January	3374.891	3713.909			
February	3658.685	4070.117			
March	4350.312	4762.228	3794.629	4182.085	3988.357
April	6873.079	6351.518			
May	14052.762	11335.411			
June	32639.896	24314.457	17855.246	14000.462	15927.854

Source: INDEC.

a/ A simple average of the CPI and the WPI.

August 1989

Table 9.4: ARGENTINA - CHANGES IN WHOLESALE AND CONSUMER PRICE INDICES: MAJOR COMPONENTS, 1970-1988

(Annual average percentage changes)

	WPI Weight	1970	1971	1972	1973	1974	1975	1976	1977	1978	1979	1980	1981	1982	1983	1984
WHOLESALE PRICES																
General Index	100.0	14.1	39.5	77.0	50.0	20.0	192.5	499.0	149.4	146.0	149.3	75.4	109.6	256.2	360.9	573.6
Domestic Goods	95.2	14.3	40.3	76.0	49.5	19.2	188.7	485.2	151.7	152.1	152.7	75.5	107.4	249.2	362.9	575.4
Domestic Agriculture	26.5	15.8	48.5	94.8	42.5	10.0	144.9	529.6	163.6	141.6	150.8	63.0	93.9	293.4	373.5	552.8
Domestic Non-agriculture	68.7	13.2	36.7	67.2	53.3	23.9	208.6	469.2	146.9	156.6	153.5	80.4	112.2	234.8	358.8	588.6
Food & Beverages	19.4	27.8	48.3	62.6	55.7	12.7	134.5	511.9	192.2	161.2	164.7	75.5	113.0	214.7	307.7	552.7
Tobacco	1.6	1.0	3.3	33.6	72.1	53.5	90.5	554.5	271.8	159.6	131.9	92.2	87.0	131.1	420.4	425.1
Textiles	9.1	1.3	33.2	75.7	55.0	29.1	176.3	417.5	154.5	152.7	150.5	75.9	120.9	279.4	419.9	538.0
Clothing	5.0	3.6	35.4	86.1	36.1	29.2	225.2	551.6	77.8	172.8	162.9	84.6	103.7	235.7	400.4	647.0
Wood	3.0	9.0	28.8	89.1	71.7	33.0	236.7	408.2	95.2	129.8	187.3	120.6	102.4	229.5	465.4	582.4
Paper	1.5	9.3	32.5	62.0	40.0	44.6	256.9	381.2	122.3	161.1	170.9	78.6	120.7	234.9	400.3	606.4
Chemicals	4.9	4.4	24.1	57.4	45.6	23.2	271.4	519.8	149.9	165.3	126.8	91.7	127.9	243.9	329.1	450.7
Oil Products	2.7	3.8	34.5	54.1	81.2	73.3	208.4	335.2	158.6	155.6	102.3	96.7	142.2	174.0	516.0	753.2
Rubber	0.9	0.8	13.7	59.4	38.4	17.1	184.0	654.4	159.1	154.1	126.9	97.4	148.6	321.5	319.9	491.8
Leather	2.2	10.0	29.0	89.2	50.3	21.7	284.3	462.4	136.6	145.3	208.1	51.8	97.7	310.4	345.3	688.9
Nonmetallic minerals	2.7	6.0	27.1	44.9	45.0	37.2	280.4	484.6	125.9	207.1	157.8	93.5	115.0	209.1	340.5	544.5
Metal, excl.machinery	6.2	7.9	28.3	68.0	51.5	33.9	370.6	429.0	126.5	137.7	129.4	84.0	103.6	277.5	323.7	500.9
Vehicles and machinery	5.8	4.3	24.0	70.3	57.8	28.9	309.5	431.7	115.5	130.8	130.5	92.4	102.5	247.1	418.0	599.1
Electric machinery	2.5	5.7	15.2	67.3	38.9	22.2	318.6	492.4	102.5	143.7	122.4	80.1	94.9	239.3	365.6	577.0
Extractive industry	1.2	8.7	36.8	60.6	39.8	28.1	497.1	424.0	129.4	190.0	132.8	75.3	111.7	214.3	520.7	603.8
Imported Goods Non-agriculture	4.8	16.6	23.1	100.0	62.7	36.9	257.5	690.4	126.2	75.9	93.0	74.5	157.7	377.1	335.7	568.9
CONSUMER PRICES																
General Index	..	13.6	34.7	58.7	60.1	29.9	170.6	444.1	176.0	175.5	159.5	100.8	104.5	164.8	343.8	626.7

	1985	1986	1987	1988
WHOLESALE PRICES a/				
General Index	664.2	63.8	122.9	412.5
Domestic Goods	650.3	64.3	122.1	409.7
Domestic Agriculture	490.2	111.5	116.3	378.3
Domestic Non-agriculture	686.0	57.9	123.1	408.4
Food, Beverages & Tobacco	785.6	71.7	144.6	402.0
Textiles, Clothing & Leather	432.0	57.2	142.8	300.2
Wood & Furniture	632.4	58.9	124.9	373.5
Paper & Products	715.0	62.4	134.4	343.5
Chemicals, Oil & Rubber Products	755.8	51.4	112.6	452.2
Non-metallic Products	714.6	50.6	107.9	410.4
Basic Metals	830.7	51.9	111.2	465.4
Metal Products, Machinery & Equip.	668.8	56.1	117.9	451.6
Imported Goods Non-agriculture	764.7	60.4	130.5	438.8
CONSUMER PRICES				
General Index	672.2	90.1	131.3	343.0

Source: INDEC.

a/ The categories for domestic non-agricultural wholesale price indices were changed in 1985.

May 1989

Distributors of World Bank Publications

ARGENTINA
Carlos Hirsch, SRL
Galeria Guemes
Florida 165, 4th Floor-Ofc. 453/465
1333 Buenos Aires

AUSTRALIA, PAPUA NEW GUINEA, FIJI, SOLOMON ISLANDS, VANUATU, AND WESTERN SAMOA
D.A. Books & Journals
648 Whitehorse Road
Mitcham 3132
Victoria

AUSTRIA
Gerold and Co.
Graben 31
A-1011 Wien

BAHRAIN
Bahrain Research and Consultancy Associates Ltd.
P.O. Box 22103
Manama Town 317

BANGLADESH
Micro Industries Development Assistance Society (MIDAS)
House 5, Road 16
Dhanmondi R/Area
Dhaka 1209

Branch offices:
156, Nur Ahmed Sarak
Chittagong 4000

76, K.D.A. Avenue
Kulna

BELGIUM
Publications des Nations Unies
Av. du Roi 202
1060 Brussels

BRAZIL
Publicacoes Tecnicas Internacionais Ltda.
Rua Peixoto Gomide, 209
01409 Sao Paulo, SP

CANADA
Le Diffuseur
C.P. 85, 1501B rue Ampère
Boucherville, Quebec
J4B 5E6

CHINA
China Financial & Economic Publishing House
8, Da Fo Si Dong Jie
Beijing

COLOMBIA
Enlace Ltda.
Apartado Aereo 34270
Bogota D.E.

COTE D'IVOIRE
Centre d'Edition et de Diffusion Africaines (CEDA)
04 B.P. 541
Abidjan 04 Plateau

CYPRUS
MEMRB Information Services
P.O. Box 2098
Nicosia

DENMARK
SamfundsLitteratur
Rosenoerns Allé 11
DK-1970 Frederiksberg C

DOMINICAN REPUBLIC
Editora Taller, C. por A.
Restauracion e Isabel la Catolica 309
Apartado Postal 2190
Santo Domingo

EL SALVADOR
Fusades
Avenida Manuel Enrique Araujo #3530
Edificio SISA, 1er. Piso
San Salvador

EGYPT, ARAB REPUBLIC OF
Al Ahram
Al Galaa Street
Cairo

The Middle East Observer
8 Chawarbi Street
Cairo

FINLAND
Akateeminen Kirjakauppa
P.O. Box 128
SF-00101
Helsinki 10

FRANCE
World Bank Publications
66, avenue d'Iéna
75116 Paris

GERMANY, FEDERAL REPUBLIC OF
UNO-Verlag
Poppelsdorfer Allee 55
D-5300 Bonn 1

GREECE
KEME
24, Ippodamou Street Platia Plastiras
Athens-11635

GUATEMALA
Librerias Piedra Santa
Centro Cultural Piedra Santa
11 calle 6-50 zona 1
Guatemala City

HONG KONG, MACAO
Asia 2000 Ltd.
Mongkok Post Office
Bute Street No. 37
Mongkok, Kowloon
Hong Kong

HUNGARY
Kultura
P.O. Box 139
1389 Budapest 62

INDIA
Allied Publishers Private Ltd.
751 Mount Road
Madras - 600 002

Branch offices:
15 J.N. Heredia Marg
Ballard Estate
Bombay - 400 038

13/14 Asaf Ali Road
New Delhi - 110 002

17 Chittaranjan Avenue
Calcutta - 700 072

Jayadeva Hostel Building
5th Main Road Gandhinagar
Bangalore - 560 009

3-5-1129 Kachiguda Cross Road
Hyderabad - 500 027

Prarthana Flats, 2nd Floor
Near Thakore Baug, Navrangpura
Ahmedabad - 380 009

Patiala House
16-A Ashok Marg
Lucknow - 226 001

INDONESIA
Pt. Indira Limited
Jl. Sam Ratulangi 37
P.O. Box 181
Jakarta Pusat

IRELAND
TDC Publishers
12 North Frederick Street
Dublin 1

ITALY
Licosa Commissionaria Sansoni SPA
Via Benedetto Fortini, 120/10
Casella Postale 552
50125 Florence

JAPAN
Eastern Book Service
37-3, Hongo 3-Chome, Bunkyo-ku 113
Tokyo

KENYA
Africa Book Service (E.A.) Ltd.
P.O. Box 45245
Nairobi

KOREA, REPUBLIC OF
Pan Korea Book Corporation
P.O. Box 101, Kwangwhamun
Seoul

KUWAIT
MEMRB Information Services
P.O. Box 5465

MALAYSIA
University of Malaya Cooperative Bookshop, Limited
P.O. Box 1127, Jalan Pantai Baru
Kuala Lumpur

MEXICO
INFOTEC
Apartado Postal 22-860
14060 Tlalpan, Mexico D.F.

MOROCCO
Societe d'Etudes Marketing Marocaine
12 rue Mozart, Bd. d'Anfa
Casablanca

NETHERLANDS
InOr-Publikaties b.v.
P.O. Box 14
7240 BA Lochem

NEW ZEALAND
Hills Library and Information Service
Private Bag
New Market
Auckland

NIGERIA
University Press Limited
Three Crowns Building Jericho
Private Mail Bag 5095
Ibadan

NORWAY
Narvesen Information Center
Bertrand Narvesens vei 2
P.O. Box 6125 Etterstad
N-0602 Oslo 6

OMAN
MEMRB Information Services
P.O. Box 1613, Seeb Airport
Muscat

PAKISTAN
Mirza Book Agency
65, Shahrah-e-Quaid-e-Azam
P.O. Box No. 729
Lahore 3

PERU
Editorial Desarrollo SA
Apartado 3824
Lima

PHILIPPINES
National Book Store
701 Rizal Avenue
P.O. Box 1934
Metro Manila

POLAND
ORPAN
Patac Kultury i Nauki
00-901 Warszawa

PORTUGAL
Livraria Portugal
Rua Do Carmo 70-74
1200 Lisbon

SAUDI ARABIA, QATAR
Jarir Book Store
P.O. Box 3196
Riyadh 11471

MEMRB Information Services
Branch offices:
Al Alsa Street
Al Dahna Center
First Floor
P.O. Box 7188
Riyadh

Haji Abdullah Alireza Building
King Khaled Street
P.O. Box 3969
Damman

33, Mohammed Hassan Awad Street
P.O. Box 5978
Jeddah

SINGAPORE, TAIWAN, MYANMAR, BRUNEI
Information Publications Private, Ltd.
02-06 1st Fl., Pei-Fu Industrial Bldg.
24 New Industrial Road
Singapore 1953

SOUTH AFRICA, BOTSWANA
For single titles:
Oxford University Press Southern Africa
P.O. Box 1141
Cape Town 8000

For subscription orders:
International Subscription Service
P.O. Box 41095
Craighall
Johannesburg 2024

SPAIN
Mundi-Prensa Libros, S.A.
Castello 37
28001 Madrid

Librería Internacional AEDOS
Consell de Cent, 391
08009 Barcelona

SRI LANKA AND THE MALDIVES
Lake House Bookshop
P.O. Box 244
100, Sir Chittampalam A. Gardiner Mawatha
Colombo 2

SWEDEN
For single titles:
Fritzes Fackboksforetaget
Regeringsgatan 12, Box 16356
S-103 27 Stockholm

For subscription orders:
Wennergren-Williams AB
Box 30004
S-104 25 Stockholm

SWITZERLAND
For single titles:
Librairie Payot
6, rue Grenus
Case postal 381
CH 1211 Geneva 11

For subscription orders:
Librairie Payot
Service des Abonnements
Case postal 3312
CH 1002 Lausanne

TANZANIA
Oxford University Press
P.O. Box 5299
Dar es Salaam

THAILAND
Central Department Store
306 Silom Road
Bangkok

TRINIDAD & TOBAGO, ANTIGUA BARBUDA, BARBADOS, DOMINICA, GRENADA, GUYANA, JAMAICA, MONTSERRAT, ST. KITTS & NEVIS, ST. LUCIA, ST. VINCENT & GRENADINES
Systematics Studies Unit
#9 Watts Street
Curepe
Trinidad, West Indies

TURKEY
Haset Kitapevi, A.S.
Istiklal Caddesi No. 469
Beyoglu
Istanbul

UGANDA
Uganda Bookshop
P.O. Box 7145
Kampala

UNITED ARAB EMIRATES
MEMRB Gulf Co.
P.O. Box 6097
Sharjah

UNITED KINGDOM
Microinfo Ltd.
P.O. Box 3
Alton, Hampshire GU34 2PG
England

URUGUAY
Instituto Nacional del Libro
San Jose 1116
Montevideo

VENEZUELA
Libreria del Este
Aptdo. 60.337
Caracas 1060-A

YUGOSLAVIA
Jugoslovenska Knjiga
P.O. Box 36
Trg Republike
YU-11000 Belgrade